289
Current Topics in Microbiology and Immunology

D.E. Griffin (Ed.)

Role of Apoptosis in Infection

With 40 Figures

Springer

Professor Dr. Diane E. Griffin, MD, PhD
Johns Hopkins University
Bloomberg School of Public Health
Department of Molecular Microbiology and Immunology
615 N. Wolfe Street
Baltimore, MD 21205
USA

e-mail: dgriffin@jhsph.edu

Cover illustration by Hilary Hurd, this volume

Library of Congress Catalog Card Number 72-152360

ISSN 0070-217X
ISBN 3-540-23006-8 Springer Berlin Heidelberg New York

Springer is a part of Springer Science+Business Media
springeronline.com

Printed in Germany

Editor: Dr. Rolf Lange, Heidelberg
Desk editor: Anne Clauss, Heidelberg
Production editor: Andreas Gösling, Heidelberg
Cover design: design & production GmbH, Heidelberg
Typesetting: Stürtz GmbH, Würzburg
Printed on acid-free paper 27/3150/ag – 5 4 3 2 1 0

Preface

Apoptosis is a regulated, energy-dependent process by which a cell self-destructs. This mechanism of programmed cell death plays an important role in normal development and control of cell numbers in mature animals. Apoptosis was initially defined by morphological criteria to describe the distinctive appearance of dying cells that developed nuclear condensation, cell shrinkage, and cytoplasmic blebbing. Initiation of the apoptotic process can come from external or internal stimuli and is highly regulated both by molecules that facilitate and by molecules that inhibit the process. Common features of apoptosis include activation of proteases and nucleases, mitochondrial membrane permeabilization, chromatin disruption, and translocation of phosphatidylserine from the inner to the outer surface of the plasma membrane. Apoptotic cells attract phagocytes that engulf the apoptotic bodies and prevent tissue damage in the region. Intense investigation of the cell death process has defined many molecular features of the pathway by which regulation and execution can be exploited by pathogens.

It can be envisioned that apoptosis as a response to an intracellular pathogen is a useful way for the host to eliminate infected cells, decreasing the likelihood of spread of the infection to neighboring cells and preventing pathogen persistence. Alternatively, the apoptotic response may be a major mechanism by which the host is harmed by a pathogen. Apoptosis can also play an important role in regulation of the immune response both by allowing cross-presentation of antigens and enhancing the T cell response and also by inducing death of inflammatory cells and suppressing the immune response.

Apoptosis can be initiated directly by contact of the pathogen with a target cell (e.g., *Yersinia, Entamoeba histolytica*), by delivery of proteases into a target cell (e.g., natural killer cells, cytotoxic T lymphocytes), or by triggering intracellular cell death signaling pathways (e.g., Sindbis virus, *Salmonella*). Apoptosis can also be initiated indirectly by induction of the expression of ligands that interact with death receptors on the cell surface (e.g., reovirus) or activating intracellular stress pathways (e.g., herpes simplex virus). For some slowly replicating intracellular pathogens inhibition of apoptosis is necessary for life cycle completion (e.g., baculovirus, *Toxoplasma gondii*). For rapidly replicating pathogens (e.g., Sindbis virus, poliovirus) delay of cell death is not necessary and pathogen growth may actually be facilitated by the apoptotic intracellular environment. Interestingly, for malaria, apoptosis of both the *Plasmodium* parasite and cells in the mosquito vector work together to enhance the likelihood of subsequent pathogen transmission by the mosquito.

The immune response to a pathogen can also be regulated to the host's or the pathogen's advantage by apoptosis. If the infected cell expresses a ligand that induces the death of macrophages, activated lymphocytes or other leukocytes (e.g., rabies virus, *Yersinia*, *Salmonella*, *T. gondii*), there will be suppression of the immune response and clearance mechanisms may be impaired. If the apoptotic bodies are engulfed by antigen-presenting cells, then the immune response will be enhanced and cross-presentation of antigens to CD8 T cells facilitated.

This volume provides examples and reviews of this wide variety of contributions of apoptosis to the pathogenesis of infectious diseases.

Diane E. Griffin

List of Contents

List of Contributors

(Their addresses can be found at the beginning of their respective chapters.)

CTMI (2005) 289:1–24

Mechanisms of Apoptosis During Reovirus Infection

P. Clarke[1, 2] (✉) · S. M. Richardson-Burns[1, 2] · R. L. DeBiasi[1, 2] · K. L. Tyler[1, 2]

[1] Department of Neurology (B 182), University of Colorado Health Sciences Center, 4200 East 9th Ave., Denver, CO 80262, USA
Penny.Clarke@uchsc.edu
[2] Denver VA Medical Center, 1055 Clermont St, Denver, CO 80220, USA

Abstract Reovirus infection has proven to be an excellent experimental system for studying mechanisms of virus-induced pathogenesis. Reoviruses induce apoptosis in a wide variety of cultured cells in vitro and in target tissues in vivo, including the heart and central nervous system. In vivo, viral infection, tissue injury, and apoptosis colocalize, suggesting that apoptosis is a critical mechanism by which disease is triggered in the host. This review examines the mechanisms of reovirus-induced apoptosis and investigates the possibility that inhibition of apoptosis may provide a novel strategy for limiting virus-induced tissue damage following infection.

Abbreviations

AIF	Apoptosis-inducing factor
CNS	Central nervous system
DcR	Decoy receptor
DD	Death domain
DISC	Death-inducing signaling complex
DN	Dominant negative
DR	Death receptor
E	Embryonic
FADDFADD	Fas-associated death domain
FMK	Fluoromethyl ketone
HTLV	Human T cell lymphoma virus
IAP	Inhibitor of apoptosis protein
IκBIκB	Inhibitor κB
ISVP	Infectious subvirion particle
JAM	Junctional adhesion molecule
JNKJNK	c-Jun N-terminal kinase
MAbs	Monoclonal antibodies
MCC	Mouse cortical cultures
MDCK	Madin-Darby canine kidney
MHC	Mouse hippocampal cultures
MEF	Mouse embryo fibroblasts
NF-κB	Nuclear factor κB
PCR	Polymerase chain reaction
PI	Postinfection
PFU	Plaque-forming unit
smacsmac	Second mitochondrion-derived activator of caspases
SMN	Survival motor neuron
T1L	Type 1 reovirus, strain Lang
T3A	Type 3 reovirus, strain Abney
T3D	Type 3 reovirus, strain Dearing
TNFTNF	Tumor necrosis factor
TNFR	TNF receptor
TRAILTRAIL	TNF-related apoptosis-inducing ligand
ts	Temperature sensitive
TUNEL	TdT-mediated dUTP nick-end labeling
UV	Ultraviolet
VarK	Variant K

1 Introduction

Reoviruses are ubiquitous, nonenveloped, cytoplasmically replicating viruses that have been isolated from a wide variety of mammalian species, including humans. In humans, reovirus has been associated with diarrheal illnesses, upper respiratory infections, hepatobiliary diseases including biliary atresia, and rare cases of central nervous system (CNS) infection. However, the viruses are still considered "orphan viruses" as they have not been definitively linked to disease. In contrast, natural and experimental infection of animals with reovirus produces a variety of diseases. The most extensively studied experimental system involves infection of neonatal mice where, depending on the viral strain and route of inoculation, reovirus infection can produce disease in a variety of organs (Tyler 1998; Tyler et al. 2001).

Reovirus infection has proven to be an excellent experimental system for studying mechanisms of virus-induced pathogenesis. Reoviruses induce apoptosis in a wide variety of cultured cells in vitro and in target tissues in vivo, including the heart and CNS (Clarke and Tyler 2003). In vivo, viral infection, tissue injury, and apoptosis colocalize, suggesting that apoptosis is a critical mechanism by which disease is triggered in the host (DeBiasi et al. 2001; Oberhaus et al. 1997; Richardson-Burns et al. 2002). This review examines the mechanisms of reovirus-induced apoptosis and investigates the possibility that inhibition of apoptosis may provide a novel strategy for limiting virus-induced tissue damage following infection.

2 Reovirus-Induced Apoptosis Is Determined by the Type 3 Reovirus S1 and M2 Gene Segments

The reovirus virion is comprised of two concentric protein capsids surrounding a genome of 10 segments of double-stranded (ds) RNA. Reovirus strains differ in their ability to induce apoptosis with prototype Type 3 (T3) reovirus strains, Dearing (T3D) and Abney (T3A), inducing significantly more apoptosis in L929 fibroblasts than the prototype Type 1 (T1) reovirus strain, Lang (T1L). The T3 S1 and M2 gene segments have been identified as the viral determinants of reovirus-induced apoptosis in L929 and Madin-Darby canine kidney (MDCK) cells and the S1 gene segment alone as a determinant of apoptosis in HeLa cells (Tyler et al. 1995, 1996; Rodgers et al. 1997; Connolly et al. 2001).

2.1 Role of the S1 Gene Segment in Reovirus-Induced Apoptosis

The S1 gene segment encodes the viral attachment protein $\sigma1$ and the nonstructural protein $\sigma1s$. $\sigma1s$ is the determinant of reovirus-induced G_2/M cell cycle arrest, an effect that results from inhibition of the G_2/M regulatory kinase $p34^{cdc2}$ and the resulting inhibition of cellular DNA synthesis (Tyler et al. 1996; Poggioli et al. 2000, 2001). In vitro, $\sigma1s$ is not required for reovirus-induced apoptosis of L929 or HEK293 cells and cell cycle arrest and inhibitors of apoptosis do not prevent reovirus-induced cell cycle arrest, suggesting that these are distinct pathways (Poggioli et al. 2000). The viral attachment protein $\sigma1$ must therefore be a determinant of reovirus-induced apoptosis.

In support of this, apoptosis can be induced at nonpermissive temperatures by a variety of reovirus temperature-sensitive (*ts*) mutants (Connolly and Dermody 2002), which are arrested at defined steps in viral replication and by ultraviolet (UV)-inactivated virus, all of which contain $\sigma1$ but lack $\sigma1s$ (Tyler et al. 1995).

In virions, the reovirus $\sigma1$ protein is a homotrimer comprised of an elongated fibrous tail, which inserts into the virion, and an externally facing globular head (Chappell et al. 2002). The heads of both reovirus T1L and T3D $\sigma1$ proteins contain a binding domain for junctional adhesion molecule (JAM), which serves as the primary reovirus receptor (Barton et al. 2001b). In addition, the fibrous tail of the T3 reovirus $\sigma1$ protein contains a domain that binds α-linked sialic acid (Chappell et al. 2000). Sialic acid binding is not required for viral entry. However, the binding of sialic acid in addition to JAM at early times after attachment can enhance both reovirus attachment and growth in some cell types (Connolly et al. 2001; Barton et al. 2001a). Virus binding to both JAM and sialic acid are required for reovirus-induced activation of nuclear factor κB (NF-κB) (see below) and apoptosis (Connolly et al. 2001).

2.2 Role of the M2 Gene Segment and Viral Disassembly in Reovirus-Induced Apoptosis

The M2 gene, which encodes the major viral outer capsid protein $\mu1/\mu1c$ also contributes to the apoptotic phenotype. In fact, incubation of infected cells with monoclonal antibodies (MAbs) directed against either the $\sigma1$ (viral attachment), $\mu1$, or $\sigma3$ (outer capsid) proteins can inhibit apoptosis. In the case of the $\sigma1$ MAbs this almost certainly reflects their capacity to inhibit viral cell attachment. However, both anti-$\mu1$ and anti-$\sigma3$ MAbs, which do not inhibit viral cell attachment but do prevent virion uncoating, can inhibit apoptosis (Virgin et al. 1994; Tyler et al. 1995, 1996; Connolly et al. 2001; Rodgers et al. 1997) .

After viral cell attachment and subsequent receptor-mediated endocytosis, reovirions are proteolytically disassembled to form infectious subvirion particles (ISVPs). This process is characterized by removal of the outer capsid protein $\sigma3$, proteolytic cleavage of $\mu1/\mu1C$, and conformational changes in $\sigma1$. Reovirus-induced apoptosis (and activation of NF-κB, see below) is blocked by inhibiting proteolysis of reovirus virions with ammonium chloride, which inhibits endosomal acidification, or E64, which inhibits cysteine-containing endocytic proteases, indicating that viral disassembly is required for apoptosis in infected cells (Connolly and Dermody 2002). Early events during viral entry, but subsequent to virus engagement of cellular receptors, thus appear to be required for reovirus-induced apoptosis, an interpretation that has subsequently been supported by experiments using *ts* mutants blocked at different stages in the reovirus replication cycle (Connolly and Dermody 2002).

3
Reovirus-Induced Apoptosis Is Mediated by Death Receptor Signaling

Tumor necrosis factor (TNF)-related apoptosis-inducing ligand (TRAIL) is a widely expressed type 2 membrane protein that was identified by its homology to Fas ligand (FasL) and TNFα. TRAIL induces apoptosis by binding to the cell surface death receptors (DRs) DR4 (also called TRAIL-R1) and DR5 (also called Apo2, TRAIL-R2, TRACK2, or KILLER). In addition, TRAIL can bind to the decoy receptors DcR-1 (for decoy receptor 1) and DcR-2, which do not transduce apoptotic signals and which can prevent the induction of apoptosis in TRAIL-treated cells. TRAIL-mediated activation of DR4 and DR5 induces DR oligomerization and the close association of their cytoplasmic death domains (DDs). The cytosolic adapter molecule FADD (for Fas-associated death domain) and pro-caspases 8 and 10 are then recruited to the receptor to form the death-inducing signaling complex (DISC), where pro-caspase cleavage generates active initiator caspases, 8 and 10.

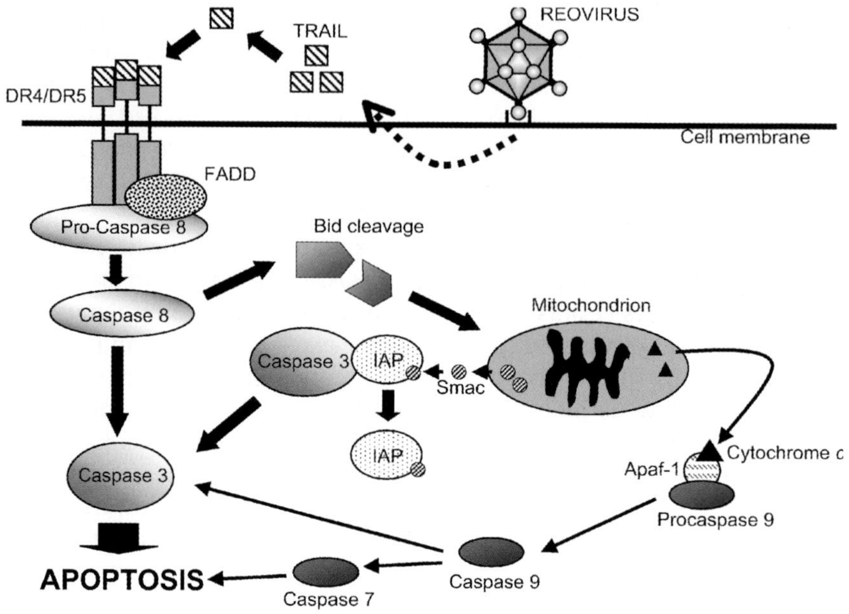

Fig. 1 Model of reovirus-induced apoptosis. After virus infection, TRAIL is released into the supernatant, by an unknown mechanism, and initiates apoptosis by binding to cell surface death receptors DR4 and DR5. TRAIL-receptor binding induces the activation of caspase 8, which in turn activates the downstream effector caspase, caspase 3. Caspase 8 also activates mitochondrial apoptotic pathways in reovirus-infected cells through the cleavage of Bid, resulting in the release of mitochondrial apoptotic factors and the amplification of death receptor apoptotic signaling. The mitochondrial release of smac, rather than cytochrome *c*, is thought to have a more significant role in reovirus induced apoptosis. (With permission from Clarke and Tyler 2003)

Soluble TRAIL receptors, anti-TRAIL antibodies, and the overexpression of DcR-1 can all inhibit reovirus-induced apoptosis in a wide variety of cells including L929 cells, human embryonic kidney (HEK293) cells, and several human cancer cell lines (Clarke et al. 2000, 2001a). This suggests that reovirus-induced apoptosis is mediated by TRAIL signaling pathways. Reovirus-infected cells release soluble TRAIL into the supernate, providing a potential mechanism for initiating both autocrine and exocrine (bystander) apoptosis (Clarke et al. 2000).

Events downstream of TRAIL receptor binding are also activated after reovirus infection. Reovirus infection thus induces the cleavage and activation of pro-caspase 8. In HEK293 cells, the activation of caspase 8 occurs in two phases. An initial phase occurs between 8 and 14 h postinfection (PI) and a later phase occurs between 24 and 34 h PI, by which time all pro-caspase 8 has been cleaved in infected cells (Kominsky et al. 2002a). Expression of dominant-negative (DN)-FADD or a peptide inhibitor of caspase 8 activity, IETD-FMK results in the inhibition of apoptosis in reovirus-infected cells, indicating that caspase 8 activity is required for reovirus-induced apoptosis (Clarke et al. 2000; Kominsky et al. 2002a). Caspase 8 activation is also required for the activation of caspase 3 in reovirus-infected cells (Kominsky et al. 2002a). Like caspase 8 activation, reovirus-induced activation of caspase 3 is biphasic, with a first phase of activation beginning around 8 h PI and a second phase of activation beginning at 24 h PI. The kinetics of caspase 8 and caspase 3 activation suggest that these two events occur in rapid succession in reovirus-infected cells. Reovirus-induced caspase 3 activity corresponds closely to the cleavage of the cellular substrate PARP, indicating that caspase 3 has biological activity in infected cells and can participate in the cleavage of cellular substrates to induce the morphological hallmarks of apoptosis (Connolly and Dermody 2002; Clarke et al. 2001a; Kominsky et al. 2002a). A model of reovirus-induced, TRAIL-mediated apoptosis is shown in Fig. 1.

4
Mitochondrial Signaling Contributes to Reovirus-Induced Apoptosis

The first indication that mitochondrial signaling pathways are involved in reovirus-induced apoptosis came from the observation that reovirus-induced apoptosis is inhibited in MDCK cells that overexpress Bcl-2 (Rodgers et al. 1997). Bcl-2 belongs to a family of proteins that contain both proapoptotic (e.g., Bax and Bak) and antiapoptotic (e.g., Bcl-2 and Bcl-xL) members. During apoptosis the interactions of pro- and anti-apoptotic Bcl-2 family proteins are disrupted, resulting in the oligomerization of Bax and Bak, the formation of a pore within the mitochondrial membrane, and the release of proapoptotic mitochondrial factors, including cytochrome *c* and a second mitochondrion-derived activator of caspases (smac; also called DIABLO). Cytochrome *c* contributes to apoptosis by activating the initiator caspase 9, which functions to activate the effector caspases. Smac contributes to apoptosis by binding to cellular inhibitor of apoptosis proteins (IAPs), thereby preventing their inhibitory actions on caspase activity (Verhagen et al. 2000).

4.1 Reovirus Induces the Cleavage of Bid

In some cell types, DR signaling activates mitochondrial apoptotic pathways through the caspase 8 (or 10)-dependent cleavage of the BH3-only Bcl-2 family protein Bid. This produces a C-terminal fragment of Bid, truncated (t) Bid, which translocates to the mitochondria and interferes with Bcl-2 protein interactions resulting in the release of cytochrome *c* and smac. Bid is cleaved in HEK293 cells after infection with reovirus (Kominsky et al. 2002a). Bid cleavage coincides with the release of cytochrome *c* from the mitochondria and the associated activation of caspase 9 (Kominsky et al. 2002a). Furthermore, both Bid cleavage and the release of cytochrome *c* are blocked in cells expressing DN-FADD and, like caspase 8 activation, Bid cleavage is biphasic. These results indicate that caspase 8 activity is required for activation of mitochondrial apoptotic pathways in reovirus-infected cells (Kominsky et al. 2002a).

4.2 Role of Smac in Reovirus-Induced Apoptosis

Smac is also released from the mitochondria of reovirus-infected cells with kinetics similar to that of the release of cytochrome *c* and, like cytochrome *c*, smac release is blocked by the overexpression of Bcl-2 (Kominsky et al. 2002a). After reovirus infection ,some cellular IAPs (survivin, cIAP1, and XIAP) undergo proteolytic cleavage and degradation (Kominsky et al. 2002b). This is prevented in cells overexpressing Bcl-2 (Kominsky et al. 2002b) and is consistent with a key role for smac in inhibiting IAP-mediated caspase inhibition (Verhagen et al. 2000). It is thought that this may be mediated through direct binding of smac with target IAPs (Verhagen et al. 2000; Chai et al. 2001; Huang et al. 2001; Riedl et al. 2001). Alternatively, IAP degradation has also been shown to represent an important apoptotic event both in mammalian cells and in *Drosophila*, where proteins with regions homologous to smac (Grim and REAPER) promote ubiquitin-mediated degradation of *Drosophila* IAP1 (Johnson et al. 2000; Deveraux et al. 1999; Yang et al. 2000; Palaga and Osborne 2002).

Smac release, rather than caspase 9 activation, plays a critical role in the mitochondrion-related augmentation of reovirus-induced, DR-mediated, apoptotic pathways (Kominsky et al. 2002b). Thus stable transfection of HEK293 cells with DN-caspase 9 (caspase 9b) only inhibits reovirus-induced activation of caspase 9, unlike Bcl-2 overexpression, which blocks all mitochondrially mediated events. In addition, caspase 9b expression does not affect reovirus-induced activation of caspase 3 or reovirus-induced PARP cleavage, suggesting that, although caspase 9 is activated in reovirus-infected cells, other pathways are necessary for effector caspase activation (Kominsky et al. 2002b).

A variety of other mitochondrial apoptotic factors have been identified, including apoptosis-inducing factor (AIF). Reovirus infection is not associated with release of AIF in HEK293 cells (Kominsky et al. 2002b). Nor does it result in disruption of mitochondrial transmembrane potential, indicating that the release of proapoptotic factors from the mitochondria after reovirus infection is selective (Kominsky et al. 2002b). This is consistent with ultrastructural studies in reovirus-infected cells that suggest that disruption of mitochondrial architecture is not a typical feature of reovirus infection.

4.3 Mitochondrial Pathways Amplify Death Receptor Apoptotic Signaling Following Reovirus Infection

As previously described, activation of both caspase 8 and Bid is biphasic (Kominsky et al. 2002a). In contrast, reovirus-induced release of smac is not biphasic, occurring just after the early phase and before the late phase of caspase 8 activation and Bid cleavage (Kominsky et al. 2002b). Overexpression of Bcl-2 inhibits the late phase of caspase 8 activation without affecting the early phase, suggesting that the late phase is mitochondrion dependent (Kominsky et al. 2002a). Together, these results are consistent with a model in which DR activation initiates reovirus apoptosis and results in early low-level activation of effector caspases. Mitochondrial events, likely initiated by Bid translocation and involving release of smac, then amplify the initial DR-initiated signal and dramatically augment effector caspase activation (Fig. 1).

In addition to caspase 3, the effector caspase 7 is also activated after reovirus infection (Kominsky et al. 2002a). This activation occurs later than the first phase of caspase 3 activity and is less robust, suggesting that caspase 7 activation may play a less critical role in reovirus-induced apoptosis than caspase 3. In addition, the observations that caspase 7 activation parallels that of caspase 9 and is not biphasic suggest that it may result from the activation of caspase 9.

5
Role of NF-κB in Reovirus-Induced Apoptosis

Nuclear factor κB (NF-κB) is a transcription factor that is normally prevented from migrating to the nucleus and binding to DNA by its association in the cytoplasm with members of the inhibitor κB (IκB) family of proteins. Site-specific phosphorylation, followed by ubiquitination and proteosomal degradation of IκB, allows for NF-κB activation. Reovirus infection transiently activates NF-κB in a variety of cell types, including L929, MDCK, and HeLa cells (Connolly et al. 2000). This activation can be detected in HeLa cells as early as 4 h PI, peaks at 10 h PI, and then declines (Connolly et al. 2000). Similarly, expression of an NF-κB-dependent luciferase reporter gene is transient in reovirus-infected cells (Connolly et al. 2000). Inhibition of NF-κB by stable overexpression of an IκB super-repressor or treatment of cells with a proteasome inhibitor that blocks IκB degradation (Z-L_sVS) inhibits reovirus-induced apoptosis (Connolly et al. 2000). Apoptosis is also inhibited in immortalized mouse embryo fibroblasts (MEFs) with targeted disruptions in the genes encoding the p50 or p65 subunits of NF-κB (Connolly et al. 2000). These results suggest, in contradistinction to many other models of apoptosis, that after reovirus infection early activation NF-κB exerts a pro- rather than antiapoptotic influence.

The regulation of TRAIL and DR expression is upregulated by NF-κB in a variety of systems, including cells undergoing apoptosis induced by human T cell lymphoma virus (HTLV-1) Tax and the chemotherapeutic agents etoposide and doxorubicin (Ravi et al. 2001; Gibson et al. 2000; Spalding et al. 2002; Rivera-Walsh et al. 2001). Studies are now under way to determine the role of NF-κB in mediating TRAIL and DR expression during reovirus-induced apoptosis.

Several lines of evidence suggest that activation of NF-κB does not completely explain the involvement of NF-κB in reovirus-induced apoptosis. First, NF-κB activation is transient and occurs before the onset of apoptosis in reovirus-infected cells (Connolly et al. 2000; Clarke et al. 2003b). Second, our preliminary experiments indicate that both T1 and T3 reoviruses activate NF-κB to a similar extent in HeLa and HEK293 cells, although T3 reoviruses induce significantly more apoptosis, suggesting that NF-κB is required, but not sufficient, for apoptosis in these cells. Third, although activation of NF-κB may be required for apoptosis in both HeLa and HEK293 cells the difference in magnitude of the response is dramatic, whereas apoptosis occurs with similar efficiency. Thus weak NF-κB activation can be detected in HEK293 cells 2–4 h after infection with T3A reovirus (Clarke et al. 2003b), whereas robust NF-κB activation is detected in HeLa cells 2–12 h PI (Connolly et al. 2000).

In addition to activating NF-κB at early times PI, reovirus has now been shown to induce a second phase of NF-κB regulation where NF-κB activity is inhibited in reovirus-infected cells at later times PI (Clarke et al. 2003a). This phase of regulation results in the transient nature of NF-κB activation in reovirus-infected cells and also inhibits stimulus-induced activation of NF-κB (Clarke et al. 2003a). Reovirus-induced inhibition of NF-κB activation in HEK293 cells is inhibited by the viral RNA synthesis inhibitor ribavirin, which is also required for efficient apoptosis in these cells and for the ability of reovirus to sensitize these cells to TRAIL-induced apoptosis (Clarke et al. 2003a). Because DR receptor-induced apoptosis is enhanced in many systems if NF-κB signaling is inhibited and because reovirus-induced apoptosis is mediated by DR signaling, it is likely that the inhibition of NF-κB at later times PI is necessary for efficient reovirus-induced apoptosis and for reovirus-induced apoptosis in TRAIL-resistant cells (Clarke et al. 2003a).

The mechanism by which NF-κB is regulated by reovirus is not fully understood. One possible activator of NF-κB in reovirus-infected cells is the calcium-dependent, papain-like, neutral cysteine protease calpain. Calpain is activated as early as 2 h after reovirus infection of L929 cells and myocardiocytes and inhibition of this activation inhibits reovirus-induced apoptosis (DeBiasi et al. 1999, 2001). In most models of apoptosis calpains act upstream of caspases, and the early onset of calpain activity in reovirus-infected cells suggests that this may also be true for reovirus-induced apoptosis. However, calpain has also been implicated in the regulation of a variety of cellular transcription factors, including NF-κB (Chen et al. 1997; Watt and Molloy 1993). The early activation of calpain after reovirus infection makes this an attractive candidate for reovirus-induced activation of NF-κB. In addition, reovirus induces the upregulation of the proapoptotic protein par-4 (DeBiasi et al. 2003), which can inhibit the phosphorylation and degradation of IκB, thereby preventing NF-κB activation (Camandola and Mattson 2000; Diaz-Meco et al. 1999). This results in the downmodulation of Bcl-2 and can result in sensitization of cells to TNF- and FasL-induced apoptosis (Diaz-Meco et al. 1999).

6
Reovirus-Induced Apoptosis Is Associated with Activation of JNK and the Transcription Factor c-Jun

Reovirus infection results in a viral strain-specific pattern activation of the c-Jun N-terminal kinase (JNK) and the JNK-associated transcription factor c-Jun (Clarke et al. 2001b). The capacity of reovirus strains to activate JNK correlates closely with their capacity to induce apoptosis (Clarke et al. 2001b). In addition, experiments using T1L × T3D reassortants indicate that the same viral gene segments that determine apoptosis induction (S1 and M2) are also key determinants of JNK activation (Clarke et al. 2001b). Furthermore, our preliminary experiments indicate that reovirus-induced apoptosis is inhibited in cells deficient in MEK kinase 1, an upstream activator of JNK in reovirus-infected cells and in cells treated with inhibitors of JNK activity. These results indicate that JNK is required for reovirus-induced apoptosis.

Our recent experiments also indicate that JNK is required for the efficient release of smac and cytochrome *c* from the mitochondria of reovirus-infected cells, suggesting that JNK promotes mitochondrial pathways of apoptosis in reovirus-infected cells (Clarke et al. 2004). Both JNK-induced phosphorylation of Bcl-2 family proteins and c-Jun-induced expression of the BH3-only protein Bim have previously been shown to promote mitochondrial apoptotic signaling. Experiments to determine the mechanism by which JNK and c-Jun influence reovirus-induced apoptosis are currently under way.

7
Reovirus-Induced Alteration in Expression of Genes with Potential Roles in Virus-Induced Apoptosis and Pathogenesis

Reovirus infection induces the activation of transcription factors NF-κB and c-Jun (Clarke et al. 2001b, 2003a; Connolly et al. 2000). This suggests that activation of specific cellular genes contributes to virus-induced cellular signaling, including apoptotic signaling, in infected cells. High-density oligonucleotide microarrays used to perform a global analysis of virus-induced cellular gene expression after reovirus infection of HEK293 cells (DeBiasi et al. 2003; Poggioli et al. 2002) showed that the expression of 24 genes related to apoptosis were altered in cells infected with the apoptosis-inducing reovirus strain T3A (Table 1). These genes encode proteins with potential roles in DR, endoplasmic reticulum

Table 1 Reovirus-induced alteration in expression of genes encoding proteins with known apoptotic involvement

Gene	GenBank accession no.[a]	Change in expression (n-fold)[b]	
		T3A	T1L
Mitochondrial signaling			
Pim-2 proto-oncogene homolog	U77735	−2.2±0.1	
Mcl-1	L08246	2.0±0.0	2.2±0.0
BAC 15E1-cytochrome *c* oxidase polypeptide	AL021546	2.1±0.0	
Par-4	U63809	2.1±0.0	
HSP-70 (heat shock protein 70 testis variant)	D85730	2.2±0.1	
BNIP-1 (Bcl-2 interacting protein)	U15172	2.3±0.2	
SMN/Btfp44/NAIP (survival motor neuron/ neuronal apoptosis inhibitor protein)	U80017	2.5±0.1	
DRAK-2	AB011421	2.8±0.2	
SIP-1	AF027150	3.0±0.2	
DP5	D83699	5.5±1.1	
Death receptor signaling			
Bcl-10	AJ006288	5.6±1.1	
PML-2	M79463	3.4±0.3	
Ceramide glucosyltransferase	D50840	4.0±1.2	
Sp 100	M60618	6.1±0.5	
ER stress-induced signaling			
ORP150	U65785	−2.4±0.2	
GADD 34	U83981	3.7±0.2	2.9±0.2
GADD 45	M60974	4.9±0.1	4.4±0.1
Proteases			
Calpain	X04366	−2.6±0.1	
Beta-4 adducin	U43959	−2.1±0.1	
Caspase 7	U67319	2.6±0.2	
Caspase 3	U13737	3.2±0.2	2.8±0.1
Undefined			
Frizzled-related protein	AF056087	−2.5±0.1	−3.3±0.5
TCBP (T cluster binding protein)	D64015	3.3±0.2	
Cug-BP/hAb50 (RNA binding protein)	U63289	6.6±1.1	

[a] GenBank accession number corresponds to the sequence from which the Affymetrix microarray U95A probe set was designed.
[b] Data are means ± standard errors of the means.

stress, and mitochondrial apoptotic signaling and cysteine proteases (caspases and calpains) (DeBiasi et al. 2003). Only five of these genes were also differentially expressed after T1L (weakly apoptotic) infection, emphasizing their potential importance in reovirus-induced apoptosis.

To date the best-characterized example of a potentially apoptosis-inducing gene identified by microarray analysis to be differentially expressed after reovirus infection is the survival motor neuron (SMN) gene. This gene was found by polymerase chain reaction (PCR) analysis to be upregulated at the transcriptional level in reovirus-infected HEK293 cells and at the translational level in the hearts of reovirus-infected baby mice (DeBiasi et al. 2003). The SMN protein has been shown to interact with Bcl-2, conferring a synergistic protective effect against Bax-induced or Fas-mediated apoptosis that has been shown to underlie the pathogenesis of spinal muscular atrophy (Sato et al. 2000; Iwahashi et al. 1997).

Changes in the expression of genes encoding proteins known to be involved in DNA repair and cell cycle regulation were also identified in this study and may also affect virus-induced pathogenesis (DeBiasi et al. 2003).

8 Reovirus Sensitizes Cells to TRAIL-Induced Apoptosis

In addition to inducing TRAIL-mediated apoptosis, reovirus infection also sensitizes cells to TRAIL-induced apoptosis by a mechanism that results in an increase in the activation of caspases 8 and 3 and is blocked by the caspase 8 inhibitor IETD-FMK (Clarke et al. 2000, 2001a). Reovirus infection and TRAIL treatment have synergistic rather than merely additive effects on apoptosis, and infection can confer TRAIL sensitivity to previously TRAIL-resistant cells as well as increasing the TRAIL sensitivity of partially resistant lines (Clarke et al. 2000, 2001a). This finding may increase the potential utility of TRAIL as an agent for cancer therapy, which is currently limited by the fact that cancer cells of all types differ in sensitivity to TRAIL-induced apoptosis.

The ability of reovirus to sensitize cells to TRAIL does not appear to reflect an increase in the expression of TRAIL receptors as assayed in several human cancer cell lines (Clarke et al. 2001a) and may instead be the result of inhibition of TRAIL-induced activation of NF-κB in reovirus-infected cells (Clarke et al. 2003a).

The ability of reovirus to sensitize cells to TRAIL also suggests that reovirus-infected cells in vivo are also susceptible to killing through the TRAIL pathway by immune cells such as natural killer and $CD4^+$ cells that bear membrane-bound TRAIL.

9 Reovirus-Induced Apoptosis in the Mouse CNS

T3 reovirus strains infect neurons within specific regions of neonatal mouse brains, producing a lethal meningoencephalitis. Viral antigen and pathology colocalize in the brain and have a predilection for the cortex, hippocampus and thalamus (Fig. 2). T3 reovirus infection also induces apoptosis in the brains of newborn mice (Oberhaus et al. 1997; Richardson-Burns et al. 2002). Thus fragmentation of DNA into oligonucleosomal-length ladders can be detected in tissue samples prepared from T3D- but not mock-infected brains at 8–9 days PI, which coincides with maximal viral growth (Oberhaus et al. 1997). The presence of apoptotic cells also correlates with areas of tissue injury and viral infection in T3-infected brain sections (Fig. 2) (Oberhaus et al. 1997; Richardson-

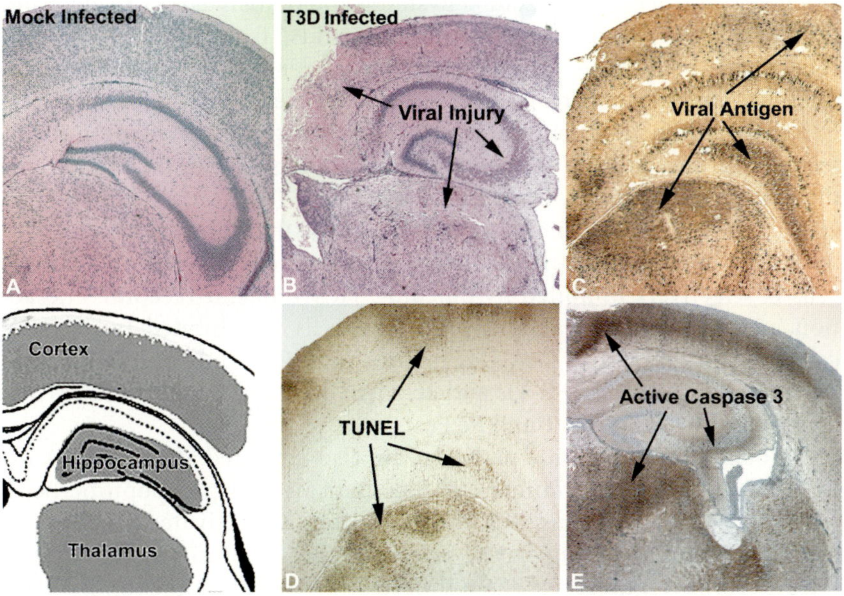

Fig. 2 Coronal sections of neonatal mouse brain 7 days after intracranial inoculation of 10,000 plaque forming units (PFU) of reovirus strain T3D or mock inoculation. Hematoxylin and eosin-stained tissue reveals marked destruction of brain tissue in the T3D-infected brain (**B**) as compared to the uninfected brain (**A**). By immunohistochemistry, serial sections of T3D-infected brain tissue were stained for viral antigen (**C**), TUNEL/apoptosis marker (**D**), and active caspase 3 (**E**). Staining for viral antigen, TUNEL, and caspase 3 were undetectable in the mock-infected brains (data not shown). (With permission from Richardson-Burns et al. 2002)

Burns et al. 2002). Most cells in infected brain regions are both TUNEL (TdT-mediated dUTP nick-end labeling)-positive (apoptotic) and reovirus antigen-positive (infected). However, there are cells in these regions that are apoptotic but antigen negative, suggesting that apoptosis occurs both as a result of direct viral infection and in uninfected "bystander" cells (Oberhaus et al. 1997). Reovirus infection in a mouse neuroblastoma-derived cell line (NB41a3) and in primary mouse cortical cultures (MCC) derived from embryonic (E20) mice is also associated with increased levels of caspase 3 activity and is blocked with the caspase 3 inhibitor DEVD-FMK (Richardson-Burns et al. 2002). Studies of reovirus infection in neuronal cultures also provides further evidence of bystander apoptosis. In both MCC and NB4 cells dual labeling with immunocytochemistry and TUNEL showed that although a great majority of infected cells were undergoing apoptosis there was also a subset of apoptotic cells that were uninfected but located in proximity to virus-infected cells (Richardson-Burns et al. 2002). Bystander apoptosis could result from the release of TRAIL, or other death ligands, from reovirus-infected cells. If this is the case, the amount of bystander apoptosis would reflect the sensitivity of the surrounding cells to the released ligand.

Reovirus infection also induces increased caspase 8 activation in infected neurons, indicating that neuronal apoptosis, like that in its epithelial cell counterparts, involves DR activation (Richardson-Burns et al. 2002). However, the ligand-receptor trigger for this activation appears to be less specific. Thus, whereas reovirus-induced apoptosis in HEK293 cells is selectively inhibited by blocking TRAIL ligand-receptor interaction, reovirus-induced apoptosis in NB4 cells is inhibited by treating cells with both soluble TRAIL receptors (Fc:DR5) and soluble TNF receptors (TNFR) (FcTNFR-1), and reovirus-induced apoptosis in MCCs is inhibited by Fc:TNFR-1 and Fc:FasL.

Mitochondrial apoptotic pathways are also activated after reovirus infection of neurons. Preliminary studies indicate that proapoptotic Bcl-2 family proteins, including Bid, Bax, and Bim, are activated in virus-infected neurons, resulting in the release of proapoptotic mitochondrial factors. However, there are again differences between mitochondrial signaling pathways activated after reovirus infection of neuronal and epithelial cells. In HEK293 cells, reovirus infection is associated with robust release of cytochrome *c* and smac and the subsequent activation of caspase 9 and inhibition of cellular IAPs. In neuronal cultures, however, our preliminary results indicate that reovirus infection results in the discordant release of smac and cytochrome *c*. Smac is released around 17 h PI and coincides with the cleavage of cellular IAPs. In contrast, cytochrome *c* release occurs only at low levels and at later times after infection, resulting in only low levels of activation of caspase 9 in these cells (Richardson-Burns et al. 2002). Consistent with these findings, the caspase 9 inhibitor Z-LEHD-FMK has little effect on reovirus-induced neuronal apoptosis, which is significantly inhibited by caspase 8 (Z-IETD-FMK), caspase 3 (Z-DEVD-FMK), or pan-caspase inhibitors.

10
Reovirus-Induced Apoptosis in the Heart

The T1L × T3D reassortant virus 8B efficiently produces myocarditis in infected neonatal mice. Similar to results seen in mouse brain, DNA extracted from the hearts of 8B-infected mice is fragmented into oligonucleosomal-length ladders, indicative of apoptosis (DeBiasi et al. 2001) and areas of TUNEL-positive cells in 8B-infected hearts correlates with areas of histological damage and reovirus antigen (DeBiasi et al. 2001). Injury to the heart following reovirus infection occurrs in the absence of an inflammatory response, also suggesting that it results from apoptotic cell death (DeBiasi et al. 2001).

Treatment of mice with the calpain inhibitor CX295 [dipeptide α-ketoamide calpain inhibitor z-Leu-aminobutyric acid-CONH(CH_2)-3-morpholine] is protective against reovirus-induced myocarditis and results in a dramatic reduction in histopathologic evidence of myocardial injury (Fig. 3), a reduction in serum creatine phosphokinase (an intracellular enzyme whose release into the serum is a quantitative marker of skeletal and cardiac muscle damage), and improved weight gain (DeBiasi et al. 2001).

Prevention of myocardial injury by apoptosis inhibitors is accompanied by a virtually complete inhibition of apoptotic myocardial cell death, strongly suggesting that virus-induced apoptosis is a key mechanism of cell death, tissue injury, and mortality in reovirus-infected mice and that inhibitors of apoptosis may prove useful in the treatment of virus-induced diseases (DeBiasi et al. 2001).

11
Apoptosis and Viral Growth

Early studies showed that there is little correlation in continuous nonneuronal cell lines between the efficiency with which reovirus strains

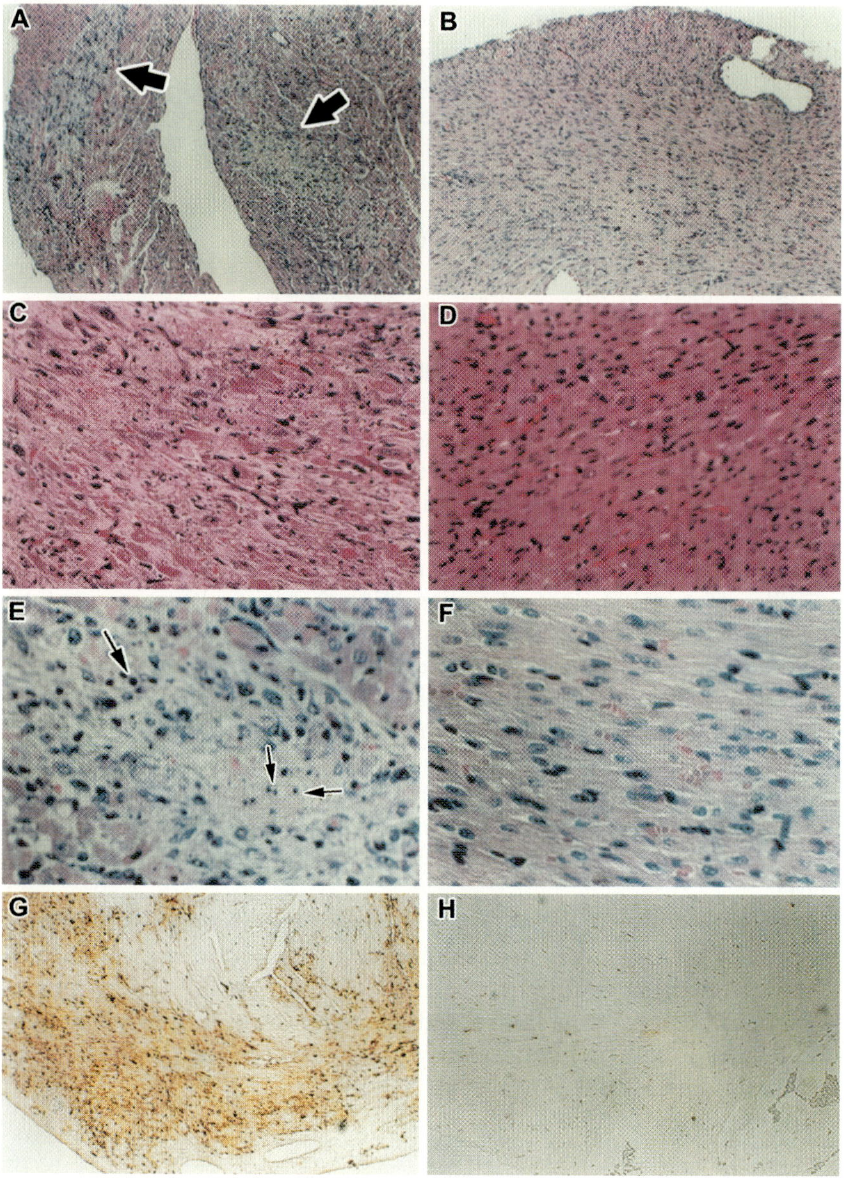

Fig. 3 Cardiac midsections from reovirus 8B-infected neonatal mice treated with the calpain inhibitor CX295 (**B**, **D**, **F**, and **H**) compared to those from inactive diluent control mice (**A**, **C**, **E**, and **G**) 7 days after intramuscular inoculation with 1,000 PFU of reovirus 8B. Hematoxylin and eosin-stained sections at an original magnification of ×25 reveal extensive focal areas of myocardial injury (*arrows*) in the control ani-

replicate and their capacity to induce apoptosis. For example, T1L and T3D grow to approximately equivalent titers in L929 fibroblasts, yet T3D induces significantly more apoptosis (Tyler et al. 1995). Similarly, T1L grows better than T3D in MDCK cells, yet again T3D induces significantly more apoptosis (Rodgers et al. 1997).

11.1 The Effects of Virus Replication and Growth on Apoptosis

In HeLa cells, viral RNA synthesis is not required for reovirus-induced apoptosis because the viral RNA synthesis inhibitor ribavirin does not prevent apoptosis (Connolly and Dermody 2002). In addition, particles lacking genomic dsRNA can induce apoptosis in HeLa cells, and *ts* reovirus mutants with mutations resulting in defects in outer capsid assembly (*ts*B352/L2 gene), and in dsRNA synthesis (*ts*D357/L1 gene, *ts*E320/S3 gene), are capable of inducing similar levels of apoptosis at both nonpermissive (39°C) and permissive (32°C) temperatures. Temperature-sensitive mutants with defects in viral core (*ts*C447/S2 gene) and outer capsid assembly (*ts*G453/S4 gene) also still induce apoptosis at 39°C, but only about half as efficiently as they do at 32°C (Connolly and Dermody 2002).

Because all these *ts* mutants undergo endosomal processing, their ability to induce apoptosis is consistent with a key role for endosomal vesicle-related events in apoptosis induction. However, the fact that assembly defects can influence the efficiency of this process suggests that additional viral growth-related factors are also involved. Furthermore, UV-inactivated virions, which are not transcriptionally active, were 100 times less apoptotic than their T3D derivative in L929 cells (Tyler et al. 1995) and ribavirin inhibits reovirus-induced apoptosis after T3A-infection of HEK293 cells (Clarke et al. 2003a). Together these results suggest

mal (**A**), which are absent in the CX295-treated animal (**B**), despite identical viral infections. Views at an original magnification of ×50 demonstrate minimal inflammatory cell infiltrate in the affected area (**C**), but myocardial architecture is dramatically disrupted, compared to that of a CX295-treated mouse (**D**). At an original magnification of ×100, nuclei with apoptotic morphology are easily seen in the control animal (**E**) as are cells with pyknotic nuclei (*long arrows*) as well as apoptotic bodies (*short arrows*). These characteristics are absent in the drug-treated mouse (**F**). TUNEL analysis of the control animal reveals extensive areas of positively staining cells in the same regions of injury (**G**) but no TUNEL-positive areas in the drug-treated mouse. (With permission from DeBiasi et al. 2001) that viral replication enhances, but is not absolutely required for, virus-induced apoptosis.

11.2 Effect of Apoptosis on Viral Growth

In several systems inhibition of apoptosis leads to a modest but reproducible decrease in viral titer in reovirus T3-infected cells. Thus a fivefold reduction in viral yield was observed in L929 cells treated with an inhibitor of calpain activity compared to untreated cells (DeBiasi et al. 1999). Similarly, in p50 or p65$^{-/-}$ immortalized fibroblasts, a two- to fivefold reduction in viral yield was seen, compared to wild-type cells (Connolly et al. 2000). Both of these treatments resulted in a significant inhibition of reovirus-induced apoptosis.

Treatment of mice with inhibitors of apoptosis also results in a reduction in viral growth. Ergo, after treatment with an inhibitor of calpain, a reduction of 0.5 $\log_{10}$ PFU/ml was observed at the site of primary replication primary (hindlimb), whereas a 0.7 $\log_{10}$ PFU/ml reduction was observed in the heart (DeBiasi et al. 2001). Our preliminary results indicate that chemical inhibition of caspase 3 produced similar effects on viral replication. However, in caspase 3-deficient mice, nearly a 2 log reduction was noted, compared to wild-type and heterozygous controls. As expected, treatment with these inhibitors and infection of caspase 3-deficient mice resulted in a marked decrease in apoptosis in infected tissues.

In a recent study, reovirus growth and apoptosis were determined after infection with reovirus variant K (VarK), which is an antigenic variant of T3D that has nearly a millionfold reduction in neurovirulence after intracerebral (IC) inoculation and a restricted pattern of CNS injury, with damage limited to the hippocampus (Richardson-Borns et al. 2004). It was found that VarK grew to similar titer as T3D in the hippocampus but had significantly lower titer in the cortex. Similarly, whereas the viruses grew to identical titers and infected the same percentage of cells in mouse primary hippocampal cultures (MHC), both the number of infected cells and the viral yield per infected cell were significantly lower for VarK than T3D in mouse primary cortical cultures (MCC). Like growth, VarK-induced apoptosis was limited to the hippocampus in vivo and to MHC in vitro. As expected (see above), growth of T3D in MCC was reduced to levels comparable to VarK after treatment of MCC with caspase inhibitors. However, also of note was the finding that induction of apoptosis in VarK-infected MCC with Fas-activating antibody significantly enhanced viral yield. These results thus suggest that the decreased growth and neurovirulence of VarK may be due to its failure to efficiently induce apoptosis in cortical neurons.

Together these results suggest that reovirus-induced apoptosis is required for maximal viral growth.

12
Conclusions and Future Directions

Significant discoveries have been made regarding the mechanisms of and the requirement of apoptosis following reovirus infection. However, much still remains to be determined. For example, we now know that many individual apoptotic signaling pathways play a role in reovirus-induced apoptosis but our understanding of the regulation of these pathways and the way in which these pathways interact is much less clear.

In addition, some apoptotic signaling events are cell type specific, making the confirmation of apoptotic signaling pathways in primary cells and infected tissues critical for our evaluation both of the role of reovirus-induced apoptosis in vivo and of the use of apoptosis inhibitors as a novel strategy for limiting virus-induced tissue damage.

Acknowledgements This work was supported by Merit and REAP grants from the Department of Veterans Affairs (KLT), RO1NS050138 from the NIH/NINDS (KLT), the Reuler-Lewin Family Professorship of Neurology (KLT), the Department of Defence/ US Army Medical Research and Material Command (DAMD 17-98-8614) (KLT) and the Ovarian Cancer Research Fund (PC).

References

Barton ES, Connolly JL, Forrest JC, Chappell JD, Dermody TS (2001a) Utilization of sialic acid as a coreceptor enhances reovirus attachment by multistep adhesion strengthening. J Biol Chem 276:2200–2211

Barton ES, Forrest JC, Connolly JL, Chappell JD, Liu Y, Schnell FJ, Nusrat A, Parkos CA, Dermody TS (2001b) Junction adhesion molecule is a receptor for reovirus. Cell 104:441–451

Camandola S, Mattson MP (2000) Pro-apoptotic action of PAR-4 involves inhibition of NF-kappaB activity and suppression of BCL-2 expression. J Neurosci Res 61:134–139

Chai J, Shiozaki E, Srinivasula SM, Wu Q, Datta P, Alnemri ES, Shi Y, Dataa P (2001) Structural basis of caspase-7 inhibition by XIAP. Cell 104:769–780

Chappell JD, Duong JL, Wright BW, Dermody TS (2000) Identification of carbohydrate-binding domains in the attachment proteins of type 1 and type 3 reoviruses. J Virol 74:8472–8479
Chappell JD, Prota AE, Dermody TS, Stehle T (2002) Crystal structure of reovirus attachment protein sigma1 reveals evolutionary relationship to adenovirus fiber. EMBO J 21:1–11
Chen F, Lu Y, Kuhn DC, Maki M, Shi X, Sun SC, Demers LM (1997) Calpain contributes to silica-induced I kappa B-alpha degradation and nuclear factor-kappa B activation. Arch Biochem Biophys 342:383–388
Clarke P, Meintzer SM, Gibson S, Widmann C, Garrington TP, Johnson GL, Tyler KL (2000) Reovirus-induced apoptosis is mediated by TRAIL. J Virol 74:8135–8139
Clarke P, Meintzer SM, Moffitt LA, Tyler KL (2003a) Two distinct phases of virus-induced nuclear factor kappa B regulation enhance tumor necrosis factor-related apoptosis-inducing ligand-mediated apoptosis in virus-infected cells. J Biol Chem 278:18092–18100
Clarke P, Meintzer SM, Spalding AC, Johnson GL, Tyler KL (2001a) Caspase 8-dependent sensitization of cancer cells to TRAIL-induced apoptosis following reovirus-infection. Oncogene 20:6910–6919
Clarke P, Meintzer SM, Wang Y, Moffitt LA, Richardson-Burns SM, Johnson GL, Tyler KL (2004) J Virol 78:13132–13138
Clarke P, Meintzer SM, Widmann C, Johnson GL, Tyler KL (2001b) Reovirus infection activates JNK and the JNK-dependent transcription factor c-Jun. J Virol 75:11275–11283
Clarke P, Tyler KL (2003) Reovirus-induced apoptosis: A minireview. Apoptosis 8:141–150
Connolly JL, Barton ES, Dermody TS (2001) Reovirus binding to cell surface sialic acid potentiates virus-induced apoptosis. J Virol 75:4029–4039
Connolly JL, Dermody TS (2002) Virion disassembly is required for apoptosis induced by reovirus. J Virol 76:1632–1641
Connolly JL, Rodgers SE, Clarke P, Ballard DW, Kerr LD, Tyler KL, Dermody TS (2000) Reovirus-induced apoptosis requires activation of transcription factor NF-kappaB. J Virol 74:2981–2989
DeBiasi RL, Clarke P, Meintzer S, Jotte R, Kleinschmidt-Demasters BK, Johnson GL, Tyler KL (2003) Reovirus-induced alteration in expression of apoptosis and DNA repair genes with potential roles in viral pathogenesis. J Virol 77:8934–8947
DeBiasi RL, Edelstein CL, Sherry B, Tyler KL (2001) Calpain inhibition protects against virus-induced apoptotic myocardial injury. J Virol 75:351–361
DeBiasi RL, Squier MK, Pike B, Wynes M, Dermody TS, Cohen JJ, Tyler KL (1999) Reovirus-induced apoptosis is preceded by increased cellular calpain activity and is blocked by calpain inhibitors. J Virol 73:695–701
Deveraux QL, Leo E, Stennicke HR, Welsh K, Salvesen GS, Reed JC (1999) Cleavage of human inhibitor of apoptosis protein XIAP results in fragments with distinct specificities for caspases. EMBO J 18:5242–5251
Diaz-Meco MT, Lallena MJ, Monjas A, Frutos S, Moscat J (1999) Inactivation of the inhibitory kappaB protein kinase/nuclear factor kappaB pathway by Par-4 expression potentiates tumor necrosis factor alpha-induced apoptosis. J Biol Chem 274:19606–19612

Gibson SB, Oyer R, Spalding AC, Anderson SM, Johnson GL (2000) Increased expression of death receptors 4 and 5 synergizes the apoptosis response to combined treatment with etoposide and TRAIL. Mol Cell Biol 20:205–212
Huang Y, Park YC, Rich RL, Segal D, Myszka DG, Wu H (2001) Structural basis of caspase inhibition by XIAP: differential roles of the linker versus the BIR domain. Cell 104:781–790
Iwahashi H, Eguchi Y, Yasuhara N, Hanafusa T, Matsuzawa Y, Tsujimoto Y (1997) Synergistic anti-apoptotic activity between Bcl-2 and SMN implicated in spinal muscular atrophy. Nature 390:413–417
Johnson DE, Gastman BR, Wieckowski E, Wang GQ, Amoscato A, Delach SM, Rabinowich H (2000) Inhibitor of apoptosis protein hILP undergoes caspase-mediated cleavage during T lymphocyte apoptosis. Cancer Res 60:1818–1823
Kominsky DJ, Bickel RJ, Tyler KL (2002a) Reovirus-induced apoptosis requires both death receptor- and mitochondrial-mediated caspase-dependent pathways of cell death. Cell Death Differ 9:926–933
Kominsky DJ, Bickel RJ, Tyler KL (2002b) Reovirus-induced apoptosis requires mitochondrial release of Smac/DIABLO and involves reduction of cellular inhibitor of apoptosis protein levels. J Virol 76:11414–11424
Oberhaus SM, Smith RL, Clayton GH, Dermody TS, Tyler KL (1997) Reovirus infection and tissue injury in the mouse central nervous system are associated with apoptosis. J Virol 71:2100–2106
Palaga T, Osborne B (2002) The 3D's of apoptosis: death, degradation and DIAPs. Nat Cell Biol 4:E149-E151
Poggioli GJ, DeBiasi RL, Bickel R, Jotte R, Spalding A, Johnson GL, Tyler KL (2002) Reovirus-induced alterations in gene expression related to cell cycle regulation. J Virol 76:2585–2594
Poggioli GJ, Dermody TS, Tyler KL (2001) Reovirus-induced sigma1s-dependent G_2/M phase cell cycle arrest is associated with inhibition of p34 (cdc2). J Virol 75:7429–7434
Poggioli GJ, Keefer C, Connolly JL, Dermody TS, Tyler KL (2000) Reovirus-induced G_2/M cell cycle arrest requires sigma1s and occurs in the absence of apoptosis. J Virol 74:9562–9570
Ravi R, Bedi GC, Engstrom LW, Zeng Q, Mookerjee B, Gelinas C, Fuchs EJ, Bedi A (2001) Regulation of death receptor expression and TRAIL/Apo2L-induced apoptosis by NF-kappaB. Nat Cell Biol 3:409–416
Richardson-Burns SM, Kominsky DJ, Tyler KL (2002) Reovirus-induced neuronal apoptosis is mediated by caspase 3 and is associated with the activation of death receptors. J Neurovirol 8:365–380
Richardson-Burns SM, Tyler KL (2004) Regional differences in viral growth and central nervous system injury correlate with apoptosis. J Virol 78:5466–5475
Riedl SJ, Renatus M, Schwarzenbacher R, Zhou Q, Sun C, Fesik SW, Liddington RC, Salvesen GS (2001) Structural basis for the inhibition of caspase-3 by XIAP. Cell 104:791–800
Rivera-Walsh I, Waterfield M, Xiao G, Fong A, Sun SC (2001) NF-kappaB signaling pathway governs TRAIL gene expression and human T-cell leukemia virus-I Tax-induced T-cell death. J Biol Chem 276:40385–40388

Rodgers SE, Barton ES, Oberhaus SM, Pike B, Gibson CA, Tyler KL, Dermody TS (1997) Reovirus-induced apoptosis of MDCK cells is not linked to viral yield and is blocked by Bcl-2. J Virol 71:2540–2546

Sato K, Eguchi Y, Kodama TS, Tsujimoto Y (2000) Regions essential for the interaction between Bcl-2 and SMN, the spinal muscular atrophy disease gene product. Cell Death Differ 7:374–383

Spalding AC, Jotte RM, Scheinman RI, Geraci MW, Clarke P, Tyler KL, Johnson GL (2002) TRAIL and inhibitors of apoptosis are opposing determinants for NF-kappaB-dependent, genotoxin-induced apoptosis of cancer cells. Oncogene 21:260–271

Tyler KL (1998) Pathogenesis of reovirus infections of the central nervous system. Curr Top Microbiol Immunol 233 Reovir.ii:93–124

Tyler KL, Clarke P, DeBiasi RL, Kominsky D, Poggioli GJ (2001) Reoviruses and the host cell. Trends Microbiol 9:560–564

Tyler KL, Squier MK, Brown AL, Pike B, Willis D, Oberhaus SM, Dermody TS, Cohen JJ (1996) Linkage between reovirus-induced apoptosis and inhibition of cellular DNA synthesis: role of the S1 and M2 genes. J Virol 70:7984–7991

Tyler KL, Squier MK, Rodgers SE, Schneider BE, Oberhaus SM, Grdina TA, Cohen JJ, Dermody TS (1995) Differences in the capacity of reovirus strains to induce apoptosis are determined by the viral attachment protein sigma 1. J Virol 69:6972–6979

Verhagen AM, Ekert PG, Pakusch M, Silke J, Connolly LM, Reid GE, Moritz RL, Simpson RJ, Vaux DL (2000) Identification of DIABLO, a mammalian protein that promotes apoptosis by binding to and antagonizing IAP proteins. Cell 102:43–53

Virgin HW, Mann MA, Tyler KL (1994) Protective antibodies inhibit reovirus internalization and uncoating by intracellular proteases. J Virol 68:6719–6729

Watt F, Molloy PL (1993) Specific cleavage of transcription factors by the thiol protease, m-calpain. Nucleic Acids Res 21:5092–5100

Yang Y, Fang S, Jensen JP, Weissman AM, Ashwell JD (2000) Ubiquitin protein ligase activity of IAPs and their degradation in proteasomes in response to apoptotic stimuli. Science 288:874–877

CTMI (2005) 289:25–56

Poliovirus, Pathogenesis of Poliomyelitis, and Apoptosis

B. Blondel[1] (✉) · F. Colbère-Garapin[1] · T. Couderc[2] · A. Wirotius[1] · F. Guivel-Benhassine[3]

[1] Laboratoire des Virus Entérotropes et Stratégies Antivirales, Institut Pasteur, 75724 Paris Cedex 15, France
bblondel@pasteur.fr
[2] Unité Postulante de Neuroimmunologie Virale, Institut Pasteur, 75724 Paris Cedex 15, France
[3] Groupe Virus et Immunité, Institut Pasteur, 75724 Paris Cedex 15, France

Abstract Poliovirus (PV) is the causal agent of paralytic poliomyelitis, an acute disease of the central nervous system (CNS) resulting in flaccid paralysis. The development of new animal and cell models has allowed the key steps of the pathogenesis of poliomyelitis to be investigated at the molecular level. In particular, it has been shown that PV-induced apoptosis is an important component of the tissue injury in the CNS of infected mice, which leads to paralysis. In this review the molecular biology of PV and the pathogenesis of poliomyelitis are briefly described, and then several models of PV-induced apoptosis are considered; the role of the cellular receptor of PV, CD155, in the modulation of apoptosis is also addressed.

Poliovirus (PV) is the causal agent of paralytic poliomyelitis, an acute disease of the central nervous system (CNS) resulting in flaccid paralysis. In addition, another neuromuscular pathology, called the post-polio syndrome, affects some poliomyelitis survivors several decades after the most severe forms of the acute disease (Dalakas 1995). A killed vaccine and an oral live attenuated vaccine were both developed in the 1950s (Salk 1955; Sabin and Boulger 1973), and subsequent massive vaccination campaigns resulted in near-total eradication of wild-type PV from industrialized countries. However, wild strains are currently still endemic in a few countries in South-East Asia and Africa. Moreover, oral polio vaccine strains are genetically unstable in vaccinees, who excrete neurovirulent vaccine-derived PV (VDPV) mutants. Consequently, there are two problems complicating the eradication of poliomyelitis, both due to the emergence of VDPV mutants. First, in very rare cases, VDPV causes chronic infections in the gut of immunodeficient individuals, in particular agammaglobulinemics. Chronically infected patients can excrete neurovirulent VDPV for up to 22 years (Martin et al. 2000; MacLennan et al. 2004). Second, there have been recent poliomyelitis outbreaks due to VDPV strains in four regions of the world (Egypt, Hispaniola, the Philippines, and Madagascar) (Centers for Disease Control and Prevention 2001, 2002; Kew et al. 2002; Rousset et al. 2003; Yang et al. 2003). These iatrogenic epidemics reveal that these particular VDPV can circulate; therefore they have been named cVDPV. All of the cVDPV responsible for these epidemics have a recombinant genome including mutated vaccine PV sequences and unidentified human enterovirus sequences belonging to the same phylogenetic cluster, cluster C, as PV (Centers for Disease Control and Prevention 2001, 2002; Kew et al. 2002; Rousset et al. 2003; Yang et al. 2003).

PV is an enterovirus belonging to the *Picornaviridae* family that is one of the most important groups of human and animal pathogens. This family also includes human hepatitis A virus, human rhinoviruses, the agents of the common cold, and foot-and-mouth disease virus. PV is classified into three serotypes (PV-1, PV-2, and PV-3). Because of its very simple structure, PV has been used as a model for studying non-retroviral RNA viruses, and consequently PV is now one of the best-characterized animal viruses. The development of new animal and cell models, together with the identification of the virus receptor CD155, has allowed the key steps of the pathogenesis of poliomyelitis to be investigated at the molecular level (reviewed by Blondel et al. 1998). In particular, the involvement of PV-induced apoptosis in CNS injury has been studied (see below).

Apoptosis is an active process of cell death that occurs in response to various stimuli, including viral infection (Roulston et al. 1999). It involves a number of distinct morphological and biochemical features, such as cell shrinkage, translocation of phosphatidylserine from the inner to the outer surface of the cell membrane, plasma membrane blebbing, chromatin condensation, loss of the inner mitochondrial transmembrane potential, and internucleosomal DNA cleavage. These changes are mediated in particular by a family of proteases called caspases (cysteine proteases with aspartate specificity) (Earnshaw et al. 1999). The apoptotic pathways leading to cell death can be generally divided into two nonexclusive signaling cascades, one involving death receptors (extrinsic pathway) and the other involving mitochondria (intrinsic pathway) (Kaufmann and Hengartner 2001).

The death receptor pathway is activated by the binding of ligand [such as tumor necrosis factor (TNF) family death ligands] to the membrane receptor. This leads to the formation of the death-inducing signaling complex (DISC),which allows caspase-8 and/or caspase-10 autoactivation followed by caspase-3 activation (Wallach 1997; Ashkenazi and Dixit 1998). Caspase-3 activation results in the activation of both critical DNA repair enzymes including poly-ADP ribose polymerase (PARP) and specific endonucleases. Ultimately, DNA cleavage and nuclear collapse occur.

Apoptosis via the mitochondrial pathway involves specific cellular stress, such as viral infection signals, that leads to the loss of the mitochondrial transmembrane potential and release of proapoptotic molecules such as cytochrome *c* from mitochondria to the cytosol (Green and Amarante-Mendes 1998; Desagher and Martinou 2000). Loss of mitochondrial transmembrane potential is regulated by members of the Bcl-2 family, which exist as heterodimers in the cell. Some, such as Bcl-2 and Bcl-XL, inhibit apoptosis, whereas others, including Bax, Bak, and Bid, induce apoptosis. Bcl-2 family proteins act on the mitochondrial voltage-dependent channel (mitochondrial porin). Bax and Bak open this channel, resulting in the release of cytochrome *c*. In the cytosol, cytochrome *c* forms a caspase-activating complex by interaction with Apaf-1 (apoptosis protease-activating factor 1) and pro-caspase-9. This event triggers caspase-9 activation and initiates the apoptotic cascade by processing executive caspase-3. However, the mitochondrial pathway does not have an absolute requirement for caspase activation, as factors such as apoptosis-inducing factor (AIF) can induce apoptosis without caspase activation (Thornberry and Lazebnik 1998).

Apoptosis may be a process exploited by the virus to spread to neighboring cells, to protect progeny virus from host immune defenses, and to avoid an inflammatory response (Teodoro and Branton 1997; O'Brien 1998).

In this review we briefly describe the molecular biology of PV and the pathogenesis of poliomyelitis, and then consider several models of PV-induced apoptosis; we will also address the role of the cellular receptor of PV, CD155, in the modulation of apoptosis.

1 Poliovirus

1.1 Structure of the Virion

The PV is composed of a single-stranded RNA genome of positive polarity surrounded by a nonenveloped icosahedral protein capsid. The mature virion is approximately 30 nm in diameter, and the three-dimensional structures of the three serotypes of PV have been determined by X-ray crystallography (Hogle et al. 1985; Filman et al. 1989; Lentz et al. 1997). The capsid consists of 60 copies of each of the four viral structural proteins, VP1 to VP4 (Fig. 1A). A deep surface depression, called the "canyon", surrounds each fivefold axis of symmetry and contains the site for cell receptor binding (Colston and Racaniello 1994, 1995; Belnap et al. 2000; He et al. 2000).

The PV RNA genome is about 7,500 nucleotides long (Fig. 1B). It is poly-adenylated at its 3′-terminus and covalently linked to a small viral protein, VPg (3B), at its 5′-terminus (for review see Racaniello 2001). It contains a long 5′ noncoding region (NCR) followed by a single large open reading frame (ORF) and a short 3′ NCR that includes the poly(A) tail. The ORF is translated to produce a 247-kDa polyprotein that is processed into three large precursors of structural (P1) and nonstructural (P2 and P3) proteins (Fig. 1B).

1.2 PV Receptor

The human PV receptor, CD155, and its simian counterparts are members of the immunoglobulin superfamily (Mendelsohn et al. 1989; Koike et al. 1990, 1992). They are related to the nectin family of adhesion molecules found at intercellular junctions (Eberle et al. 1995a; Lopez et al. 1995; Takahashi et al. 1999). CD155 is predicted to contain three extracellular Ig-like domains in the order V–C2–C2, followed by a transmembrane region and a short cytoplasmic tail. There are four isoforms of

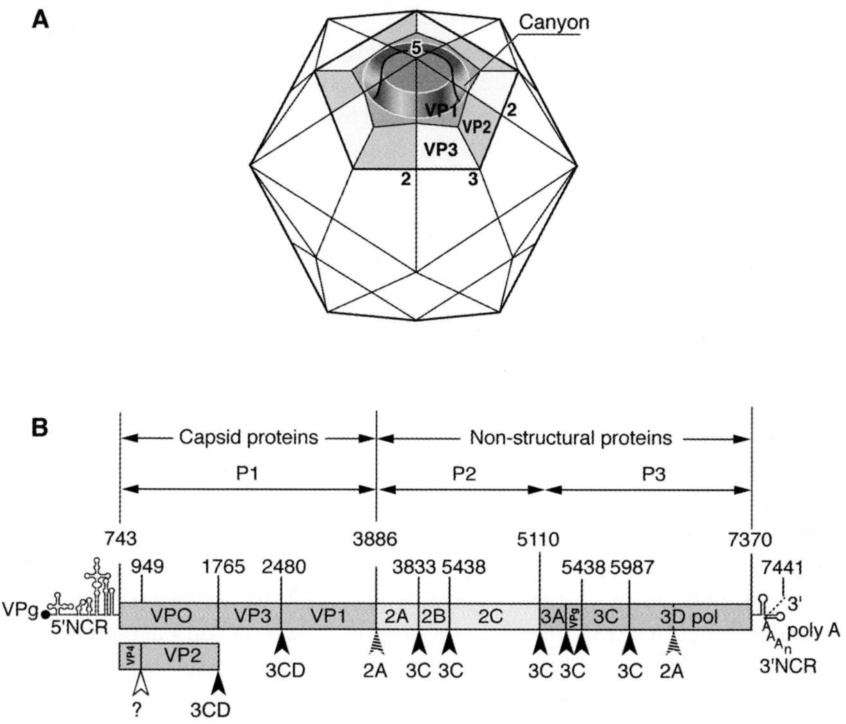

Fig. 1 A Schematic structure of the PV capsid. The two-, three- and fivefold axes of symmetry and the positions of capsid proteins VP1, VP2, and VP3 are indicated for one protomer. Five molecules of VP1 surround the fivefold axis of symmetry, whereas VP2 and VP3 alternate around the threefold axis of symmetry; VP4 is exclusively internal. The depression surrounding the fivefold axis, called the canyon, is formed by residues of VP1, VP2, and VP3, and contains the site for cell receptor binding. (Adapted from Hogle et al. 1985). **B** Genetic organization of PV-1/Mahoney. The 5′ and 3′ noncoding regions, indicated as *5′NCR* and *3′NCR*, respectively, flank the single open-reading frame, encoding the polyprotein, which is shown as an elongated rectangle. The protein precursors P1, P2, and P3, are designated by *arrows* above the genome. The viral proteins are indicated in the *rectangles*. The small viral protein VPg is covalently linked to the 5′ end of the RNA genome. Proteolytic cleavages occur between the amino acid pairs Asn-Ser, Gln-Gly, and Tyr-Gly, as indicated by *empty*, *solid*, and *cross-hatched arrowheads*, respectively. The cleavage sites of proteases 2A, 3CD, and 3C are shown. The mechanism of cleavage of the precursor VP0 giving VP4 and VP2 is not known. (Adapted from Kitamura et al. 1981)

CD155: Two of them (β and γ), lacking the transmembrane domain, are secreted, and the other two (α and δ) are membrane-bound proteins that can serve as PV receptors. The α and δ isoforms differ by the length of the intracytoplasmic part of the protein (Koike et al. 1990). The binding site for PV has been mapped to the N-terminal, V-like, extracellular Ig domain (domain 1) (Koike et al. 1991a; Aoki et al. 1994; Bernhardt et al. 1994; Morrison et al. 1994; Belnap et al. 2000; He et al. 2000). In human epithelial HeLa cells, two different bases (G and A) have been found at nucleotide position 199 in the mRNA encoding CD155 (Mendelsohn et al. 1989; Koike et al. 1990). Consequently, amino acid position 67, within domain 1 of CD155, is either Ala or Thr. Both of these CD155 forms have been found in humans (Karttunen et al. 2003; Saunderson et al. 2004), suggesting that CD155 with a Thr residue at amino acid position 67 is an allelic form of CD155.

In epithelial cells, the cytoplasmic domain of CD155 binds the μ1B subunit of the clathrin adaptor complex, and this interaction is responsible for sorting the CD155α isotype protein to basolateral membranes (Ohka et al. 2001). In contrast, CD155δ isotype is sorted to both the basolateral and apical surfaces (Ohka et al. 2001). These observations are consistent with CD155 being preferentially expressed on basolateral surfaces of human enterocytes, although it is also expressed on apical and lateral surfaces (Iwasaki et al. 2002). Similarly, CD155 expression is stronger on the basal side of M cells than on the apical side (Iwasaki et al. 2002).

The ectodomain of CD155 establishes cell-matrix contacts by interaction with vitronectin (Lange et al. 2001), as does the integrin α_v β3, and both proteins colocalize in microdomains on transfected mouse fibroblasts (Mueller and Wimmer 2003). CD155 also binds to nectin-3 ectodomain (Fabre et al. 2002). This interaction depends on dimerization of CD155, promoted by cell type-specific factors (Mueller and Wimmer 2003). Thus there may be *trans*-interaction between the bona fide cell-cell adherens type adhesion system (cadherin/nectin) and the cell-matrix adhesion system (integrin/CD155) by virtue of their nectin-3 and CD155 components, respectively (Mueller and Wimmer 2003).

In the CNS, expression of CD155 is activated by the secreted morphogen sonic hedgehog protein (Solecki et al. 2002). Both the expression of CD155 and vitronectin production are associated with regions of the CNS involved in the differentiation of motor neurons during embryonic development (Martinez-Moralez et al. 1997; Gromeier et al. 2000b). The short cytoplasmic domain of CD155 interacts with Tctex-1, a light chain subunit of dynein motor complex (Ohka and Nomoto 2001; Mueller et

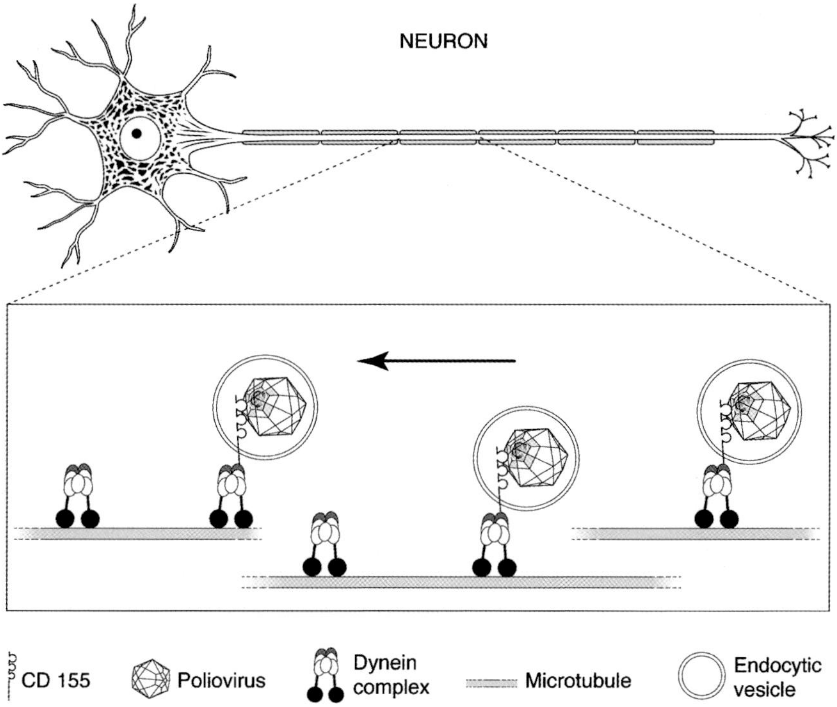

Fig. 2 Model of retrograde axonal transport of PV. A neuron, with an enlarged portion of the axon, is shown. The cytoplasmic domain of CD155 interacts with the light chain Tctex-1 (in *gray*) of the dynein motor complex, and the virus enclosed in endocytic vesicles is transported along microtubules by fast retrograde axonal transport. (Adapted from Ohka and Nomoto 2001; Mueller et al. 2002)

al. 2002; Ohka et al. 2004). This interaction is thought to mediate retrograde axonal transport of endocytic vesicles containing CD155. This mechanism could therefore be responsible for the fast retrograde axonal transport of PV-CD155 complexes (Fig. 2).

CD155 (and nectin-2) also specifically induce NK cell activation by interacting with DNAM-1 (the leukocyte adhesion molecule DNAX accessory molecule-1), also called CD226 (Bottino et al. 2003). The surface expression of CD155 (or nectin-2) in cell transfectants results in DNAM-1-dependent enhancement of interleukin (IL)-2-activated T and NK cell-mediated cytotoxicity (Bottino et al. 2003; Tahara-Hanaoka et al. 2004). In addition, it has recently been shown that CD96, also named Tactile (for T cell-activated increased late expression), is another NK cell receptor for CD155 (Fuchs et al. 2004). CD96 promotes NK cell adhesion to CD155-expressing cells and stimulates activated NK cell cytotoxicity (Fuchs et al. 2004). NK cells have thus evolved a dual receptor system to recognize nectin-2 and nectin-like (CD155) molecules. CD155 is highly expressed in certain human tumors, including colorectal carcinoma (Masson et al. 2001) and malignant gliomas (Gromeier et al. 2000a). It is possible that this receptor is essential for NK cell recognition of tumors (Fuchs et al. 2004).

Despite these numerous data, the cellular role of CD155, in particular in the CNS, is not completely elucidated.

1.3 Viral Cycle

In vitro, PV multiplies exclusively in primate cell lines (either human or simian). The viral cycle of PV (Fig. 3) proceeds entirely in the cytoplasm of the host cell (Racaniello 2001). It is one of the fastest known viral cycles, lasting approximately 8 h at 37°C in cell culture.

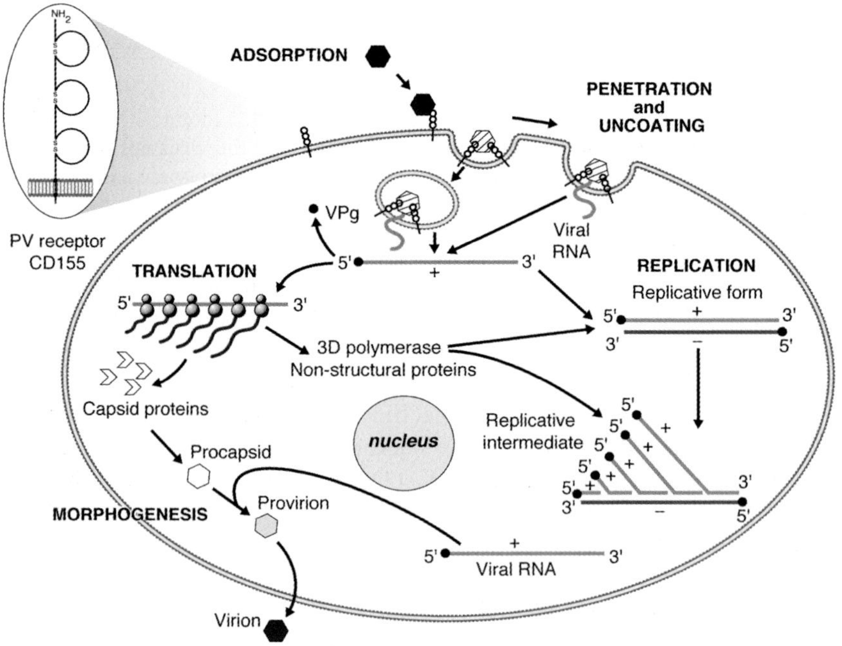

Fig. 3 Viral cycle of PV

The initial event of the viral cycle is attachment of the virion to the receptor, CD155: This receptor is only found on the surface of primate cells. PV binding to CD155 destabilizes the virion and induces the conformational modifications required for RNA release (Hogle 2002). The uncoating intermediates may be VP4-free particles, named A particles, because they are the major cell-associated form and remain infectious under particular conditions (Hogle 2002; Hogle and Racaniello 2002). Other uncoating intermediates have recently been proposed: VP4-containing particles such as those accumulating following interaction of PV with soluble CD155 receptor forms at high concentration (Duncan et al. 1998; Duncan and Colbère-Garapin 1999; Pelletier et al. 2003).

After PV RNA has been released into the cytoplasm of infected cells, translation of PV RNA is initiated by the binding of ribosomes to a highly structured region in the 5′ NCR, called the internal ribosomal entry site (IRES) (Ehrenfeld and Teterina 2002). Efficient IRES-dependent translation of PV RNA requires IRES-specific cellular factors as well as canonical initiation factors (Jackson 2002). Translation produces the large polyprotein, which is processed co- and posttranslationally by viral proteases 3C, its precursor 3CD, and 2A. This releases the structural and nonstructural proteins responsible for the proteolytic activities, RNA synthesis, and biochemical and structural changes that occur in the infected cell.

Viral RNA replicates on the surface of membranous vesicles that bud from various host cell organelles (Carrasco et al. 2002; Egger et al. 2002). RNA is replicated by the viral RNA-dependent RNA polymerase 3Dpol, but most of the other PV nonstructural proteins and cellular factors are also involved in viral RNA synthesis. RNA replication starts with the formation of a complementary negative-stranded RNA molecule, which serves as the template for the synthesis of progeny positive-stranded viral RNAs (Xiang et al. 1997; Andino et al. 1999; Gromeier et al. 1999; Paul 2002).

The formation of viral particles seems to be coupled to RNA synthesis (Hellen and Wimmer 1995; Ansardi et al. 1996; Nugent et al. 1999). VP0 (the precursor of VP2 and VP4), VP1, and VP3 aggregate with the viral RNA to form the provirion (Hogle 2002; Hogle and Racaniello 2002). During the last step of virus assembly, VP0 is cleaved by an unknown mechanism to give VP2 and VP4. Once assembled, the virions accumulate in the cytoplasm of infected cells in the form of crystalline inclusions, which are liberated by the bursting of vacuoles at the cell surface (Dunnebacke et al. 1969; Bienz et al. 1973). Vectorial release has been described in polarized human intestinal epithelial cells (Tucker et al.

1993). The massive release of new progeny virions occurs during cell lysis (Lwoff et al. 1955).

1.4
Effect of PV Replication on the Host Cell

As PV infection progresses in vitro, the host cell undergoes substantial metabolic and morphological changes commonly referred to as cytopathic effects (Haller and Semler 1995; Schlegel et al. 1996; Carrasco et al. 2002).

Early in the infectious cycle, PV proteases mediate the shutoff of both host cell transcription and translation (Dasgupta et al. 2002; Kuechler et al. 2002; Zamora et al. 2002). Both genetic and biochemical studies have shown that the virus-encoded protease 3C cleaves the transcription factors (TF) and is directly responsible for the shutoff of host cell transcription (Dasgupta et al. 2002). Four sequence-specific DNA binding pol II TF are either cleaved or degraded in PV-infected cells: the TATA-binding protein (TBP), the cyclic AMP-responsive element binding protein (CREB), the Octamer-binding factor (Oct-1), and the transcriptional activator p53 (Clark and Dasgupta 1990; Clark et al. 1993; Yalamanchili et al. 1996, 1997a,b; Weidman et al. 2001). Similarly, the pol III factor TFIIIC, which interacts with pol III promoters, as well as the 110-kDa TBP-associated factor (TAF 110), a subunit of the pol I factor SL-1, are also cleaved in PV-infected cells (Rubinstein and Dasgupta 1989; Clark et al. 1991; Rubinstein et al. 1992). The 2A and 3C proteases of PV cleave or induce cleavage of factors involved in translation initiation, such as eIF4G, and poly(A)-binding protein (PABP), respectively, thus inhibiting host cap-dependent mRNA translation (Etchison et al. 1982; Kräusslich et al. 1987; Gradi et al. 1998; Joachims et al. 1999; Novoa and Carrasco 1999; Kuechler et al. 2002; Zamora et al. 2002; Kuyumcu-Martinez et al. 2004).

In addition, nonstructural proteins have large effects on host intracellular membrane structure and function. Protein 2C induces membrane vesiculation (Cho et al. 1994; Aldabe and Carrasco 1995; Teterina et al. 1997), and proteins 2B and 3A are each sufficient to inhibit protein traffic through the host secretory pathway (Doedens and Kirkegaard 1995; Doedens et al. 1997; Dodd et al. 2001; Neznanov et al. 2001). Of note, 3A limits IL-6, IL-8, and β-interferon secretion (Dodd et al. 2001) and inhibits TNF receptor (Neznanov et al. 2001, 2002) and MHCI-dependent antigen presentation (Deitz et al. 2000) at the cell surface in PV-infected cells. The native immune response and inflammation could thus be reduced during PV infection.

Recent studies have also shown that PV infection can cause inhibition of nuclear-cytoplasmic trafficking, leading to accumulation of nuclear proteins in cytoplasm (Belov et al. 2000; Gustin and Sarnow 2001; Gustin 2003). Nuclear proteins, including the La autoantigen (Meerovitch et al. 1989), Sam 68 (McBride et al. 1996), and nucleolin (Waggoner and Sarnow 1998), accumulate in the cytoplasm of infected cells and interact with either viral RNA or proteins encoded by the virus. The cytoplasmic retention of La in PV-infected cells may, at least in part, be due to truncation of La by the 3C protease, resulting in the loss of the nuclear location sequence (NLS) (Shiroki et al. 1999). In the cytoplasm, La stimulates IRES-mediated PV translation (Meerovitch et al. 1989).

Finally, PV infection can trigger the development of apoptosis, as described below.

2
Pathogenesis of Poliomyelitis and Post-Polio Syndrome

The term poliomyelitis is derived from the Greek words *polios*, meaning "gray," and *myelos*, meaning "marrow," referring to the fact that the disease results from the destruction of neurons in the gray matter of the anterior horn of the spinal cord.

Humans are the only natural host of PV, and infections with PV result in poliomyelitis in only 1%–2% of cases. The virus is transmitted mainly via the fecal-oral route or via droplets from the pharynx. After oral ingestion (Fig. 4), it infects the oropharynx and the gut, where it causes few if any symptoms. Virus is excreted in the oropharyngeal secretions and in the stool for several weeks (for review see Minor 1997; Gromeier et al. 1999; Ohka and Nomoto 2001; Pallansch and Roos 2001). PV multiplies extensively in the tonsils and Peyer's patches (Bodian 1955), but the target cell where initial multiplication occurs is still unidentified. In humans and some Old World monkeys, sensitivity to oral infection correlates with CD155 expression in the intestinal epithelium, including the follicle-associated epithelium and microfold M cells of Peyer's patches, and in follicular dendritic cells and B cells of germinal centers within Peyer's patches (Iwasaki et al. 2002). It has been demonstrated that PV particles adhere specifically to and are endocytosed by human intestinal M cells, suggesting that M cells in humans are the site of PV penetration of the intestinal epithelial barrier (Sicinski et al. 1990). In agreement

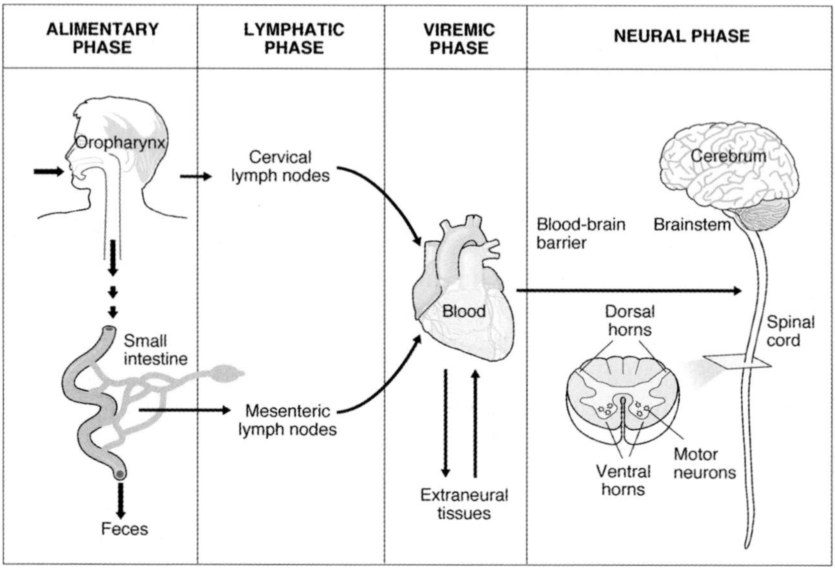

Fig. 4 Poliomyelitis pathogenesis

with this hypothesis, PV transcytosis is much more efficient in a cell culture model of enterocytes containing M-like cells than in enterocytes cultured alone (Ouzilou et al. 2002). Direct delivery of virions by M cells to underlying mucosa-associated lymphoid tissue implicates immune cells as candidates for PV transport to mesenteric lymph nodes. The ability of PV to replicate in mononuclear phagocytic cells from the blood (Freistadt et al. 1993; Eberle et al. 1995b; Freistadt and Eberle 2000) suggests that resident phagocytic cells in the tonsils and the Peyer's patches may harbor the initial rounds of PV replication. However, translocated PV may also infect enterocytes through their basal, CD155-covered face. The vectorial release of PV from the apical surface of human intestinal epithelial cells, toward the lumen, is consistent with dissemination of progeny virus into the gut (Tucker et al. 1993).

The virus is released from lymphoid tissues into the bloodstream. In a minority of cases, this results in viral spread to other tissues, amplifying the viremia. The circulating virus invades the CNS probably through the blood-brain barrier and, at least in mice, this does not require CD155 (Yang et al. 1997). Infected mononuclear phagocytic cells that shuttle the blood-brain barrier may introduce PV into the CNS (Freistadt et al. 1993).

Retrograde axonal transport has also been suspected to have occurred in PV-inoculated primates (Bodian 1955), in humans in a series of cases of iatrogenic poliomyelitis known as the "Cutter incident" (Nathanson and Langmuir 1963), and in children receiving intramuscular injection within a few weeks after oral vaccination ("provocation" poliomyelitis) (Strebel et al. 1995). This pathway has found strong support in experiments with Tg-CD155 mice (Ren and Racaniello 1992; Gromeier and Wimmer 1998; Ohka et al. 1998). Thus PV may enter the CNS by the retrograde axonal pathway (Bodian 1955; Nathanson and Langmuir 1963; Ren and Racaniello 1992; Gromeier and Wimmer 1998; Ohka et al. 1998; Ohka and Nomoto 2001; Gromeier and Nomoto 2002). It has been proposed that, while the virus remains attached to the receptor, the cytoplasmic tail of CD155 interacts with Tctex-1, the latter mediating retrograde transport of the endocytic vesicle via the dynein complex to the motor neuron cell body (Ohka and Nomoto 2001; Mueller et al. 2002; Ohka et al. 2004) (Fig. 2).

In the CNS, the main target cell of PV is the motor neuron of the spinal cord and the brain stem (Bodian 1955), and PV probably travels by cell-to-cell spread (Ponnuraj et al. 1998). The destruction of motor neurons, a consequence of PV replication, results in paralysis. Studies with mice transgenic for the PV receptor molecule CD155 (Tg-CD155) have shown that the specific tropism for motor neurons is a consequence of their expression of CD155. When CD155 is expressed under the control of a heterologous promoter in transgenic mice, the pathology induced by PV is substantially modified: PV can infect glial and ependymal cells in addition to the neurons (Ida-Hosonuma et al. 2002). In this model, paralysis of the limbs is rarely observed and mice survive without showing subclinical abnormality: The immune response may be induced by PV-infected nonneuronal cells to clear the virus before it can infect neurons (Ida-Hosonuma et al. 2002).

After decades of clinical stability following acute paralytic poliomyelitis, many patients develop a disease called post-polio syndrome, involving slowly progressive muscle weakness (Dalakas 1995). The presence of PV RNA sequences or PV-related RNA (Leon-Monzon and Dalakas 1995; Muir et al. 1995; Leparc-Goffart et al. 1996) and of anti-PV IgM antibodies (Sharief et al. 1991) suggests that this syndrome may be a consequence of PV persistence. Consistent with this hypothesis, PV can establish persistent infections in human cell cultures of neuronal origin (Colbère-Garapin et al. 1989, 2002; Pavio et al. 1996).

Monkeys and Tg-CD155 mice are susceptible to wild strains of all three PV serotypes after intracerebral or intraspinal inoculation (Ren et al. 1990; Koike et al. 1991b). In Tg-CD155 mice, infection of the CNS is usually extensive and most often results in fatal poliomyelitis. Only a small number of mouse-adapted PV strains are able to induce poliomyelitis in non-Tg-CD155 mice. We have isolated and characterized a PV mutant pathogenic for mice, PV-1/Mah-T1022I (Couderc et al. 1993, 1996). Interestingly, this mutant induces paralytic poliomyelitis in Swiss mice that is, as in humans, not always lethal. With this model, we have shown that PV RNA and capsids persist in the CNS throughout the life of animals (Destombes et al. 1997) and that PV RNA replication in the CNS of persistently infected mice is restricted, mainly as a consequence of inhibition of plus-strand RNA synthesis (Girard et al. 2002).

These animal models have been used to study principally the neurological phase of poliomyelitis, but mice are not suitable for study of the digestive phase after oral infection. However, excretion of PV in the feces has also been described after intraperitoneal injection of PV in Tg-CD155 mice (Buisman et al. 2000; Boot et al. 2003). Recently, Tg-CD155 mouse models allowing infection by PV through the mucosal route have been reported: These animals develop paralysis after intranasal inoculation (Crotty et al. 2002; Nagata et al. 2004).

3 Poliovirus and Apoptosis

Cell damage in the CNS in response to virus infection can involve apoptosis (for exhaustive review see Griffin and Hardwick 1999; Roulston et al. 1999; Levine 2002). This has been illustrated in vivo with a number of human and murine neurotropic RNA viruses including HIV (Petito and Roberts 1995), HTLV-1 (Umehara et al. 1994), reovirus (Oberhaus et al. 1997), rabies virus (Jackson and Rossiter 1997; Baloul and Lafon 2003), measles virus (Manchester et al. 1999; Evlashev et al. 2000), dengue virus (Desprès et al. 1998), Sindbis virus (Levine et al. 1993; Lewis et al. 1996), and Theiler murine encephalomyelitis virus (Tsunoda et al. 1997), a member of the Picornaviridae family. Three other picornaviruses, Coxsackievirus B3 (Carthy et al. 1998), HAV (Brack et al. 1998), and enterovirus 71 (Kuo et al. 2002; Li et al. 2002), induce apoptosis in cell cultures.

3.1
PV-Induced Apoptosis in Nerve Cells In Vivo and Ex Vivo

As stated above, PV-induced paralysis results from the destruction of motor neurons, but the process leading to the death of neurons was unknown until recently.

Work with mouse models (Tg-CD155 and non-Tg mice) has shown that PV-infected motor neurons in the spinal cord die by apoptosis (Fig. 5): The extent of apoptosis correlates with viral load, and its onset coincides with that of paralysis (Girard et al. 1999). Moreover, CNS injury may be enhanced by apoptosis in uninfected nonneuronal cells, probably glial or inflammatory cells, contiguous to the PV-infected neurons (Girard et al. 1999). PV-induced apoptosis is therefore an important component of the tissue injury in the CNS of infected mice that leads to paralysis.

To investigate PV-induced apoptosis in nerve cells, cultures of mixed mouse primary nerve cells from the cerebral cortex of Tg-CD155 mice have been developed (Couderc et al. 2002). These cultures contain all three main cell types of the CNS, i.e., neurons, astrocytes, and oligodendrocytes. All of these cell types are susceptible to PV infection, and viral replication leads to the DNA fragmentation characteristic of apoptosis (Couderc et al. 2002). Furthermore, PV-induced apoptosis in this model

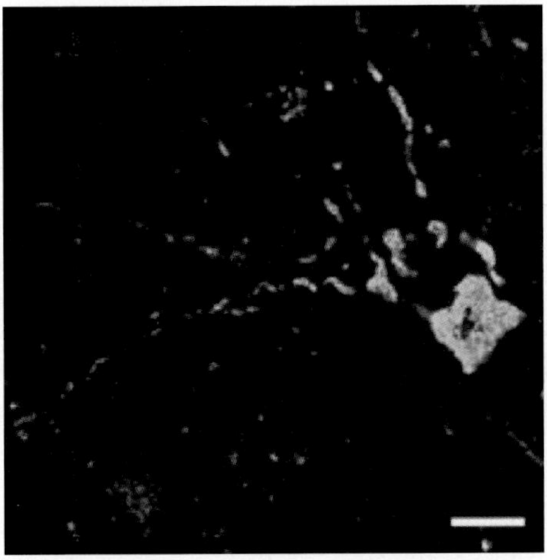

Fig. 5 PV-infected motor neurons (immunofluorescence labeling for viral antigen, *white*) with an apoptotic nucleus (terminal deoxynucleotidyltransferase-mediated dUTP nick end labeling, *dark*) in the mouse spinal cord. *Bar*, 20 μm

involves caspases. Interestingly, in contrast to the observations in vivo, mixed primary nerve cell cultures harbor PV-infected glial cells in addition to PV-infected neurons. This confirms previous data obtained in human primary fetal brain cells (Pavio et al. 1996), and the discrepancy could be due to a difference in the expression of CD155 ex vivo and in vivo. Indeed, it was shown recently that glial and ependymal cells expressing CD155 in a new Tg-CD155 mouse model became susceptible to PV infection (Ida-Hosonuma et al. 2002). Cultures of mixed mouse primary nerve cells could serve as a model for investigations of the molecular mechanisms of PV-induced apoptosis in nerve cells and lead to the identification of specific neuronal pathways.

3.2
PV-Induced Apoptosis In Vitro

PV can trigger apoptosis in human tissue cultures of colon carcinoma cells (Caco-2) and promonocytic cells (U937) (Ammendolia et al. 1999; Lopez-Guerrero et al. 2000; Calandria and Lopez-Guerrero 2002). If enterocytes and monocytic cells are indeed infected by PV during viral infection in humans, PV-induced apoptosis in these two target cells could allow the virus to reach the CNS with only a limited activation of immune inflammatory responses.

In a subline of human epithelial cells (HeLa), Tolskaya et al. showed that no apoptotic reaction occurs after productive PV infection. In contrast, apoptosis can develop after nonpermissive infection with various PV mutants (guanidine-sensitive, guanidine-dependent, or temperature-sensitive mutants) (Tolskaya et al. 1995). In permissive conditions, apoptosis induced by apoptotic inducers, such as metabolic inhibitors, is suppressed by PV infection (Tolskaya et al. 1995; Koyama et al. 2001). Therefore, PV appears to encode two separate functions, which have opposite effects: one triggering and the other suppressing apoptosis. Agol et al. proposed a model in which during the productive PV infection, the commitment of cells switches in the middle of the viral cycle from apoptosis to an antiapoptotic state and cytopathic effect (Agol et al. 2000). Analysis of specific apoptotic pathways in this model suggested that early PV-induced apoptosis results in translocation of cytochrome *c* from mitochondria to the cytosol and thus activates caspase-9 and caspase-3 (Belov et al. 2003). The apoptotic signal may then be amplified by a loop, which includes activation of Bid, a proapoptotic member of the Bcl-2 family (Belov et al. 2003). The subsequent suppression of the apoptotic program in this subline of HeLa cells may be due, at least in part, to aberrant processing and degradation of pro-caspase-9 (Belov et al. 2003).

Expression of either PV protease 2A or 3C is sufficient to trigger apoptosis (Barco et al. 2000; Goldstaub et al. 2000), suggesting that PV proteases are able to activate an endogenous cell suicide program. Similarly, the 2A and 3C proteases encoded by another neurotropic enterovirus, enterovirus 71, induce cell apoptosis (Kuo et al. 2002; Li et al. 2002). Apoptosis induced by 3C seems to depend on the caspase pathway (Barco et al. 2000), whereas 2A-induced apoptosis may be caspase independent (Goldstaub et al. 2000). When elucidating the exact molecular mechanism used by 3C to provoke apoptosis it should be remembered that this protease cleaves a variety of host proteins, including transcription factors (Clark et al. 1991, 1993; Carrasco et al. 2002) and the cytoskeletal protein MAP4 (Joachims et al. 1995). Cleavage of these proteins could thus trigger apoptosis via a transcriptional inhibition mechanism or cytoskeleton changes, respectively. Protease 2A is involved in the cleavage of eIF4GI, eIF4GII, and PABP, which are factors involved in cellular protein synthesis activity (Etchison et al. 1982; Joachims et al. 1999; Goldstaub et al. 2000). Consequently, the PV protease 2A may induce apoptosis by arresting cap-dependent translation of particular cellular mRNAs encoding proteins that are required for maintaining cellular viability. Alternatively, the protease 2A may induce apoptosis by allowing preferential cap-independent translation of cellular mRNAs encoding proteins that induce apoptosis. Proteases 2A and 3C may also trigger the apoptotic process via the cleavage of other unidentified cellular substrates.

Double-stranded PV RNAs generated during viral replication may also trigger apoptosis in HeLa cells, under conditions of permissive PV infection, through activation of interferon, 2′-5′-oligoadenylate synthetase (OAS), and RNase L (Castelli et al. 1997, 1998a,b).

PV-mediated suppression of apoptosis may involve other mechanisms in addition to pro-caspase-9 degradation. Expression of 3A specifically suppresses the host secretory pathway, resulting notably in the loss from the cell surface of TNF receptor and receptors for various other cytokines (Dodd et al. 2001; Neznanov et al. 2001, 2002). This causes a decrease in cell sensitivity to these cytokines and thus may be another indirect mechanism preventing apoptosis. Furthermore, although expression of PV protease 3C induces apoptosis (Barco et al. 2000), it may also be able to delay or prevent apoptosis as it mediates the degradation of the transcriptional activator and tumor suppressor p53 (Weidman et al. 2001).

3.3 CD155 and Apoptosis

PV can establish persistent infections in cells of neural origin (neuroblastoma cells) and in human fetal brain cell cultures (Colbère-Garapin et al. 1989, 2002; Pavio et al. 1996). During persistent infection in human neuroblastoma IMR-32 cells, specific mutations are selected in CD155 that affect domain 1, the binding site for PV (Pavio et al. 2000). These mutations include the Ala 67→Thr substitution, corresponding to a switch from one allelic form of the PV receptor to the other form previously identified. This mutated form of CD155 is not expressed in parental IMR-32 cells. A role for CD155 in cell cytopathic effects has been suggested, with mutated forms of CD155 generated by site-directed mutagenesis affecting residues in domain 1 (Morrison et al. 1994). To investigate the effect of the mutations selected during persistent infection on cell susceptibility to PV-induced cytopathic effects, the two forms of PV receptors, mutated ($CD155_{Thr67}$), and nonmutated ($CD155_{IMR}$), have been expressed independently in murine LM cells lacking the CD155 gene. Interestingly, although virus adsorption and viral growth were identical in the two cell lines, PV-induced cytopathic effects appeared later in cells expressing the mutated form, $CD155_{Thr67}$ (LM-$CD155_{Thr67}$) than in cells expressing the nonmutated form $CD155_{IMR}$ (LM-$CD155_{IMR}$) (Pavio et al. 2000).

Analysis of the PV-induced death of LM-$CD155_{Thr67}$ or LM-$CD155_{IMR}$ cells showed that PV infection triggers the DNA fragmentation characteristic of apoptosis in both LM-cell lines, but at a lower level in LM-$CD155_{Thr67}$ cells than in LM-$CD155_{IMR}$ cells (Gosselin et al. 2003). Intrinsic (mitochondrial dysfunction) and extrinsic (death receptor mediated) apoptotic pathways have been investigated in both cell lines. Mitochondrial dysfunction was analyzed by detecting the translocation of cytochrome *c* from mitochondria and by measuring the activation of caspases-3 and -9. Levels of cytochrome *c* release and caspase-9 and caspase-3 activation were lower in PV-infected LM-$CD155_{Thr67}$ cells than in LM-$CD155_{IMR}$ cells (Gosselin et al. 2003). The death receptor-mediated pathway was studied by measuring the activation of caspase-8 and caspase-10: In both cell lines, activation of caspases-8 and -10 paralleled that of caspase-9 and caspase-3. Thus it appeared that the two main apoptotic signaling pathways are simultaneously activated in response to PV infection. However, this observation does not exclude possible cross-talk between the intrinsic and the extrinsic apoptotic cascades. Indeed, caspase-8 can generate a truncated form of the Bid protein, which then translocates to mitochondria and initiates the mitochondrial pathway (Kaufmann and Hengartner 2001). Moreover, some of the caspases activated by the mitochondrial pathway may also activate caspase-8 in a feedback loop (Viswanath et al. 2001).

Altogether, these data indicate that the level of PV-induced apoptosis is lower in those cells, expressing mutated $CD155_{Thr67}$, selected during persistent PV infection of IMR-32 cells than in cells expressing $CD155_{IMR}$. As stated above, expression of 2A and 3C PV proteases is sufficient to induce cell apoptosis (Barco et al. 2000; Goldstaub et al. 2000). Thus CD155 may be an additional cellular factor involved in the modulation of PV-induced apoptosis. One attractive possibility is that an apoptotic signal is triggered by the PV-CD155 interaction and modulated according to the CD155 allelic form. The triggering of apoptosis by the binding of a virus to a receptor, and also at a postbinding entry step, has been demonstrated for other viruses, including reovirus (Tyler et al. 2001), Sindbis virus (Jan and Griffin 1999), and HIV (Banda et al. 1992). However, no data are currently available in support of the notion that an apoptotic signal is directly transduced via CD155. Alternatively, a molecule interacting with CD155 may transduce this signal. One candidate is CD44, the major receptor for hyaluronic acid, which is physically associated with CD155 and is able to induce apoptotic signals (Shepley and Racaniello 1994; Freistadt and Eberle 1997; Foger et al. 2000; Okamoto et al. 2001). Another candidate is the Bcl2-interacting mediator of cell death protein (Bim) (Adams and Cory 1998): Bim has a proapoptotic effect by interacting with the antiapoptotic factor Bcl-2. In the CNS, Bim is expressed in neurons of the gray matter but not in glial cells, and it is involved in a variety of neuronal cell death models (Putcha et al. 2001; Sanchez and Yuan 2001). Interestingly, the cytoplasmic domain of CD155 interacts with the dynein complex (Pallansch and Roos 2001) that keeps Bim inactivated at the level of the cytoskeleton (Puthalakath et al. 1999). Thus after interaction between the PV capsid and CD155, it is possible that Bim is released from the cytoskeleton such that it can then neutralize Bcl-2, thereby leading to apoptosis.

4 Conclusion

During recent years, the effort invested in studying PV-induced apoptosis has increased, and this has revealed the complexity of the apoptotic process in PV-infected cells. In the CNS, PV-induced apoptosis seems to be a significant factor in the pathogenesis of poliomyelitis, and it would be interesting to identify the molecular mechanisms of this apoptotic process in neurons. New ex vivo cellular models will allow the analysis of apoptotic pathways triggered by PV infection in nerve cells. In vitro studies using various cellular models have shown that PV can either trigger or, on the contrary, suppress the development of apoptosis according to the conditions of infection and to the host cell type. Several viral proteins are good candidates for involvement in the apoptotic process in PV-infected cells. Proteases 2A and 3C are able to trigger apoptosis, whereas protein 3A can inhibit it. Protease 3C can also prevent apoptosis. However, most of these data were obtained with individual protein expression systems and the situation is more complex in the context of cellular infection. PV interaction with CD155 could also trigger apoptosis either via CD155 itself or more probably via another molecule interacting with CD155. However, the molecular details of the apoptotic pathways leading to cell death following PV/CD155 interaction remain to be determined. PV-induced apoptosis is less extensive in cells expressing mutated $CD155_{Thr67}$ selected during persistent PV infection of IMR-32 cells than in cells expressing nonmutated $CD155_{IMR}$. Thus CD155 may also be a cellular factor involved in the modulation of PV-induced apoptosis. This modulation could be profitable to the virus: If apoptosis is not induced correctly or not induced at all in a context where it should be, it may allow the establishment of a persistent infection. Thus it would be interesting to determine which CD155 allelic forms are expressed in patients developing poliomyelitis and post-polio syndrome. It was recently shown that the frequency of $CD155_{Thr67}$ was significantly higher in patients with amyotrophic lateral sclerosis or progressive muscular atrophy than in controls (Saunderson et al. 2004).

Acknowledgements The authors are very grateful to Laurent Blondel for his precious help with the figures. This work was supported by grants from the Institut Pasteur and the Association Française contre les Myopathies (contract N° 6932 and 7290).

References

Adams JM, Cory S (1998) The Bcl-2 protein family: arbiters of cell survival. Science 281:1322–1326

Agol VI, Belov GA, Bienz K, Egger D, Kolesnikova MS, Romanova LI, Sladkova LV, Tolskaya EA (2000) Competing death programs in poliovirus-infected cells: commitment switch in the middle of the infectious cycle. J Virol 74:5534–5541

Aldabe R, Carrasco L (1995) Induction of membrane proliferation by poliovirus proteins 2C and 2BC. Biochem. Biophys. Res Commun 206:64–76
Ammendolia MG, Tinari A, Calcabrini A, Superti F (1999) Poliovirus infection induces apoptosis in CaCo-2 cells. J Med Virol 59:122–129
Andino R, Böddeker N, Silvera D, Gamarnik A (1999) Intracellular determinants of picornavirus replication. Trends Microbiol 7:76–82
Ansardi D, Porter D, Anderson M, Morrow C (1996) Poliovirus assembly and encapsidation of genomic RNA. Adv Virus Res 46:1–68
Aoki J, Koike S, Ise I, Satoyoshida Y, Nomoto A (1994) Amino acid residues on human poliovirus receptor involved in interaction with poliovirus. J Biol.-Chem.269:8431–8438
Ashkenazi A, Dixit VM (1998) Death receptors: signaling and modulation. Science 281:1305–1308
Baloul L, Lafon M (2003) Apoptosis and rabies virus neuroinvasion. Biochimie 85:777–788
Banda NK, Bernier J, Kurahara DK, Kurrle R, Haigwood N, Sekaly RP, Finkel T (1992) Crosslinking CD4 by human immunodeficiency virus gp120 primes T cells for activation-induced apoptosis. J. Exp. Med. 176:1099–1106
Barco A, Feduchi E, Carrasco L (2000) Poliovirus protease 3Cpro kills cells by apoptosis. Virology 266:352–360
Belnap DM, McDermott BM, Filman DJ, Cheng NQ, Trus BL, Zuccola HJ, Racaniello VR, Hogle JM, Steven AC (2000) Three-dimensional structure of poliovirus receptor bound to poliovirus. Proc. Natl. Acad. Sci. USA 97:73–78
Belov GA, Evstafieva AG, Rubtsov YP, Mikitas OV, Vartapetian AB, Agol VI (2000) Early alteration of nucleocytoplasmic traffic induced by some RNA viruses. Virology 275:244–248
Belov GA, Romanova LI, Tolskaya EA, Kolesnikova MS, Lazebnik YA, Agol VI (2003) The major apoptotic pathway activated and suppressed by poliovirus. J Virol 77:45–56
Bernhardt G, Harber J, Zibert A, de Crombrugghe M, Wimmer E (1994) The poliovirus receptor: Identification of domains and amino acid residues critical for virus binding. Virology 203:344–356
Bienz K, Egger D, Wolff DA (1973) Virus replication, cytopathology, and lysosomal enzyme response of mitotic and interphase HEp-2 cells infected with poliovirus. J Virol 11:565–574
Blondel B, Duncan G, Couderc T, Delpeyroux F, Pavio N, Colbère-Garapin F (1998) Molecular aspects of poliovirus biology with a special focus on the interactions with nerve cells. J Neurovirol 4:1–26
Bodian D (1955) Viremia, invasiveness, and the influence of injections. Ann N Y Acad Sci 61:877–882
Boot HJ, Kasteel DT, Buisman AM, Kimman TG (2003) Excretion of wild-type and vaccine-derived poliovirus in the feces of poliovirus receptor-transgenic mice. J Virol 77:6541–6545
Bottino C, Castriconi R, Pende D, Rivera P, Nanni M, Carnemolla B, Cantoni C, Grassi J, Marcenaro S, Reymond N, Vitale M, Moretta L, Lopez M, Moretta A (2003) Identification of PVR (CD155) and Nectin-2 (CD112) as cell surface ligands for the human DNAM-1 (CD226) activating molecule. J Exp Med 198:557–567

Brack K, Frings W, Dotzauer A, Vallbracht A (1998) A cytopathogenic, apoptosis-inducing variant of hepatitis A virus. J Virol 72:3370–3376

Buisman AM, Sonsma JA, Kimman TG, Koopmans MP (2000) Mucosal and systemic immunity against poliovirus in mice transgenic for the poliovirus receptor: the poliovirus receptor is necessary for a virus-specific mucosal IgA response. J Infect Dis 181:815–823

Calandria C, Lopez-Guerrero JA (2002) Poliovirus modulates Bcl-xl expression in the human U937 promonocytic cell line. Arch Virol 147:2445–2452

Carrasco L, Guinea R, Irurzun A, Barco A (2002) Effects of viral replication on cellular membrane metabolism and function. In: Semler BL, Wimmer E (eds) Poliovirus receptors and cell entry. ASM Press, Washington DC, pp 337–354

Carthy CM, Granville DJ, Watson KA, Anderson DR, Wilson JE, Yang D, Hunt DWC, McManus BM (1998) Caspase activation and specific cleavage of substrates after coxsackievirus B3-induced cytopathic effect in HeLa cells. J Virol 72:7669–7675

Castelli J, Wood KA, Youle RJ (1998a) The 2-5A system in viral infection and apoptosis. Biomed Pharmacother 52:386–390

Castelli JC, Hassel BA, Maran A, Paranjape J, Hewitt JA, Li XL, Hsu YT, Silverman RH, Youle RJ (1998b) The role of 2′–5′ oligoadenylate-activated ribonuclease L in apoptosis. Cell Death Differ 5:313–320

Castelli JC, Hassel BA, Wood KA, Li XL, Amemiya K, Dalakas MC, Torrence PF, Youle RJ (1997) A study of the interferon antiviral mechanism—apoptosis activation by the 2-5a system. J Exp Med 186:967–972

Centers for Disease Control and Prevention (2001) Acute flaccid paralysis associated with circulating vaccine-derived poliovirus—Philippines, 2001. Morbid Mortal Wkly Rep 50:874–875

Centers for Disease Control and Prevention (2002) From the Centers for Disease Control and Prevention. Acute flaccid paralysis associated with circulating vaccine-derived poliovirus—Philippines, 2001. JAMA 287:311

Cho MW, Teterina N, Egger D, Bienz K, Ehrenfeld E (1994) Membrane rearrangement and vesicle induction by recombinant poliovirus 2C and 2BC in human cells. Virology 202:129–145

Clark ME, Dasgupta A (1990) A transcriptionally active form of TFIIIC is modified in poliovirus-infected HeLa cells. Mol Cell Biol 10:5106–5113

Clark ME, Hammerle T, Wimmer E, Dasgupta A (1991) Poliovirus proteinase-3C converts an active form of transcription factor-IIIC to an inactive form: a mechanism for inhibition of host cell polymerase-III transcription by poliovirus. EMBO J 10:2941–2947

Clark ME, Lieberman PM, Berk AJ, Dasgupta A (1993) Direct cleavage of human TATA-binding protein by poliovirus protease 3C in vivo and in vitro. Mol Cell Biol 13:1232–1237

Colbère-Garapin F, Christodoulou C, Crainic R, Pelletier I (1989) Persistent poliovirus infection of human neuroblastoma cells. Proc. Natl Acad Sci USA 86:7590–7594

Colbère-Garapin F, Pelletier I, Ouzilou L (2002) Persistent infections by picornaviruses. In: Semler BL, Wimmer E (eds) Molecular biology of picornaviruses. ASM Press, Washington DC, pp 437–448

Colston E, Racaniello VR (1994) Soluble receptor-resistant poliovirus mutants identify surface and internal capsid residues that control interaction with the cell receptor. EMBO J 13:5855–5862

Colston EM, Racaniello VR (1995) Poliovirus variants selected on mutant receptor-expressing cells identify capsid residues that expand receptor recognition. J Virol 69:4823–4829

Couderc T, Delpeyroux J, Le Blay H, Blondel B (1996) Mouse adaptation determinants of poliovirus type 1 enhance viral uncoating. J Virol 70:305–312

Couderc T, Guivel-Benhassine F, Calaora V, Gosselin AS, Blondel B (2002) An ex vivo murine model to study poliovirus-induced apoptosis in nerve cells. J Gen Virol 83:1925–1930

Couderc T, Hogle J, Le Blay H, Horaud F, Blondel B (1993) Molecular characterization of mouse-virulent poliovirus type 1 Mahoney mutants: involvement of residues of polypeptides VP1 and VP2 located on the inner surface of the capsid protein shell. J Virol 67:3808–3817

Crotty S, Hix L, Sigal LJ, Andino R (2002) Poliovirus pathogenesis in a new poliovirus receptor transgenic mouse model: age-dependent paralysis and a mucosal route of infection. J Gen Virol 83:1707–1720

Dalakas MC (1995) The post-polio syndrome as an evolved clinical entity. Definition and clinical description. In: Dalakas MC, Bartfeld H, Kurland LT (eds) The post-polio syndrome, vol 753. The New York Academy of Sciences, New York, pp 68–80

Dasgupta A, Yalamanchili P, Clark ME, Kliewer S, Fradkin L, Rubinstein S, Das S, Shen Y, Weidman MK, Banerjee R, Datta U, Igo M, Kundu P, Barat B, Berk AJ (2002) Effects of picornavirus proteinases on host cell transcription. In: Semler BL, Wimmer E (eds) Molecular biology of picornaviruses. ASM Press, Washington DC, pp 321–335

Deitz SB, Dodd DA, Cooper S, Parham P, Kirkegaard K (2000) MHC I-dependent antigen presentation is inhibited by poliovirus protein 3A. Proc Natl Acad Sci USA 97:13790–13795

Desagher S, Martinou JC (2000) Mitochondria as the central control point of apoptosis. Trends Cell Biol 10:369–377

Desprès P, Frenkiel MP, Ceccaldi PE, Duarte dos Santos C, Deubel V (1998) Apoptosis in the mouse central nervous system in response to infection with mouse-neurovirulent dengue viruses. J Virol 72:823–829

Destombes J, Couderc T, Thiesson D, Girard S, Wilt SG, Blondel B (1997) Persistent poliovirus infection in mouse motoneurons. J Virol 71:1621–1628

Dodd DA, Giddings TH, Kirkegaard K (2001) Poliovirus 3A protein limits interleukin-6 (IL-6), IL-8, and beta interferon secretion during viral infection. J Virol 75:8158–8165

Doedens JR, Giddings TH, Kirkegaard K (1997) Inhibition of endoplasmic reticulum-to-golgi traffic by poliovirus protein 3A—genetic and ultrastructural analysis. J Virol 71:9054–9064

Doedens JR, Kirkegaard K (1995) Inhibition of cellular protein secretion by poliovirus proteins 2B and 3A. EMBO J 14:894–907

Duncan G, Colbère-Garapin F (1999) Two determinants in the capsid of a persistent type 3 poliovirus exert different effects on mutant virus uncoating. J Gen Virol 80: 2601–2605

Duncan G, Pelletier I, Colbère-Garapin F (1998) Two amino acid substitutions in the type 3 poliovirus capsid contribute to the establishment of persistent infection in HEp-2c cells by modifying virus-receptor interactions. Virology 241:14–29

Dunnebacke TH, Levinthal JD, Williams RC (1969) Entry and release of poliovirus as observed by electron microscopy of cultured cells. J Virol 4:504–513

Earnshaw W, Martins L, Kaufmann S (1999) Mammalian caspases: structure, activation, substrates, and functions during apoptosis. Annu Rev Biochem 68:383–424

Eberle F, Dubreuil P, Mattei MG, Devilard E, Lopez M (1995a) The human PRR2 gene, related to the human poliovirus receptor gene (PVR), is the true homolog of the murine MPH gene. Gene 159:267–272

Eberle KE, Nguyen VT, Freistadt MS (1995b) Low levels of poliovirus replication in primary human monocytes: possible interactions with lymphocytes. Arch Virol 140:2135–2150

Egger D, Gosert R, Bienz K (2002) Role of cellular structures in viral RNA replication. In: Semler BL, Wimmer E (eds) Molecular biology of picornaviruses. ASM Press, Washington DC, pp 247–253

Ehrenfeld E, Teterina N (2002) Initiation of translation of picornavirus RNAs: structure and function of internal ribosome entry site. In: Semler BL, Wimmer E (eds) Molecular biology of picornaviruses. ASM Press, Washington DC, pp 159–169

Etchison D, Milburn SC, Edery I, Sonenberg N, Hershey JW (1982) Inhibition of HeLa cell protein synthesis following poliovirus infection correlates with the proteolysis of a 220,000-dalton polypeptide associated with eucaryotic initiation factor 3 and a cap binding protein complex. J Biol Chem 257:14806–14810

Evlashev A, Moyse E, Valentin H, Azocar O, Trescol-Biemont MC, Marie JC, Rabourdin-Combe C, Horvat B (2000) Productive measles virus brain infection and apoptosis in CD46 transgenic mice. J Virol 74:1373–1382

Fabre S, Reymond N, Cocchi F, Menotti L, Dubreuil P, Campadelli-Fiume G, Lopez M (2002) Prominent role of the Ig-like V domain in trans-interactions of nectins. Nectin 3 and nectin 4 bind to the predicted C-C′-C″-D beta-strands of the nectin1V domain. J Biol Chem 277:27006–27013

Filman DJ, Syed R, Chow M, Macadam AJ, Minor PD, Hogle JM (1989) Structural factors that control conformational transitions and serotype specificity in type 3 poliovirus. EMBO J 8:1567–1579

Foger N, Marhaba R, Zoller M (2000) CD44 supports T cell proliferation and apoptosis by apposition of protein kinases. Eur J Immunol 30:2888–2899

Freistadt MS, Eberle KE (1997) Physical association between CD155 and CD44 in human monocytes. Mol Immunol 34:1247–1257

Freistadt MS, Eberle KE (2000) Hematopoietic cells from CD155-transgenic mice express CD155 and support poliovirus replication ex vivo. Microbial Pathogenesis 29:203–212

Freistadt MS, Fleit HB, Wimmer E (1993) Poliovirus receptor on human blood cells: a possible extraneural site of poliovirus replication. Virology 195:798–803

Fuchs A, Cella M, Giurisato E, Shaw AS, Colonna M (2004) Cutting edge: CD96 (tactile) promotes NK cell-target cell adhesion by interacting with the poliovirus receptor (CD155). J Immunol 172:3994–3998
Girard S, Couderc T, Destombes J, Thiesson D, Delpeyroux F, Blondel B (1999) Poliovirus induces apoptosis in the mouse central nervous system. J Virol 73:6066–6072
Girard S, Gosselin AS, Pelletier I, Colbère-Garapin F, Couderc T, Blondel B (2002) Restriction of poliovirus RNA replication in persistently infected nerve cells. J Gen Virol 83:1087–1093
Goldstaub D, Gradi A, Bercovitch Z, Grosmann Z, Nophar Y, Luria S, Sonenberg N, Kahana C (2000) Poliovirus 2A protease induces apoptotic cell death. Mol Cell Biol 20:1271–1277
Gosselin AS, Simonin Y, Guivel-Benhassine F, Rincheval V, Vayssiere JL, Mignotte B, Colbère-Garapin F, Couderc T, Blondel B (2003) Poliovirus-induced apoptosis is reduced in cells expressing a mutant CD155 selected during persistent poliovirus infection in neuroblastoma cells. J Virol 77:790–798
Gradi A, Svitkin YV, Imataka H, Sonenberg N (1998) Proteolysis of human eukaryotic translation initiation factor eIF4GII, but not eIF4GI, coincides with the shutoff of host protein synthesis after poliovirus infection. Proc Natl Acad Sci USA 95:11089–11094
Green DR, Amarante-Mendes GP (1998) The point of no return: mitochondria, caspases, and the commitment to cell death. Results Probl Cell Differ 24:45–61
Griffin DE, Hardwick JM (1999) Perspective: virus infections and the death of neurons. Trends Microbiol 7:155–160
Gromeier M, Lachmann S, Rosenfeld MR, Gutin PH, Wimmer E (2000a) Intergeneric poliovirus recombinants for the treatment of malignant glioma. Proc. Natl Acad Sci USA 97:6803–6808
Gromeier M, Nomoto A (2002) Determinants of poliovirus pathogenesis. In: Semler BL, Wimmer E (eds) Molecular biology of picornaviruses. ASM Press, Washington DC, pp 367–379
Gromeier M, Solecki D, Patel DD, Wimmer E (2000b) Expression of the human poliovirus receptor/CD155 gene during development of the central nervous system: Implications for the pathogenesis of poliomyelitis. Virology 273:248–257
Gromeier M, Wimmer E (1998) Mechanism of injury-provoked poliomyelitis. J Virol 72:5056–5060
Gromeier M, Wimmer E, Gorbalenya AE (1999) Genetics, pathogenesis and evolution of picornaviruses. In: Domingo E, Webster RG, Holland JJ (eds) Origin and evolution of viruses. Academic Press, New York, pp 287–343
Gustin KE (2003) Inhibition of nucleo-cytoplasmic trafficking by RNA viruses: targeting the nuclear pore complex. Virus Res 95:35–44
Gustin KE, Sarnow P (2001) Effects of poliovirus infection on nucleo-cytoplasmic trafficking and nuclear pore complex composition. EMBO J 20:240–249
Haller A, Semler B (1995) Translation and host cell shutoff. In: Rotbart HA (ed) Human enterovirus infection. ASM Press, Washington, DC, pp 113–133
He YN, Bowman VD, Mueller S, Bator CM, Bella J, Peng XH, Baker TS, Wimmer E, Kuhn RJ, Rossmann MG (2000) Interaction of the poliovirus receptor with poliovirus. Proc Natl Acad Sci USA 97:79–84

Hellen C, Wimmer E (1995) Enterovirus structure and assembly. In: Rotbart HA (ed) Human enterovirus infection. ASM Press, Washington, DC, pp 155–174
Hogle JM (2002) Poliovirus cell entry: common structural themes in viral cell entry pathways. Annu Rev Microbiol 56:677–702
Hogle JM, Chow M, Filman DJ (1985) Three dimensional structure of poliovirus at 2.9 Å resolution. Science 229:1358–1365
Hogle JM, Racaniello VR (2002) Poliovirus receptors and cell entry. In: Semler BL, Wimmer E (eds) Molecular biology of picornaviruses. ASM Press, Washington DC, pp 71–83
Ida-Hosonuma M, Iwasaki T, Taya C, Sato Y, Li J, Nagata N, Yonekawa H, Koike S (2002) Comparison of neuropathogenicity of poliovirus in two transgenic mouse strains expressing human poliovirus receptor with different distribution patterns. J Gen Virol 83:1095–1105
Iwasaki A, Welker R, Mueller S, Linehan M, Nomoto A, Wimmer E (2002) Immunofluorescence analysis of poliovirus receptor expression in Peyer's patches of humans, primates, and CD155 transgenic mice: implications for poliovirus infection. J Infect Dis 186:585–592
Jackson AC, Rossiter JP (1997) Apoptosis plays an important role in experimental rabies virus infection. J Virol 71:5603–5607
Jackson R (2002) Proteins involved in the function of picornavirus internal ribosomal entry sites. In: Semler BL, Wimmer E (eds) Molecular biology of picornaviruses. ASM Press, Washington DC, pp 171–183
Jan J-T, Griffin DE (1999) Induction of apoptosis by Sindbis virus occurs at cell entry and does not require virus replication. J Virol 73:10296–10302
Joachims M, Harris KS, Etchison D (1995) Poliovirus protease 3C mediates cleavage of microtubule-associated protein 4. Virology 211:451–461
Joachims M, Van Breugel PC, Lloyd RE (1999) Cleavage of poly(A)-binding protein by enterovirus proteases concurrent with inhibition of translation in vitro. J Virol 73:718–727
Karttunen A, Poyry T, Vaarala O, Ilonen J, Hovi T, Roivainen M, Hyypia T (2003) Variation in enterovirus receptor genes. J Med Virol 70:99–108
Kaufmann SH, Hengartner MO (2001) Programmed cell death: alive and well in the new millennium. Trends Cell Biol. 11:526–534
Kew O, Morris-Glasgow V, Landaverde M, Burns C, Shaw J, Garib Z, Andre J, Blackman E, Freeman CJ, Jorba J, Sutter R, Tambini G, Venczel L, Pedreira C, Laender F, Shimizu H, Yoneyama T, Miyamura T, van Der Avoort H, Oberste MS, Kilpatrick D, Cochi S, Pallansch M, de Quadros C (2002) Outbreak of poliomyelitis in Hispaniola associated with circulating type 1 vaccine-derived poliovirus. Science 296:356–359
Kitamura N, Semler BL, Rothberg PG, Larsen GR, Adler CJ, Dorner AJ, Emini EA, Hanecak R, Lee JJ, Van der Werf S, Anderson CW, Wimmer E (1981) Primary structure, gene organization and polypeptide expression of poliovirus RNA. Nature 291:547–553
Koike S, Horie H, Ise I, Okitsu A, Yoshida M, Iizuka N, Takeuchi K, Takegami T, Nomoto A (1990) The poliovirus receptor protein is produced both as membrane-bound and secreted forms. EMBO J 9:3217–3224

Koike S, Ise I, Nomoto A (1991a) Functional domains of the poliovirus receptor. Proc Natl Acad Sci USA 88:4104–4108

Koike S, Ise I, Sato Y, Yonekawa H, Gotoh O, Nomoto A (1992) A 2nd gene for the African green monkey poliovirus receptor that has no putative N-glycosylation site in the functional N-terminal immunoglobulin-like domain. J Virol 66:7059–7066

Koike S, Taya C, Kurata T, Abe S, Ise I, Yonekawa H, Nomoto A (1991b) Transgenic mice susceptible to poliovirus. Proc Natl Acad Sci USA 88:951–955

Koyama AH, Irie H, Ueno F, Ogawa M, Nomoto A, Adachi A (2001) Suppression of apoptotic and necrotic cell death by poliovirus. J Gen Virol 82:2965–2972

Kräusslich HG, Nicklin MJ, Toyoda H, Etchison D, Wimmer E (1987) Poliovirus proteinase 2A induces cleavage of eukaryotic initiation factor 4F polypeptide p220. J Virol 61:2711–2718

Kuechler E, Seipelt J, Liebig H-D, Sommergruber W (2002) Picornavirus proteinase-mediated shutoff of host cell translation: direct cleavage of a cellular initiation factor. In: Semler BL, Wimmer E (eds) Molecular biology of picornaviruses. ASM Press, Washington DC, pp 301–311

Kuo RL, Kung SH, Hsu YY, Liu WT (2002) Infection with enterovirus 71 or expression of its 2A protease induces apoptotic cell death. J Gen Virol 83:1367–1376

Kuyumcu-Martinez NM, Van Eden ME, Younan P, Lloyd RE (2004) Cleavage of poly(A)-binding protein by poliovirus 3C protease inhibits host cell translation: a novel mechanism for host translation shutoff. Mol Cell Biol 24:1779–1790

Lange R, Peng X, Wimmer E, Lipp M, Bernhard G (2001) The poliovirus receptor CD155 mediates cell-to-matrix contacts by specifically binding to vitronectin. Virology 285:218–227

Lentz KN, Smith AD, Geisler SC, Cox S, Buontempo P, Skelton A, Demartino J, Rozhon E, Schwartz J, Girijavallabhan V, Oconnell J, Arnold E (1997) Structure of poliovirus type 2 Lansing complexed with antiviral agent sch48973—comparison of the structural and biological properties of the three poliovirus serotypes. Structure 5:961–978

Leon-Monzon ME, Dalakas MC (1995) Detection of poliovirus antibodies and poliovirus genome in patients with post-polio syndrome (PPS). In: Dalakas MC, Bartfeld H, Kurland LT (eds) The post-polio syndrome, vol 753. The New York Academy of Sciences, New York, pp 208–218

Leparc-Goffart I, Julien J, Fuchs F, Janatova I, Aymard M, Kopecka H (1996) Evidence of presence of poliovirus genomic sequences in cerebrospinal fluid from patients with postpolio syndrome. J Clin Microbiol 34:2023–2026

Levine B (2002) Apoptosis in viral infections of neurons: a protective or pathologic host response? Curr Top Microbiol Immunol 265:95–118

Levine B, Huang Q, Isaacs JT, Reed JC, Griffin DE, Hardwick JM (1993) Conversion of lytic to persistent alphavirus infection by the bcl-2 cellular oncogene. Nature 361:739–742

Lewis J, Wesselingh SL, Griffin DE, Hardwick JM (1996) Alphavirus-induced apoptosis in mouse brains correlates with neurovirulence. J Virol 70:1828–1835

Li ML, Hsu TA, Chen TC, Chang SC, Lee JC, Chen CC, Stollar V, Shih SR (2002) The 3C protease activity of enterovirus 71 induces human neural cell apoptosis. Virology 293:386–395

Lopez M, Eberle F, Mattei MG, Gabert J, Birg F, Bardin F, Maroc C, Dubreuil P (1995) Complementary DNA characterization and chromosomal localization of a human gene related to the poliovirus receptor-encoding gene. Gene 155:261–265

Lopez-Guerrero JA, Alonso M, Martin-Belmonte F, Carrasco L (2000) Poliovirus induces apoptosis in the human U937 promonocytic cell line. Virology 272:250–256

Lwoff A, Dulbecco R, Vogt M, Lwoff M (1955) Kinetics of the release of poliomyelitis virus from single cells. Virology 1:128–139

MacLennan C, Dunn G, Huissoon AP, Kumararatne DS, Martin J, O'Leary P, Thompson RA, Osman H, Wood P, Minor P, Wood DJ, Pillay D (2004) Failure to clear persistent vaccine-derived neurovirulent poliovirus infection in an immunodeficient man. Lancet 363:1509–1513

Manchester M, Eto DS, Oldstone MB (1999) Characterization of the inflammatory response during acute measles encephalitis in NSE-CD46 transgenic mice. J Neuroimmunol 96:207–217

Martin J, Dunn G, Hull R, Patel V, Minor PD (2000) Evolution of the Sabin strain of type 3 poliovirus in an immunodeficient patient during the entire 637-day period of virus excretion. J Virol 74:3001–3010

Martinez-Moralez JR, Barbas JA, Marti E, Bovolenta P, Edgar D, Rodriguez-Tébar A (1997) Vitronectin is expressed in the ventral region of the neural tube and promotes the differentiation of motor neurons. Development 124:5139–5147

Masson D, Jarry A, Baury B, Blanchardie P, Laboisse C, Lustenberger P, Denis MG (2001) Overexpression of the CD155 gene in human colorectal carcinoma. Gut 49:236–240

McBride AE, Schlegel A, Kirkegaard K (1996) Human protein Sam68 relocalization and interaction with poliovirus RNA polymerase in infected cells. Proc Natl Acad Sci USA 93:2296–2301

Meerovitch K, Pelletier J, Sonenberg N (1989) A cellular protein that binds to the 5′-noncoding region of poliovirus RNA : implications for internal translation initiation. Genes Dev 3:1026–1034

Mendelsohn CL, Wimmer E, Racaniello VR (1989) Cellular receptor for poliovirus: molecular cloning, nucleotide sequence and expression of a new member of the immunoglobulin superfamily. Cell 56:855–865

Minor P (1997) Poliovirus. In: Nathanson N (ed) Viral pathogenesis. Lippincott-Raven, Philadelphia, pp 555–574

Morrison ME, He YJ, Wien MW, Hogle JM, Racaniello VR (1994) Homolog-scanning mutagenesis reveals poliovirus receptor residues important for virus binding and replication. J Virol 68:2578–2588

Mueller S, Cao X, Welker R, Wimmer E (2002) Interaction of the poliovirus receptor CD155 with the dynein light chain Tctex-1 and its implication for poliovirus pathogenesis. J Biol Chem 277:7897–7904

Mueller S, Wimmer E (2003) Recruitment of nectin-3 to cell-cell junctions through trans-heterophilic interaction with CD155, a vitronectin and poliovirus receptor that localizes to alpha(v)beta3 integrin-containing membrane microdomains. J Biol Chem 278:31251–31260

Muir P, Nicholson F, Sharief MK, Thompson EJ, Cairns NJ, Lantos P, Spencer GT, Kaminski HJ, Banatvala JE (1995) Evidence for persistent enterovirus infection of the central nervous system in patients with previous paralytic poliomyelitis. In: Dalakas MC, Bartfeld H, Kurland LT (eds) The post-polio syndrome, vol 753. The New York Academy of Sciences, New York, pp 219–232

Nagata N, Iwasaki T, Ami Y, Sato Y, Hatano I, Harashima A, Suzaki Y, Yoshii T, Hashikawa T, Sata T, Horiuchi Y, Koike S, Kurata T, Nomoto A (2004) A poliomyelitis model through mucosal infection in transgenic mice bearing human poliovirus receptor, TgPVR21. Virology 321:87–100

Nathanson N, Langmuir AD (1963) The Cutter incident. Am J Hyg 78:16–81

Neznanov N, Chumakov KP, Ullrich A, Agol VI, Gudkov AV (2002) Unstable receptors disappear from cell surface during poliovirus infection. Med Sci Monit 8: BR391–BR396

Neznanov N, Kondratova A, Chumakov KM, Angres B, Zhumabayeva B, Agol VI, Gudkov AV (2001) Poliovirus protein 3A inhibits tumor necrosis factor (TNF)-induced apoptosis by eliminating the TNF receptor from the cell surface. J Virol 75:10409–10420

Novoa I, Carrasco L (1999) Cleavage of eukaryotic translation initiation factor 4G by exogenously added hybrid proteins containing poliovirus 2A(pro) in HeLa cells: Effects on gene expression. Mol Cell Biol 19:2445–2454

Nugent CI, Johnson KL, Sarnow P, Kirkegaard K (1999) Functional coupling between replication and packaging of poliovirus replicon RNA. J Virol 73:427–435

O'Brien V (1998) Viruses and apoptosis. J Gen Virol 79: 1833–1845

Oberhaus SM, Smith RL, Clayton GH, Dermody TS, Tyler KL (1997) Reovirus infection and tissue injury in the mouse central nervous system are associated with apoptosis. J Virol 71:2100–2106

Ohka S, Matsuda N, Tohyama K, Oda T, Morikawa M, Kuge S, Nomoto A (2004) Receptor (CD155)-dependent endocytosis of poliovirus and retrograde axonal transport of the endosome. J Virol 78:7186–7198

Ohka S, Nomoto A (2001) Recent insights into poliovirus pathogenesis. Trends Microbiol 9:501–506

Ohka S, Ohno H, Tohyama K, Nomoto A (2001) Basolateral sorting of human poliovirus receptor alpha involves an interaction with the mu1B subunit of the clathrin adaptor complex in polarized epithelial cells. Biochem. Biophys. Res Commun 287:941–948

Ohka S, Yang W-X, Terada E, Iwasaki K, Nomoto A (1998) Retrograde transport of intact poliovirus through the axon via the fast transport system. Virology 250:67–75

Okamoto I, Kawano Y, Murakami D, Sasayama T, Araki N, Miki T, Wong AJ, Saya H (2001) Proteolytic release of CD44 intracellular domain and its role in the CD44 signaling pathway. J Cell Biol 155:755–762

Ouzilou L, Caliot E, Pelletier I, Prevost MC, Pringault E, Colbère-Garapin F (2002) Poliovirus transcytosis through M-like cells. J Gen Virol 83:2177–2182

Pallansch M, Roos R (2001) Enteroviruses: polioviruses, coxsackieviruses, echoviruses, and newer enteroviruses. In: Knipe DM, Howley PM (eds) Fields Virology, vol 1. Lippincott Williams and Wilkins, Philadelphia, pp 723–775

Paul AV (2002) Possible unifying mechanisms of picornavirus genome replication. In: Semler BL, Wimmer E (eds) Molecular biology of picornaviruses. ASM Press, Washington DC, pp 227–246

Pavio N, Buc-Caron M-H, Colbère-Garapin F (1996) Persistent poliovirus infection of human fetal brain cells. J Virol 70:6395–6401

Pavio N, Couderc T, Girard S, Sgro JY, Blondel B, Colbère-Garapin F (2000) Expression of mutated receptors in human neuroblastoma cells persistently infected with poliovirus. Virology 274:331–342

Pelletier I, Ouzilou L, Arita M, Nomoto A, Colbère-Garapin F (2003) Characterization of the poliovirus 147S particle: new insights into poliovirus uncoating. Virology 305:55–65

Petito CK, Roberts B (1995) Evidence of apoptotic cell death in HIV encephalitis. Am J Pathol 146:1121–1130

Ponnuraj EM, John TJ, Levin MJ, Simoes EAF (1998) Cell-to-cell spread of poliovirus in the spinal cord of bonnet monkeys (*Macaca radiata*). J Gen Virol 79:2393–2403

Putcha GV, Moulder KL, Golden JP, Bouillet P, Adams JA, Strasser A, Johnson EM (2001) Induction of BIM, a proapoptotic BH3-only BCL-2 family member, is critical for neuronal apoptosis. Neuron 29:615–628

Puthalakath H, Huang DC, O'Reilly LA, King SM, Strasser A (1999) The proapoptotic activity of the Bcl-2 family member Bim is regulated by interaction with the dynein motor complex. Mol Cell 3:287–296

Racaniello VR (2001) Picornaviridae: the viruses and their replication. In: Knipe DM, Howley PM (eds) Fields Virology, vol 1. Lippincott Williams and Wilkins, Philadelphia, pp 685–722

Ren R, Racaniello VR (1992) Poliovirus spreads from muscle to the central nervous system by neural pathways. J Infect Dis 166:747–752

Ren RB, Costantini F, Gorgacz EJ, Lee JJ, Racaniello VR (1990) Transgenic mice expressing a human poliovirus receptor: a new model for poliomyelitis. Cell 63:353–362

Roulston A, Marcellus R, Branton PE (1999) Virus and apoptosis. Annu Rev Microbiol 53:577–628

Rousset D, Rakoto-Andrianarivelo M, Razafindratsimandresy R, Randriamanalina B, Guillot S, Balanant J, Mauclère P, Delpeyroux F (2003) Emergence of recombinant vaccine-derived poliovirus in Madagascar. Emerging Infect Dis 9:885–887

Rubinstein SJ, Dasgupta A (1989) Inhibition of rRNA synthesis by poliovirus: specific inactivation of transcription factors. J Virol 63:4689–4696

Rubinstein SJ, Hammerle T, Wimmer E, Dasgupta A (1992) Infection of HeLa cells with poliovirus results in modification of a complex that binds to the rRNA promoter. J Virol 66:3062–3068

Sabin AB, Boulger LR (1973) History of Sabin attenuated poliovirus oral live vaccine strains. J Biol Stand 1:115–118

Salk JE (1955) Consideration in the preparation and use of poliomyelitis virus vaccine. JAMA 1548:1239–1248

Sanchez I, Yuan J (2001) A convoluted way to die. Neuron 29:563–566

Saunderson R, Yu B, Trent RJ, Pamphlett R (2004) A polymorphism in the poliovirus receptor gene differs in motor neuron disease. Neuroreport 15:383–386

Schlegel A, Giddings TH, Ladinsky MS, Kirkegaard K (1996) Cellular origin and ultrastructure of membranes induced during poliovirus infection. J Virol 70:6576–6588

Sharief MK, Hentges MR, Ciardi M (1991) Intrathecal immune response in patients with the post-polio syndrome. N Engl J Med 325:749–755

Shepley MP, Racaniello VR (1994) A monoclonal antibody that blocks poliovirus attachment recognizes the lymphocyte homing receptor CD44. J Virol 68:1301–1308

Shiroki K, Isoyama T, Kuge S, Ishii T, Ohmi S, Hata S, Suzuki K, Takasaki Y, Nomoto A (1999) Intracellular redistribution of truncated La protein produced by poliovirus 3Cpro-mediated cleavage. J Virol 73:2193–2200

Sicinski P, Rowinski J, Warchol JB, Jarzabek Z, Gut W, Szczygiel B, Bielicki K, Koch G (1990) Poliovirus type 1 enters the human host through intestinal M cells. Gastroenterology 98:56–58

Solecki DJ, Gromeier M, Mueller S, Bernhardt G, Wimmer E (2002) Expression of the human poliovirus Receptor/CD155 gene is activated by sonic-hedgehog. J Biol Chem 277:25697–25702

Strebel PM, Ion-Nedelcu N, Baughman AL, Sutter RW, Cochi SL (1995) Intramuscular injections within 30 days of immunization with oral poliovirus vaccine—a risk factor for vaccine-associated paralytic poliomyelitis. N Engl J Med 332:500–506

Tahara-Hanaoka S, Shibuya K, Onoda Y, Zhang H, Yamazaki S, Miyamoto A, Honda S, Lanier LL, Shibuya A (2004) Functional characterization of DNAM-1 (CD226) interaction with its ligands PVR (CD155) and nectin-2 (PRR-2/CD112). Int Immunol 16:533–538

Takahashi K, Nakanishi H, Miyahara M, Mandai K, Satoh K, Satoh A, Nishioka H, Aoki J, Nomoto A, Mizoguchi A, Takai Y (1999) Nectin/PRR: an immunoglobulin-like cell adhesion molecule recruited to cadherin-based adherens junctions through interaction with Afadin, a PDZ domain-containing protein. J Cell Biol 145:539–549

Teodoro JG, Branton PE (1997) Regulation of apoptosis by viral gene products. J Virol 71:1739–1746

Teterina NL, Gorbalenya AE, Egger D, Bienz K, Ehrenfeld E (1997) Poliovirus 2C protein determinants of membrane binding and rearrangements in mammalian cells. J Virol 71:8962–8972

Thornberry NA, Lazebnik Y (1998) Caspases: enemies within. Science 281:1312–1316

Tolskaya EA, Romanova L, Kolesnikova MS, Ivannikova TA, Smirnova EA, Raikhlin NT, Agol VI (1995) Apoptosis-inducing and apoptosis-preventing functions of poliovirus. J Virol 69:1181–1189

Tsunoda I, Kurtz CIB, Fujinami RS (1997) Apoptosis in acute and chronic central nervous system disease induced by Theiler's murine encephalomyelitis virus. Virology 228:388–393

Tucker SP, Thornton CL, Wimmer E, Compans RW (1993) Vectorial release of poliovirus from polarized human intestinal epithelial cells. J Virol 67:4274–4282

Tyler KL, Clarke P, DeBiasi RL, Kominsky D, Poggioli GJ (2001) Reoviruses and the host cell. Trends Microbiol 9:560–564

Umehara F, Nakamura A, Izumo S, Kubota R, Ijchi S, Kashio N, Hashimoto K-I, Usuku K, Sato E, Osame M (1994) Apoptosis of T lymphocytes in the spinal cord lesions in HTLV-I-associated myelopathy: a possible mechanism to control viral infection in the central nervous system. J Neuropathol Exp Neurol 53:617–624

Viswanath V, Wu Y, Boonplueang R, Chen S, Stevenson FF, Yantiri F, Yang L, Beal MF, Andersen JK (2001) Caspase-9 activation results in downstream caspase-8 activation and bid cleavage in 1-methyl-4-phenyl-1,2,3,6-tetrahydropyridine-induced Parkinson's disease. J Neurosci 21:9519–9528

Waggoner S, Sarnow P (1998) Viral ribonucleoprotein complex formation and nucleolar-cytoplasmic relocalization of nucleolin in poliovirus-infected cells. J Virol 72:6699–6709

Wallach D (1997) Cell death induction by TNF: a matter of self control. Trends Biochem Sci 22:107–109

Weidman MK, Yalamanchili P, Ng B, Tsai W, Dasgupta A (2001) Poliovirus 3C protease-mediated degradation of transcriptional activator p53 requires a cellular activity. Virology 291:260–271

Xiang W, Paul AV, Wimmer E (1997) RNA signals in entero- and rhinovirus genome replication. Semin Virol 8:256–273

Yalamanchili P, Datta U, Dasgupta A (1997a) Inhibition of host cell transcription by poliovirus: cleavage of transcription factor CREB by poliovirus-encoded protease 3Cpro. J Virol 71:1220–1226

Yalamanchili P, Harris K, Wimmer E, Dasgupta A (1996) Inhibition of basal transcription by poliovirus: a virus- encoded protease (3Cpro) inhibits formation of TBP-TATA box complex in vitro. J Virol 70:2922–2929

Yalamanchili P, Weidman K, Dasgupta A (1997b) Cleavage of transcriptional activator Oct-1 by poliovirus encoded protease 3Cpro. Virology 239:176–185

Yang C, Naguib T, Yang SJ, Nasr E, Jorba J, Ahmed N, Campagnoli R, van der Avoort H, Shimizu H, Yoneyama T, Miyamura T, Pallansch M, Kew O (2003) Circulation of endemic type 2 vaccine-derived poliovirus in Egypt from 1983 to 1993. J Virol 77:8366–8377

Yang W-X, Terasaki T, Shiroki K, Ohka S, Aoki J, Tanabe S, Nomura T, Terada E, Sugiyama Y, Nomoto A (1997) Efficient delivery of circulating poliovirus to the central nervous system independently of poliovirus receptor. Virology 229:421–428

Zamora M, Marissen WE, Lloyd RE (2002) Poliovirus-mediated shutoff of host translation: an indirect effect. In: Semler BL, Wimmer E (eds) Molecular biology of picornaviruses. ASM Press, Washington DC, pp 313–320

CTMI (2005) 289:57–78

Neuronal Cell Death in Alphavirus Encephalomyelitis

D. E. Griffin

W. Harry Feinstone Department of Molecular Microbiology and Immunology, Johns Hopkins Bloomberg School of Public Health, 615 N. Wolfe St, Rm E5132, Baltimore, MD 21205, USA
dgriffin@jhsph.edu

Abstract Alphaviruses are mosquito-borne, enveloped, plus-strand RNA viruses that cause a spectrum of diseases in humans that include fever, rash, arthritis, meningitis, and encephalomyelitis. Sindbis virus (SINV) is the prototype alphavirus, causes encephalomyelitis in mice, and provides a model system for studying the pathogenesis of alphavirus-induced neurological disease. Major target cells for SINV infection in the central nervous system (CNS) are neurons, and both host and viral factors determine the fate of infected neurons. Young animals are most susceptible to fatal disease. This correlates with the ability of SINV to induce apoptosis in immature neurons. In vitro, apoptotic death of neuroblastoma cells can be induced by fusion of the virus envelope with the endosomal membrane and does not require infectious virus. This fusion process activates acid sphingomyelinase that cleaves sphingomyelin to release ceramide, an initiator of apoptosis. Within an hour, poly(ADP-ribose) polymerase is activated, and this is followed by release of cytochrome *c* and activation of effector caspases. SINV-induced cell death can be delayed or prevented by treatment with antioxidants or caspase inhibitors and by intracellular expression of Bcl-2, Beclin-1, or protease inhibitors. Older animals survive infection unless infected with a neurovirulent strain of SINV. In these mice, anterior horn motor neurons die by a primarily necrotic process that is influenced by excitotoxic amino acids and inflammation, whereas hippocampal neurons can be either apoptotic or necrotic. Death also occurs in uninfected neurons in the vicinity of infected neurons and can be delayed or prevented by treatment with glutamate receptor antagonists.

Abbreviations

AIF	Apoptosis-inducing factor
AMPAAMPA	α-Amino-3-hydroxy-5-methyl-4-isoxazole propionic acid
aSMase	Acidic sphingomyelinase
CAD	Caspase-activated DNase
CNS	Central nervous system
IM	Inner membrane
MMP	Mitochondrial membrane permeabilization
NAD	β-Nicotinamide adenine dinucleotide
NMDANMDA	*N*-methyl-D-aspartate
OM	Outer membrane
PARPPARP - poly(ADP-ribose) polymerase	Poly(ADP-ribose) polymerase
ROSROS - reactive oxygen species	Reactive oxygen species
SINV	Sindbis virus
SM	Sphingomyelin
TNF	Tumor necrosis factor

1 Introduction

Arthropod-borne viruses are the most important causes of acute encephalitis worldwide. The mosquito-borne flaviviruses Japanese encephalitis virus and West Nile virus are widespread and cause severe encephalomyelitis in Asia, Europe, and North America (Burke and Monath 2003). Alphaviruses cause a spectrum of diseases in humans that include fever, rash, arthritis, meningitis, and encephalomyelitis (Griffin 2003). Sindbis virus (SINV) is the most widespread of the alphaviruses and is found in Europe, Africa, the Asian subcontinent, and Australia. SINV is related to western equine encephalitis virus (Hahn et al. 1988), which has caused widespread outbreaks of encephalitis in North America (Longshore et al. 1956) and to Mayaro virus, an emerging cause of rash and arthritis in South America (Tesh et al. 1999). Of the encephalitic alphaviruses eastern equine encephalomyelitis virus has a high mortality in all age groups (Calisher 1994) and Venezuelan equine encephalitis virus can cause infection by the respiratory route (Danes et al. 1973; Lennette and Koprowski 1943) and has been weaponized for use in biowarfare.

SINV, the prototype alphavirus, causes encephalomyelitis in mice and provides a model system for studying the pathogenesis of alphavirus-induced neurological disease. Major target cells for SINV infection in the central nervous system (CNS) are neurons, and both host and viral factors determine the fate of infected neurons. Young animals are most susceptible to fatal disease, and SINV induces apoptosis in immature neurons, as well as in a variety of cultured cell lines, including neuroblastoma cells.

Many viruses induce apoptosis in infected tissue culture cells, but studies of virus-induced apoptosis in vivo are more limited. Survival of neurons that are infected by neurotropic viruses in vivo is determined by the maturity of the infected neuron and the virulence of the infecting virus. Encephalitis is more likely to complicate infection with a potentially neurotropic virus, and the disease is more likely to be fatal when infection occurs in young children or the elderly. In mice, many neurotropic viruses cause more severe disease and replicate to a higher titer in newborn mice than in older weanling and adult mice and there is often a sharp decrease in susceptibility to fatal infection sometime during the first 2 weeks of life (Johnson and Johnson 1968; Johnson et al. 1972; Ogata et al. 1991; Walder and Bradish 1975). During this period, neuronal axogenesis and synaptogenesis are being completed and neurons are becoming independent of trophic factors for survival. Decreased susceptibility to fatal disease reflects an intrinsic property of maturing neurons rather than a change in innate or adaptive immune responses to the virus (Oliver et al. 1997). The cellular factors that correlate with this increased resistance of neurons to apoptosis have not been clearly identified.

Virus-induced apoptosis of immature neurons in vivo has been documented for several neurotropic viruses including alphaviruses (Allsopp et al. 1998; Lewis et al. 1996), flaviviruses (Despres et al. 1998), reoviruses (Oberhaus et al. 1997), and bunyaviruses (Pekosz et al. 1996). Newborn mice often die rapidly after infection, coincident with increasing levels of virus in the CNS (Johnson et al. 1972). As during development, virus-induced neuronal death is often characterized by features of apoptosis, such as nuclear condensation, cytoplasmic blebbing, and cell shrinkage. In older mice, both the animals and their neurons are more likely to survive infection. However, apoptosis has been observed in neurons of older animals infected with poliovirus (Girard et al. 1999), Venezuelan equine encephalitis virus (Jackson and Rossiter 1997), and Theiler's murine encephalomyelitis virus (Tsunoda et al. 1997), but for a number of other viruses neurovirulent for adult mice or rats, some populations of dying infected neurons are not apoptotic (Hase et al. 1990; Havert et al. 2000; Sammin et al. 1999).

2 Overview of Cell Death

2.1 Mechanisms of Cell Death

Understanding of the biology and mechanisms of cell death is rapidly improving. Although apoptotic and necrotic forms of cell death are morphologically separable and have been considered distinct processes, it is now clear that they constitute a continuum with many overlapping features (Leist et al. 1997; Leist and Jaattela 2001; Yuan et al. 2003). In addition, a third type of cell death linked to the autophagy pathway first described as a means of cellular recycling of the components of intracellular organelles during periods of stress or starvation has recently been recognized (Yu et al. 2004; Yuan et al. 2003) and is characterized by the development of double-membrane vesicles that fuse with lysosomes (Danial and Korsmeyer 2004; Yuan et al. 2003). All of these types of cell death can affect neurons.

Apoptosis is an energy-dependent, regulated process through which a cell self-destructs and is characterized by cell shrinkage, redistribution of phosphatidylserine to the outer surface of the plasma membrane, membrane blebbing, chromatin condensation, nuclear fragmentation, and, eventually, segmentation of the cell into membrane-bound apoptotic bodies that are engulfed by neighboring cells. This phagocytic disposal of dead and dying cells prevents release of cell contents and induction of inflammation. Necrosis is characterized by cell swelling and plasma membrane rupture and may induce an inflammatory response. The apoptotic process requires ATP and is converted into a necrotic process if cellular levels of ATP fall below a critical level (Haeberlein 2004). Cells then die by necrosis with ion pump failure, swelling of the cell, and rupture of the plasma membrane (Eguchi et al. 1997; Leist et al. 1997).

Cell death can be initiated externally at the cell membrane through death receptors that are in the tumor necrosis factor (TNF) receptor family (e.g., TNFR-1, Fas, DR-4/5) and assemble a death-inducing signaling complex in the cytoplasm to activate caspases 8 and 10. Caspase 8 can execute the apoptotic process by direct activation of caspases 3 and 7 or through a mitochondrial amplification pathway that results in activation of caspase 3 (Danial and Korsmeyer 2004). Caspases are a family of cysteine proteases that are synthesized as inactive proenzymes (zymogens) and require proteolytic removal of the prodomain for activation. Caspases are generally divided into the initiator caspases (e.g., 8 and 9), which have long prodomains and are autocatalytically processed and activated within a multimeric complex, and the effector caspases (e.g., 3 and 7), which have short prodomains and are processed and activated by the initiator caspases. Effector caspases degrade cellular proteins and activate endonucleases that ultimately kill a cell.

Cell death can also be initiated internally through cell stress/signaling pathways that originate in the endoplasmic reticulum. Both externally and internally initiated pathways can converge to induce selective mitochondrial membrane permeabilization (MMP). Mitochondria have two well-defined compartments: the matrix, surrounded by the inner membrane (IM), and the intermembrane space, between the IM and the outer membrane (OM). The IM is folded into numerous cristae and contains the protein complexes of the electron transport chain, ATP synthase, and adenine nucleotide translocator. It is almost impermeable, and the respiratory chain creates a measurable electrochemical gradient ($\Delta\Psi$). Apoptotic OM permeabilization leads to release of proteins that are normally confined to the intermembrane space: cytochrome *c*, pro-caspases 3 and 9, adenylate kinase 2, smac, endoG, omi, and apoptosis-inducing factor (AIF). Factors that control MMP are Bcl-2 family proteins, intracellular levels of Ca^{2+}, lipid mediators such as ceramide, GM3 ganglioside, and palmitate, reactive oxygen species (ROS), nitric oxide, and a variety of viral proteins (Haeberlein 2004; Kroemer and Reed 2000).

The Bcl-2 family has many members that regulate MMP, and they are subdivided into the antiapoptotic multidomain proteins (e.g., Bcl-2, Bcl-x_L, Mcl-1, Bcl-w), the proapoptotic multidomain proteins (e.g., Bax, Bak) and the proapoptotic "BH3 domain only" proteins (e.g., Bid, Bim, Dp5/Hrk, Bad). In classic apoptosis the process is initiated when Bid is activated by caspase 8-mediated cleavage (tBid) and together with Bax or Bak begins formation of pores in the OM. Pore formation can be blocked or slowed by the antiapoptotic proteins Bcl-2 and Bcl-x_L. Once pores are formed, intermembrane proteins are released into the cytoplasm. Once in the cytoplasm, cytochrome *c* and pro-caspase 9 function together with Apaf-1 in the cytoplasm to form the "apoptosome." In the apoptosome, activation of pro-caspase 9, an ATP-dependent process, results in cleavage and activation of pro-caspase 3 that leads to cleavage of a variety of cellular proteins. Caspase 3-mediated cleavage can destroy or activate cellular proteins. It is the activation of caspase-activated DNase (CAD) that degrades chromatin into the oligonucleosomal fragments that characterize late stages of the apoptotic process (Fig. 1).

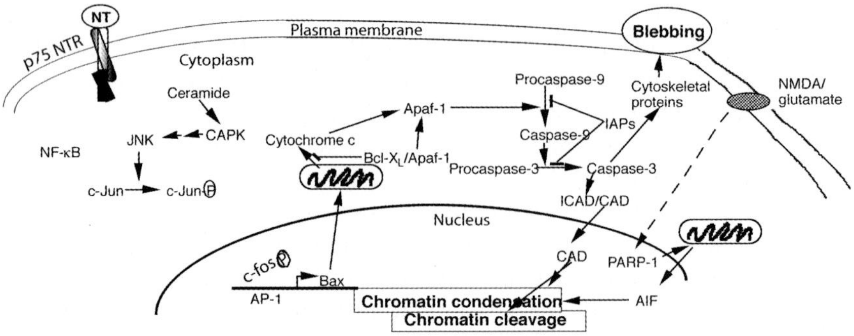

Fig. 1 Overview of the intracellular pathways that can be involved in induction of neuronal cell death. *AIF*, apoptosis-inducing factor; *CAD*, caspase-activated DNase; *CAPK*, ceramide-activated protein kinase; *IAP*, inhibitor of apoptosis protein; *NTR*, neurotrophin receptor

2.2 Neuronal Cell Death

Death of immature neurons is an important part of the normal development of the nervous system (Yuan and Yankner 2000). It is estimated that approximately half of the neurons produced during development die during the processes of migration and synapse formation necessary for generation of the mature nervous system (Burek and Oppenheim 1998). This developmental death is presumed to result from the failure of immature neurons to receive cellular signals from trophic factors required for survival and serves to match the numbers of innervating neurons to the size of their target cell populations (Becker and Bonni 2004). The process by which neurons die during development is apoptosis (Haeberlein 2004).

In immature neurons the apoptotic process is tightly controlled by the balance of survival and death signals. The Bcl-2, caspase, and Apaf-1 protein families constitute the core apoptotic machinery in neurons, as well as other types of cells. Apoptosis can be induced by withdrawal of neurotrophic growth factors and a complicated interaction of different neurotrophins with high- and low-affinity receptors, such as p75 NTR, a member of the TNFR family of death receptors (Fig. 1). As neurons mature, survival becomes independent of the presence of previously essential growth factors (Yuan and Yankner 2000) and survival signaling pathways predominate. Important survival signaling pathways are the Ras-MAPK and PI3K-Akt pathways (Yuan et al. 2003). Survival is presumably a result of changes in intracellular proteins during maturation that alter receptor expression, intracellular signaling, and the responses of the cell to withdrawal of a trophic factor.

Among the unique features of mature neurons is the fully differentiated, postmitotic state of the cells and the resultant inability of the host to replace neurons that are lost. Although many neurons die during developmental modeling of the nervous system, once neurons have been selected for inclusion in the nervous system the host has a vested interest in preventing neuronal death. The cellular factors that correlate with this increased resistance of neurons to apoptosis have not been clearly identified. Expression of the antiapoptotic proteins Bcl-2 and Bcl-x_L are actually at their highest levels in the CNS during embryogenesis rather than after development is complete. With maturation, expression of Bcl-x_L persists and is found in many populations of adult neurons (Krajewski et al. 1994) whereas Bcl-2 expression decreases (Merry et al. 1994). Therefore, expression of other cellular factors, such as the bifunctional apoptosis inhibitor (Roth et al. 2003), is likely to be involved.

Death of mature neurons can result from extrinsic or intrinsic insults and may be caspase dependent or caspase independent, apoptotic or necrotic (Leist and Jaattela 2001). Reactivation of cell cycle machinery in differentiated postmitotic neurons often accompanies apoptotic cell death in ways that are not yet fully understood (Becker and Bonni 2004). A common extrinsic mechanism for inducing death of mature neurons is excessive release of glutamate and resultant glutamate receptor activation and neuronal excitotoxicity (Andersson et al. 1993; Choi et al. 1987; Olney 1969). Excessive excitotoxic neurotransmitter release can lead to apoptotic or necrotic neuronal cell death depending on the maturity of the neurons, intensity of the injury, levels of intracellular Ca^{2+}, and mitochondrial function (Ankarcrona et al. 1995; Lee et al. 1999). An excess of glutamate can overstimulate one or more of the three types of glutamate receptors present on neurons, *N*-methyl-D-aspartate (NMDA), α-amino-3-hydroxy-5-methyl-4-isoxazole propionic acid (AMPA), and kainate, resulting in the influx of Ca^{2+}, Na^{+}, and/or Zn^{2+} ions through channels gated by these receptors. Excess intracellular Ca^{2+} can activate phospholipases, oxidases, proteases, and phosphatases that contribute to cell death (Lee et al. 1999; Nicotera and Orrenius 1998). Activated proteases include the calpains, a family of cytosolic, Ca^{2+}-activated cysteine proteases. Calpain in turn can activate cathepsins, cytoplasmic aspartyl proteases synthesized as zymogens. Calpain- and cathepsin-mediated degradation of cytoplasmic proteins is probably responsible for the

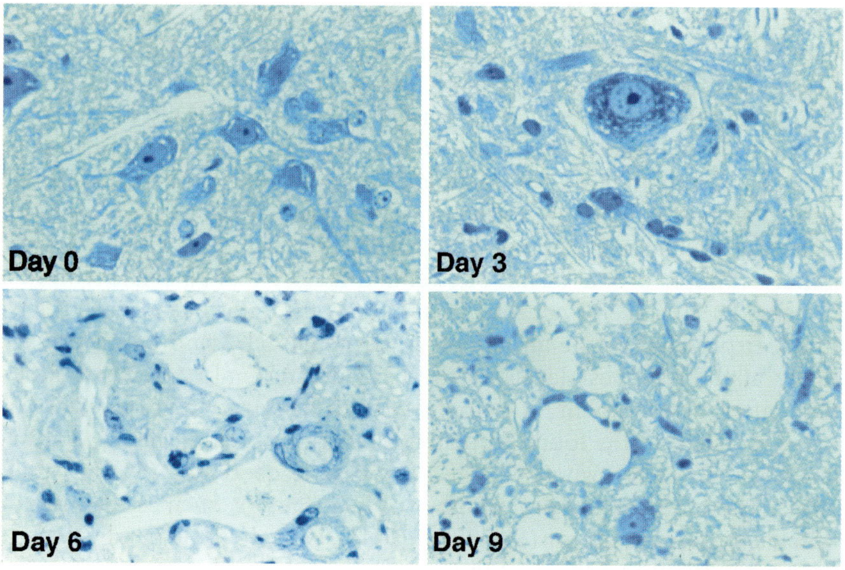

Fig. 2 Appearance of lumbar spinal cord motor neurons at various times after NSV infection of weanling C57BL/6 mice. Sections of spinal cords from the L4–L5 level of an uninfected 4-week-old mouse (**a**) and mice infected for 3 (**b**), 6 (**c**), and 9 (**d**) days. Three days after infection, some motor neurons showed mild swelling and cytoplasmic vacuolization (**b**). Six days after infection, many motor neurons were severely swollen and pale (**c**). Nine days after infection, nuclear and cytoplasmic membranes were no longer discernable, leaving empty holes in the ventral horn of the spinal cord (**d**). (Havert et al. 2000)

translucent appearance of the cytoplasm often found in necrotic cell death (Yuan et al. 2003) (Fig. 2).

NMDA receptors are highly permeable to Ca^{2+}, and treatment of neurons with NMDA receptor antagonists protects cells from glutamate-induced excitotoxic death. Therefore, this subtype of receptor has been thought to be most important in mediating excitotoxic death of mature neurons (Choi et al. 1998). However, the importance of AMPA receptors, which are generally impermeable to Ca^{2+}, for inducing excitotoxic death of certain populations of neurons has recently been appreciated. AMPA receptors are ligand-gated heteromeric channels assembled from various combinations of four subunits, GluR-1 through -4. The GluR-2 subunit is responsible for limiting Ca^{2+} permeability of AMPA-type glutamate receptors. Neurological insults can downregulate expression of mRNA encoding GluR-2 and increase the potential for Ca^{2+} influx through

AMPA receptors (Gorter et al. 1997). AMPA receptor-mediated excitotoxicity is postulated to contribute to neuronal damage in ischemia, seizures, amyotrophic lateral sclerosis, and Alzheimer disease (Buchan et al. 1991; Gill and Lodge 1997; Nellgard and Wieloch 1992; Sherdown et al. 1993; Xue et al. 1994).

A key factor in control of the mode of neuronal death may be the protein-modifying enzyme poly(ADP-ribose) polymerase (PARP). PARP catalyzes the transformation of β-nicotinamide adenine dinucleotide (NAD^+) into nicotinamide and poly(ADP-ribose) and the successive transfer of ADP-ribose units to a variety of proteins including PARP itself to produce linear or branched homopolymers. These polymers are rapidly degraded by poly(ADP-ribose) glycohydrolase. The most abundant protein in the PARP family of ADP-ribosylating proteins is PARP-1 (1 molecule for every 1,000 bp DNA) (Ha and Snyder 2000). Under homeostatic conditions, PARP-1 participates in genome repair, DNA replication, and the regulation of transcription (Satoh and Lindahl 1992). PARP is activated in response to DNA strand breaks and has been considered a relatively late participant in the apoptotic process. Massive PARP-1 activation can deplete the cell of NAD^+ and ATP, ultimately leading to energy failure (Zhang et al. 1994). PARP-1 cleavage by caspase 3 or 7 is used as a marker for apoptosis, and inactivation of this enzyme may help to prevent energy depletion and necrosis (Ha and Snyder 2000).

PARP-1 can also be activated early in the cell death process in the absence of DNA strand breaks (Homburg et al. 2000; Simbulan-Rosenthal et al. 1998). Activation of PARP-1 in neurons treated with NMDA occurs in the absence of DNA damage and is followed by translocation of AIF from the mitochondrial intermembrane space to the nucleus (Yu et al. 2002) (Fig. 1). Ceramide and other cell death signaling molecules also induce translocation of AIF to the cytoplasm and then to the nucleus (Daugas et al. 2000). AIF is a flavoprotein that normally resides in the mitochondrial intermembrane space (like cytochrome *c*, pro-caspase 3, and other apoptotic modulators) and is important for normal mitochondrial function. However, when released from the intermembrane space, AIF is a powerful trigger for caspase-independent cell death (Lorenzo et al. 1999; Pieper et al. 2000; Yuan et al. 2003). AIF induces peripheral chromatin condensation and relocation of phosphatidylserine into the outer leaflet of the plasma membrane. AIF also triggers release of cytochrome *c* and caspase activation, but caspase inhibition does not abrogate AIF-dependent cell death (Lorenzo et al. 1999; Pieper et al. 2000), although it does inhibit oligonucleosomal DNA fragmentation (Daugas et al. 2000).

3 Alphavirus Encephalomyelitis

3.1 Background on Sindbis Virus

SINV is an enveloped plus-strand RNA virus that replicates entirely in the cytoplasm. The virion has three major structural proteins: capsid, which surrounds the genomic RNA, and two surface glycoproteins, E1 that mediates fusion and E2 that mediates attachment. To initiate infection, SINV attaches to an unknown cellular receptor and is taken up by receptor-mediated endocytosis. The low pH of the endosome induces a conformational change in the E1-E2 heterodimer that results in fusion of the virion envelope with the endosomal membrane and release of the capsid and genomic RNA into the cytoplasm. Replication occurs on modified endosomal membranes. Virus is assembled and released at the plasma membrane. Cellular changes induced by virus replication include shutdown of host protein and RNA synthesis and loss of membrane potential due to decreased function of Na^+,K^+-ATPase, the sodium pump (Strauss and Strauss 1994).

Strains of SINV vary in virulence, but all replicate primarily in neurons and differ in their efficiency of neuronal replication and in their ability to induce neuronal cell death. A very neurovirulent strain of SINV is the neuroadapted strain NSV, developed by passage of the prototype SINV AR339 strain in mouse brain (Griffin and Johnson 1977). Whereas SINV AR339 causes fatal encephalomyelitis in newborn mice, but not adult mice, NSV also causes the death of adult mice after intracerebral or intranasal inoculation (Jackson et al. 1988; Thach et al. 2000).

3.2 SINV-Induced Death of Immature Neurons

Most strains of SINV cause rapidly fatal encephalitis in newborn mice. Mice die 2–4 days after infection, and infected neurons show evidence of apoptosis (Lewis et al. 1996). Neuroblastoma cells can serve as an in vitro model for SINV infection of immature neurons and undergo SINV-induced apoptosis. Cell death is triggered at the cell membrane by fusion

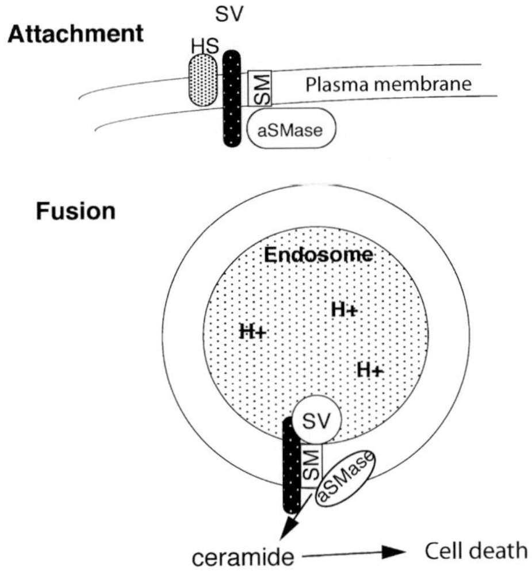

Fig. 3 Schematic diagram of the induction of cell death during SINV entry into neuroblastoma cells. SINV binds to the plasma membrane and undergoes receptor-mediated endocytosis. The lowered pH of the endosome triggers a conformational change in the E1-E2 heterodimer that leads to E1-mediated fusion. During the process of fusion acidic sphingomyelinase (*aSMase*) in the membrane is activated, sphingomyelin (*SM*) is cleaved, and ceramide is released

of the viral envelope with the endosomal membrane (Fig. 3). Virus replication is not required. Fusion-induced apoptosis is blocked by agents that prevent endosomal acidification and thereby prevent the conformational changes in E1 and E2 required for fusion (Jan and Griffin 1999). Alphavirus fusion requires cholesterol and sphingomyelin (SM) in the target membrane (Kielian and Helenius 1984; Wilschut et al. 1995). SINV fusion induces the activation of acidic sphingomyelinase (aSMase), present in the outer leaflet of the plasma membrane. SMase degrades SM to release ceramide, an intracellular initiator of apoptosis. Later in infection, Mg^{2+}-dependent neutral SMase is activated, resulting in sustained production of ceramide (Jan et al. 2000). Overexpression of acid ceramidase decreases the level of ceramide in infected cells and inhibits alphavirus-induced cell death (Jan et al. 2000; Ubol et al. 1996). Thus ceramide is an important initiator of SINV-induced apoptosis in neuroblastoma cells.

This pathway may be linked to a TNFR signaling pathway that has been implicated in SINV-induced death of PC-12 rat pheochromocytoma cells (Sarid et al. 2001) and activates the apoptotic pathway by in-

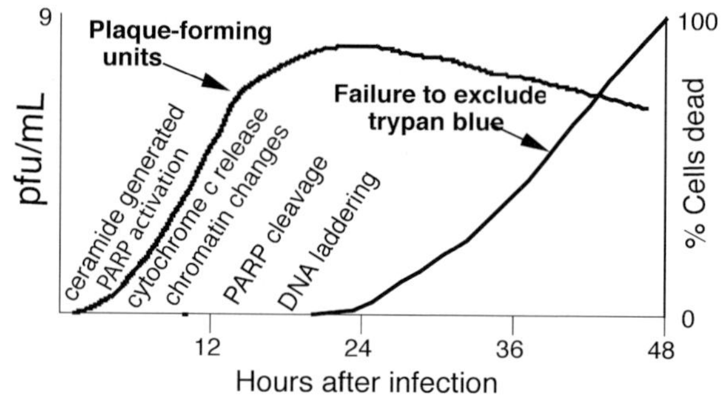

Fig. 4 Time course of intracellular changes induced by virus infection as related to virus production and failure to exclude trypan blue in N18 neuroblastoma cells. Changes begin early after infection, are progressive, and do not inhibit virus production

creasing intracellular ceramide through hydrolysis of SM (Kolesnick and Golde 1994). Members of the TNFR family trigger the assembly of the death-inducing signaling complex that results in activation of caspase 8. Expression of Crm A, a poxvirus serpin that inhibits caspase 8, protects against SINV-induced cell death and fatal disease in newborn mice (Nava et al. 1998).

In addition to activation of SMase and release of ceramide, another early event in SINV-induced apoptosis is activation of PARP-1. PARP-1 is activated within an hour after initiation of virus infection, well before caspase 3 activation, which is not detected until 6–10 h after infection (Fig. 4) (Nargi-Aizenman et al. 2002; Ubol et al. 1996). Caspase activation and SINV-induced apoptosis are delayed in cells lacking PARP-1, suggesting an important early role for this enzyme in SINV-induced apoptosis (Nargi-Aizenman et al. 2002; Ubol et al. 1996). After PARP-1 activation there is consumption of NAD^+ and release of cytochrome *c*. Caspases are involved in downstream events, as evidenced by the cleavage of a number of caspase 3 target proteins and the ability of z-VAD-fmk, a broad-spectrum caspase inhibitor, to inhibit SINV-induced apoptosis (Grandgirard et al. 1998; Jan et al. 2000; Nava et al. 1998). Activated caspase 3 cleaves PARP-1 and terminates ADP ribosylation of proteins.

Intrinsic mechanisms are also probably involved in SINV-induced apoptosis of immature neurons. An intrinsic mechanism for inducing death of mature neurons is oxidative stress. Increased ROS can result from exposure to excitotoxic amino acids or production of an excess of unfolded or aggregated proteins. ROS lead to cell damage and ultimately to mitochondrial dysfunction and cell death. Thiol antioxidants such as *N*-acetylcysteine and pyrrolidine dithiocarbamate potently inhibit SINV-induced apoptosis of neuroblastoma cells. This does not appear to act through control of NF-κB expression (Lin et al. 1995, 1998). The apoptotic process in SINV-infected neuroblastoma cells is not Ca^{2+} dependent. In cycling cells SINV-induced apoptosis results in loss of cells primarily from the G_0/G_1 population, suggesting that cells may be particularly susceptible to SINV-induced apoptosis during the G_1 phase of the cell cycle (Ubol et al. 1996). A role for cell cycle proteins is further suggested by the fact that inhibition of Ras signaling in PC-12 cells inhibits both cell proliferation and SINV-induced apoptosis (Joe et al. 1996).

Bcl-2 family member proteins are involved in regulating SINV-induced neuronal cell death because cells overexpressing Bcl-2 are relatively resistant to apoptosis induced by SINV and other alphaviruses (Levine et al. 1993; Scallan et al. 1997; Ubol et al. 1994). However, interpretation is complicated because Bcl-2 also inhibits virus replication by an unknown mechanism and virus-induced caspase activation can inactivate Bcl-2 (Grandgirard et al. 1998). SINV can be engineered to express exogenous proteins from a second subgenomic promoter. Overexpression of Bcl-2 or, paradoxically, Bax partially protects newborn mice from fatal SINV infection (Levine et al. 1996; Lewis et al. 1999; Pal et al. 1997). Ability of the Bcl-2-interacting protein beclin-1 to protect against fatal SINV-induced disease in newborn mice potentially links SV-induced neuronal death to the autophagy pathway (Liang et al. 1998).

The role of specific viral proteins in induction of apoptosis in immature neurons is unclear. Cell death occurs late after infection if only the nonstructural proteins are synthesized, but it is not clear that this is an apoptotic process (Frolov and Schlesinger 1994). Amino acid changes in nonstructural protein-2 are important determinants of cytopathic effect, but noncytopathic viruses are deficient in shutoff of host protein synthesis, potentially accounting for the effect (Agapov et al. 1998; Frolova et al. 2002). Cells infected with viruses expressing the structural proteins die more rapidly than cells infected with replicons expressing only the nonstructural proteins. Overexpression of the transmembrane domains of the E1 or E2 glycoproteins can induce cell death, implicating these proteins in the apoptotic process (Joe et al. 1998).

3.3 SINV-Induced Death of Mature Neurons

NSV induces fatal encephalomyelitis in adult C57BL/6 mice that is accompanied by hindlimb paralysis. A number of amino acid changes in the E1 and E2 glycoproteins contribute to the increased virulence of NSV for older mice, but the most important is the substitution of histidine for glutamine at position 55 of the E2 glycoprotein (Tucker et al. 1993). This change increases efficiency of SINV infection of neurons and is rapidly selected during persistent SINV infection of the CNS (Levine and Griffin 1993; Tucker et al. 1997). In the hippocampus, a mixture of morphologically apoptotic and necrotic pyramidal neurons are present in NSV-infected mice. The onset of paralysis parallels morphological changes in motor neuron cell bodies in the lumbar spinal cord and in motor neuron axons in ventral nerve roots, many of which are eventually lost. However, the loss of spinal cord motor neurons is not apoptotic, as judged by morphological criteria and lack of detectable activated caspase 3 (Havert et al. 2000) (Fig. 2). These cells are not protected by overexpression of Bcl-2 or Bax, further suggesting that the mode of death is not apoptotic (Kerr et al. 2002).

An in vitro culture system of SINV-infected primary cortical neurons and time-lapse imaging revealed evidence of both apoptosis and necrosis (Nargi-Aizenman and Griffin 2001). Surprisingly, uninfected as well as infected neurons died in these cultures, suggesting bystander neuronal death. Because glutamate excitotoxicity is a recognized cause of neuronal death, and can be mediated through different types of glutamate receptors on neurons, the role of excitotoxicity was explored. Antagonists of the NMDA subtype of glutamate receptor (MK801 and APV) protected cultured neurons from SINV-induced death without affecting virus replication, suggesting that SINV activates neurotoxic pathways that result in NMDA receptor stimulation and damage to infected and uninfected neurons (Nargi-Aizenman and Griffin 2001).

MK801 was then used to determine whether glutamate receptor antagonists could protect NSV-infected mice from fatal encephalomyelitis. Hippocampal neurons were protected from death, but the mice were not. However, when mice were treated with an AMPA receptor antagonist (GYKI-52466) neurons in the hippocampus and the lumbar spinal cord were protected and the mice were also protected from paralysis and death (Nargi-Aizenman et al. 2004) (Fig. 5). Protection was not due to altered virus replication because treatment did not affect virus distribution or peak titers and actually delayed virus clearance. These results

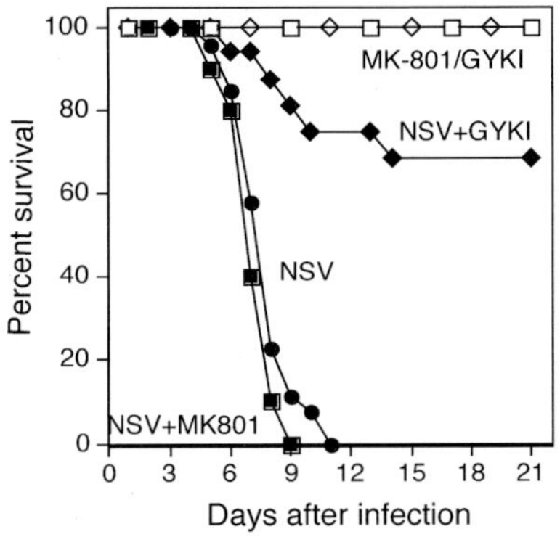

Fig. 5 Effect of glutamate receptor antagonists on NSV-induced paralysis and death of weanling mice. Four-week-old C57BL/6 mice were infected with NSV and treated with diluent, the NMDA receptor antagonist MK-801, or the AMPA receptor antagonist GYKI-52466. (Nargi-Aizenman et al. 2004)

provide evidence that NSV infection activates neurotoxic pathways that result in aberrant glutamate receptor stimulation and neuronal damage. Furthermore, AMPA receptor-mediated motor neuron death is an important contributor to paralysis and mortality in acute alphavirus-induced encephalomyelitis.

Further evidence that NSV-induced death of mature neurons differs with the population of neurons infected comes from study of NSV overexpression of Bcl-2 and Bax. These two members of the Bcl-2 family protected hippocampal neurons from death, but not lumbar spinal cord motor neurons, and the animals developed levels of paralysis similar to those of control animals (Kerr et al. 2002). Likewise, 4-week-old $Bax^{-/-}$ mice were not protected from NSV-induced paralysis or motor neuron death. Thus divergent cell death pathways are activated in different target populations of neurons during NSV infection of weanling mice and these pathways are not necessarily the same as the pathways activated by SINV-infection of immature neurons.

4 Conclusion

Most strains of SINV cause fatal encephalomyelitis in young animals by inducing apoptosis of immature neurons. This is associated with efficient virus replication and rapid death of the animals and can be modeled in vitro through the study of undifferentiated neurons and neuroblastoma cells. Viral fusion with the endosomal cell membrane activates aSMase, releasing ceramide and initiating the apoptotic cascade. As neurons mature they become more resistant to induction of apoptosis, but neurovirulent strains of SINV can still cause cell death. Death of mature motor neurons tends to be necrotic rather than apoptotic, and bystander cell death contributes to paralysis and death of the neurons. Death of mature neurons can be inhibited by treatment with drugs that prevent excitotoxic cell damage.

Acknowledgments This work was supported by research grant NS-18596 from the National Institutes of Health. The contributions to research from the author's laboratory of many individuals including Philippe DesPres, Michael Havert, Marie Hardwick, David Irani, Jia-Tsrong Jan, Beth Levine, Jennifer Nargi-Aizenman, and Sukathida Ubol are gratefully acknowledged.

References

Agapov EV, Frolov I, Lindenbach BD, Pragai BM, Schlesinger S, Rice CM (1998) Noncytopathic Sindbis virus RNA vectors for heterologous gene expression. Proc Natl Acad Sci USA 95:12989–12994

Allsopp TE, Scallan M, Williams A, Fazakerley JK (1998) Virus infection induces neuronal apoptosis: a comparison with trophic factor withdrawal. Cell Death Differ 5:50–59

Andersson T, Schwarcz R, Love A, Kristensson K (1993) Measles virus-induced hippocampal neurodegeneration in the mouse: a novel, subacute model for testing neuroprotective agents. Neurosci Lett 154:109–112

Ankarcrona M, Dypbukt J, Bonfoco E, Zhivotovsky B, Orrenius S, Lipton S, Nicotera P (1995) Glutamate-induced neuronal death: a succession of necrosis or apoptosis depending on mitochondrial function. Neuron 15:961–973

Becker EB, Bonni A (2004) Cell cycle regulation of neuronal apoptosis in development and disease. Prog Neurobiol 72:1–25

Buchan A, Li H, Cho S, Pulsinelli W (1991) Blockade of the AMPA receptor prevents CA1 hippocampal injury following severe but transient forebrain ischemia in adult rats. Neurosci Lett 132:255–258

Burek MJ, Oppenheim RW (1998) Cellular interactions that regulate programmed cell death in the developing vertebrate nervous system. In: Koliatsos VE, Ratan RR, eds., editors. Cell Death and Diseases of the Nervous System. 1998:145–179.

Burke DS, Monath TP (2003) Flaviviruses. In: Knipe DM, Howley PM, Griffin DE, Lamb RA, Martin MA, Roizman B et al., editors. Fields Virology. Philadelphia: Lippincott Williams and Wilkins 2003:1043–1126.

Calisher CH (1994) Medically important arboviruses of the United States and Canada. Clin Microbiol Rev 7:89–116

Choi D, Koh J, Peters S (1998) Pharmacology of glutamate neurotoxicity in cortical cell culture: attenuation by NMDA antagonists. J Neurosci 8:185–196

Choi D, Maulucci-Gedde M, Kriegstein A (1987) Glutamate neurotoxicity in cortical cell culture. J Neurosci 7:357–368

Danes L, Kufner J, Hruskova J, Rychterova V (1973) The role of the olfactory route on infection of the respiratory tract with Venezuelan equine encephalomyelitis virus in normal and operated *Macaca* rhesus monkeys. I. Results of virological examination. Acta Virol 17:50–56

Danial NN, Korsmeyer SJ (2004) Cell death: critical control points. Cell 116:205–219

Daugas E, Susin SA, Zamzami N, Ferri KF, Irinopoulou T, Larochette N, Prevost M, Leber B, Andrews D, Penninger J, Kroemer G (2000) Mitochondrio-nuclear translocation of AIF in apoptosis and necrosis. FASEB J 14:729–739

Despres P, Frenkiel MP, Ceccaldi PE, dos Santos CD, Deubel V (1998) Apoptosis in the mouse central nervous system in response to infection with mouse-neurovirulent dengue viruses. J Virol 72:823–829

Eguchi Y, Shimizu S, Tsujimoto Y (1997) Intracellular ATP levels determine cell death fate by apoptosis or necrosis. Cancer Res 57:1835–1840

Frolov I, Schlesinger S (1994) Comparison of the effects of Sindbis virus and Sindbis virus replicons on host cell protein synthesis and cytopathogenicity in BHK cells. J Virol 68:1721–1727

Frolova EI, Fayzulin RZ, Cook SH, Griffin DE, Rice CM, Frolov I (2002) Roles of nonstructural protein nsP2 and alpha/beta interferons in determining the outcome of sindbis virus infection. J Virol 76:11254–11264

Gill R, Lodge D (1997) Pharmacology of AMPA antagonists and their role in neuroprotection. Int Rev Neurobiol 40:197–232

Girard S, Couderc T, Destombes J, Thiesson D, Delpeyroux F, Blondel B (1999) Poliovirus induces apoptosis in the mouse central nervous system. J Virol 73:6066–6072

Gorter J, Petrozzino J, Aronica E, Rosenbaum D, Opitz T, Bennett M, Connor J, Zukin R (1997) Global ischemia induces downregulation of Glur2 mRNA and increases AMPA receptor-mediated Ca^{2+} influx in hippocampal CA1 neurons of gerbil. J Neurosci 17:6179–6188

Grandgirard D, Studer E, Monney L, Belser T, Fellay I, Borner C, Michel MR (1998) Alphaviruses induce apoptosis in Bcl-2-overexpressing cells: evidence for a caspase-mediated, proteolytic inactivation of Bcl-2. EMBO J 17:1268–1278

Griffin DE, Johnson RT (1977) Role of the immune response in recovery from Sindbis virus encephalitis in mice. J Immunol 118:1070–1075

Griffin DE (2003) Alphaviruses. In: Knipe DM, Howley PM, Griffin DE, Lamb RA, Martin MA, Roizman B et al., editors. Fields Virology. Philadelphia: Lippincott Williams & Wilkins, 2003:917–962.

Ha HC, Snyder SH (2000) Poly (ADP-ribose) polymerase-1 in the nervous system. Neurobiol Dis 7:225–239

Haeberlein SL (2004) Mitochondrial function in apoptotic neuronal cell death. Neurochem Res 29:521–530

Hahn CS, Lustig S, Strauss EG, Strauss JH (1988) Western equine encephalitis virus is a recombinant virus. Proc Natl Acad Sci USA 85:5997–6001

Hase T, Summers PL, Dubois DR (1990) Ultrastructural changes of mouse brain neurons infected with Japanese encephalitis virus. Intl J Exp Pathol 71:493–505

Havert MB, Schofield B, Griffin DE, Irani D (2000) Activation of divergent neuronal cell death pathways in different target cell populations during neuroadapted Sindbis virus infection of mice. J Virol 74:5352–5356

Homburg S, Visochek L, Moran N, Dantzer F, Priel E, Asculai E, Schwartz D, Rotter V, Dekel N, Cohen-Armon M (2000) A fast signal-induced activation of poly (ADP-ribose) polymerase: a novel downstream target of phospholipase C. J Cell Biol 150:293–307

Jackson AC, Moench TR, Trapp BD, Griffin DE (1988) Basis of neurovirulence in Sindbis virus encephalomyelitis of mice. Lab Invest 58:503–509

Jackson AC, Rossiter J (1997) Apoptotic cell death is an important cause of neuronal injury in experimental Venezuelan equine encephalitis virus infection of mice. Acta Neuropathol 93:349–353

Jan J-T, Chatterjee SB, Griffin DE (2000) Sindbis virus entry into cells triggers apoptosis by activating sphingomyelinase leading to the release of ceramide. J Virol 74:6425–6432

Jan J-T, Griffin DE (1999) Induction of apoptosis by Sindbis virus occurs at cell entry and does not require virus replication. J Virol 73:10296–10302

Joe AK, Ferrari G, Jiang HH, Liang XH, Levine B (1996) Dominant inhibitory ras delays Sindbis virus-induced apoptosis in neuronal cells. J Virol 70:7744–7751

Joe AK, Foo H, Kleeman L, Levine B (1998) The transmembrane domains of Sindbis virus envelope glycoproteins induce cell death. J Virol 72:3935–3943

Johnson KP, Johnson RT (1968) California encephalitis. II. Studies of experimental infection in the mouse. J Neuropathol Exp Neurol 27:390–400

Johnson RT, McFarland HF, Levy SE (1972) Age-dependent resistance to viral encephalitis: studies of infections due to Sindbis virus in mice. J Infect Dis 125:257–262

Kerr DA, Larsen T, Cook SH, Fannjiang YR, Choi E, Griffin DE, Hardwick JM, Irani DN (2002) BCL-2 and BAX protect adult mice from lethal Sindbis virus infection but do not protect spinal cord motor neurons or prevent paralysis. J Virol 76:10393–10400

Kielian MC, Helenius A (1984) Role of cholesterol in fusion of Semliki Forest virus with membranes. J Virol 52:281–283

Kolesnick R, Golde DW (1994) The sphingomyelin pathway in tumor necrosis factor and interleukin-1 signaling. Cell 77:325–328

Krajewski S, Krajewska M, Shabaik A, Wang HG, Irie S, Fong L, Reed JC (1994) Immunohistochemical analysis of in vivo patterns of Bcl-X expression. Cancer Res 54:5501–5507

Kroemer G, Reed JC (2000) Mitochondrial control of cell death. Nature Med 6:513–519

Lee J, Zipfel G, Cho D (1999) The changing landscape of ischaemic brain injury mechanisms. Nature 399:A7–A14

Leist M, Jaattela M (2001) Four deaths and a funeral: from caspases to alternative mechanisms. Mol Cell Biol 2:1–10

Leist M, Single B, Castoldi AF, Kuhnle S, Nicotera P (1997) Intracellular adenosine triphosphate (ATP) concentration: a switch in the decision between apoptosis and necrosis. J Exp Med 185:1481–1486

Lennette EH, Koprowski H (1943) Human infection with Venezuelan equine encephalomyelitis virus: a report on eight cases of infection acquired in the laboratory. JAMA 13:1088–1095

Levine B, Goldman JE, Jiang HH, Griffin DE, Hardwick JM (1996) Bcl-2 protects mice against fatal alphavirus encephalitis. Proc Natl Acad Sci USA 93:4810–4815

Levine B, Griffin DE (1993) Molecular analysis of neurovirulent strains of Sindbis virus that evolve during persistent infection of SCID mice. J Virol 67:6872–6875

Levine B, Huang Q, Isaacs JT, Reed JC, Griffin DE, Hardwick JM (1993) Conversion of lytic to persistent alphavirus infection by the Bcl-2 cellular oncogene. Nature 361:739–742

Lewis J, Oyler GA, Ueno K, Fannjiang Y-R, Chau BN, Vornov J, Korsmeyer SJ, Zou S, Hardwick JM (1999) Inhibition of virus-induced neuronal apoptosis by Bax. Nat Med 5:832–835

Lewis J, Wesselingh SL, Griffin DE, Hardwick JM (1996) Alphavirus-induced apoptosis in mouse brains correlates with neurovirulence. J Virol 70:1828–1835

Liang X-H, Kleeman LK, Jiang H-H, Gordon G, Goldman JE, Berry G, Herman B, Levine B (1998) Protection against fatal Sindbis virus encephalitis by Beclin, a novel Bcl-2 interacting protein. J Virol 72:8586–8596

Lin K-I, DiDonato JA, Hoffman A, Hardwick JM, Ratan RR (1998) Suppression of steady-state, but not stimulus-induced NF-κB activity inhibits alphavirus-induced apoptosis. J Cell Biol 141:1479–1487

Lin K-I, Lee SH, Narayanan R, Baraban JM, Hardwick JM, Ratan RR (1995) Thiol agents and Bcl-2 identify an alphavirus-induced apoptotic pathway that requires activation of the transcription factor NF-kappa B. J Cell Biol 131:1–14

Longshore WA, Stevens IM, Hollister AC, Gittelsohn A, Lennette EH (1956) Epidemiologic observations on acute infectious encephalitis in California with special reference to the 1952 outbreak. Am J Hyg 63:69-86

Lorenzo HK, Susin SA, Penninger J, Kroemer G (1999) Apoptosis inducing factor (AIF): a phylogenetically old, caspase-independent effector of cell death. Cell Death Differ 6:516–524

Merry DE, Veis DJ, Hickey WF, Korsmeyer SJ (1994) Bcl-2 protein expression is widespread in the developing nervous system and retained in the adult PNS. Development 120:301–311

Nargi-Aizenman JL, Griffin DE (2001) Sindbis virus-induced neuronal death is both necrotic and apoptotic and is ameliorated by *N*-methyl-D-aspartate receptor antagonists. J Virol 75:7114–7121

Nargi-Aizenman JL, Havert MB, Zhang M, Irani DN, Rothstein JD, Griffin DE (2004) Glutamate receptor protect from virus-induced neural degeneration prevented by antagonists. Ann Neurol 55:541–549

Nargi-Aizenman JL, Simbulan-Rosenthal CM, Kelly TA, Smulson ME, Griffin DE (2002) Rapid activation of poly (ADP-ribose) polymerase contributes to Sindbis virus and staurosporine-induced apoptotic cell death. Virology 293:164–171

Nava VE, Rosen A, Veliuona MA, Clem RJ, Levine B, Hardwick JM (1998) Sindbis virus induces apoptosis through a caspase-dependent, CrmA-sensitive pathway. J Virol 72:452–459

Nellgard B, Wieloch T (1992) Postischemic blockade of AMPA but not NMDA receptors mitigates neuronal damage in the rat brain following transient severe cerebral ischemia. J Cereb Blood Flow Metab 12:2–11

Nicotera P, Orrenius S (1998) The role of calcium in apoptosis. Cell Calcium 23:173–180

Oberhaus SM, Smith RL, Clayton GH, Dermody TS, Tyler KL (1997) Reovirus infection and tissue injury in the mouse central nervous system are associated with apoptosis. J Virol 71:2100–2106

Ogata A, Nagashima K, Hall WW, Ichikawa M, Kimura-Kuroda J, Yasui K (1991) Japanese encephalitis virus neurotropism is dependent on the degree of neuronal maturity. J Virol 65:880–886

Oliver KR, Scallan MF, Dyson H, Fazakerley JK (1997) Susceptibility to a neurotropic virus and its changing distribution in the developing brain is a function of CNS maturity. J Neurovirol 3:38–48

Olney J (1969) Brain lesions, obesity, and other disturbances in mice treated with monosodium glutamate. Science 164:719–721

Pal R, Garzino-Demo A, Markham PD, Burns J, Brown M, Gallo RC, DeVico AL (1997) Inhibition of HIV-1 infection by the β-chemokine MDC. Science 278:695–698

Pekosz A, Phillips J, Pleasure D, Merry D, Gonzalez-Scarano F (1996) Induction of apoptosis by La Crosse virus infection and role of neuronal differentiation and human bcl-2 expression in its prevention. J Virol 70:5329-5335

Pieper AA, Blackshaw S, Clements EE, Brat DJ, Krug DK, White AJ, Pinto-Garcia P, Favit ACJR, Snyder SH, Verma A (2000) Poly (ADP-ribosyl)ation basally activated by DNA strand breaks reflects glutamate-nitric oxide neurotransmission. Proc Natl Acad Sci USA 97:1845–1850

Roth W, Kermer P, Krajewska M, Welsh K, Davis S, Krajewski S, Reed JC (2003) Bifunctional apoptosis inhibitor (BAR) protects neurons from diverse cell death pathways. Cell Death Differ 10:1178–1187

Sammin DJ, Butler D, Atkins GJ, Sheahan BJ (1999) Cell death mechanisms in the olfactory bulb of rats infected intranasally with Semliki Forest virus. Neuropathol Appl Neurobiol 25:236–243

Sarid R, Ben Moshe T, Kazimirsky G, Weisberg S, Appel E, Kobiler D, Lustig S, Brodie C (2001) vFLIP protects PC-12 cells from apoptosis induced by Sindbis virus: implications for the role of TNF-alpha. Cell Death Differ 8:1224–1231

Satoh MS, Lindahl T (1992) Role of poly(ADP-ribose) formation in DNA repair. Nature 356:356–358

Scallan MF, Allsopp TE, Fazakerley JK (1997) Bcl-2 acts early to restrict Semliki Forest virus replication and delays virus-induced programmed cell death. J Virol 71:1583–1590

Sherdown M, Suzdak P, Nordholm L (1993) AMPA, but not NMDA, receptor antagonism is neuroprotective in gerbil global ischaemia, even when delayed 24 h. Eur J Pharmacol 236:347–353

Simbulan-Rosenthal CM, Rosenthal DS, Iyer S, Boulares AH, Smulson ME (1998) Transient poly (ADP-ribosyl)ation of nuclear proteins and role of poly (ADP-ribose) polymerase in the early stages of apoptosis. J Biol Chem 273:13703–13712

Strauss JH, Strauss EG (1994) The alphaviruses: gene expression, replication and evolution. Microbiol Rev 58:491–562

Tesh RB, Watts DM, Russell KL, Damodaran C, Calampa C, Cabezas C, Ramirez G, Vasquez B, Hayes CG, Rossi CA, Powers AM, Hice CL, Chandler LJ, Cropp CB, Karabatsos N, Roehrig JT, Gubler DJ (1999) Mayaro virus disease: an emerging mosquito-borne zoonosis in tropical South America. Clin Infect Dis 28:67–73

Thach D, Kimura T, Griffin DE (2000) Differences between C57BL/6 and BALB/cBy mice in mortality and virus replication after intranasal infection with neuroadapted Sindbis virus. J Virol 74:6156–6161

Tsunoda I, Kurtz CIB, Fujinami RS (1997) Apoptosis in acute and chronic central nervous system disease induced by Theiler's murine encephalomyelitis virus. Virology 228:388–393

Tucker PC, Lee SH, Bui N, Martinie D, Griffin DE (1997) Amino acid changes in the Sindbis virus E2 glycoprotein that increase neurovirulence improve entry into neuroblastoma cells. J Virol 71:6106–6112

Tucker PC, Strauss EG, Kuhn RJ, Strauss JH, Griffin DE (1993) Viral determinants of age-dependent virulence of Sindbis virus in mice. J Virol 67:4605–4610

Ubol S, Park S, Budihardjo I, Desnoyers S, Montrose MH, Poirier GG, Kaufmann SH, Griffin DE (1996) Temporal changes in chromatin, intracellular calcium, and poly(ADP-ribose) polymerase during Sindbis virus-induced apoptosis of neuroblastoma cells. J Virol 70:2215–2220

Ubol S, Tucker PC, Griffin DE, Hardwick JM (1994) Neurovirulent strains of alphavirus induce apoptosis in Bcl-2-expressing cells; role of a single amino acid change in the E2 glycoprotein. Proc Natl Acad Sci USA 91:5202–5206

Walder R, Bradish CJ (1975) Venezuelan equine encephalomyelitis virus (VEEV): Strain differentiation and specification of virulence markers. J Gen Virol 26:265–275

Wilschut J, Corver J, Nieva J-L, Bron R, Moesby L, Reddy KC, Bittman R (1995) Fusion of Semliki Forest virus with cholesterol-containing liposomes at low pH: a specific requirement for sphingolipids. Mol Membr Biol 12:143–149

Xue D, Huang Z, Barnes K, Lesiuk H, Smith K, Buchan A (1994) Delayed treatment with AMPA, but not NMDA, antagonists reduces neocortical infarction. J Cereb Blood Flow Metab 14:251–261

Yu L, Alva A, Su H, Dutt P, Freundt E, Welsh S, Baehrecke EH, Lenardo MJ (2004) Regulation of an ATG7-beclin 1 program of autophagic cell death by caspase-8. Science 304:1500–1502

Yu S, Wang H, Poitras MF, Coombs C, Bowers WJ, Federoff HJ, Poirier GG, Dawson TM, Dawson VL (2002) Mediation of poly (ADP-ribose) polymerase-1-dependent cell death by apoptosis-inducing factor. Science 297:259–263

Yuan J, Lipinski M, Degterev A (2003) Diversity in the mechanisms of neuronal cell death. Neuron 40:401–413

Yuan J, Yankner BA (2000) Apoptosis in the nervous system. Nature 407:802–809

Zhang J, Dawson VL, Dawson TM, Snyder SH (1994) Nitric oxide activation of poly (ADP-ribose) synthetase in neurotoxicity. Science 263:687–689

CTMI (2005) 289:79–112

HSV-Induced Apoptosis in Herpes Encephalitis

L. Aurelian

Virology/Immunology Laboratories, University of Maryland,
Bressler, Room 4-023, 655 West Baltimore Street, Baltimore, MD 21201, USA
laurelia@umaryland.edu

Abstract HSV triggers and blocks apoptosis in cell type-specific fashion. This review discusses present understanding of the role of apoptosis and signaling cascades in neuronal pathogenesis and survival and summarizes present findings relating to the modulation of these strictly balanced processes by HSV infection. Underscored are the findings that HSV-1, but not HSV-2, triggers apoptosis in CNS neurons and causes encephalitis in adult subjects. Mechanisms responsible for the different outcomes of infection with the two HSV serotypes are described, including the contribution of viral antiapoptotic genes, notably the HSV-2 gene ICP10PK. Implications for the potential use of HSV vectors in future therapeutic developments are discussed.

1 Introduction

Herpes simplex viruses types 1 (HSV-1) and 2 (HSV-2) are a major cause of worldwide morbidity (Fisman et al. 2002). Neurovirulence and the ability to establish latency in sensory neurons are characteristic properties of both HSV serotypes. However, despite an overall DNA homology of 47%–50%, HSV-1 and -2 are biologically distinct, notably in their interaction with neurons in the central nervous system (CNS), where HSV-1 but not HSV-2, causes encephalitis (Sauerbrei et al. 2000). Viral genes that control neurovirulence and latency, their mechanism of action, and their interaction with cellular genes are still poorly understood. Emerging evidence implicates a complex cross-talk between programmed cell death (apoptosis), survival, and immune pathways in virus-induced neurological disorders. This review provides a brief description of the present findings, with particular emphasis on apoptotic cascades, their modulation by HSV infection, and the implications for future therapeutic developments.

2 Apoptotic Cascades

Apoptosis is a tightly regulated, irreversible process that results in cell death in the absence of inflammation. It differs from necrosis, an established mechanism of virus-induced cell death, in that it involves active cell participation. Apoptotic stimuli induce mitochondrial release of cytochrome *c* (Cyt c), apoptosis-inducing factor (AIF), Smac/Diablo, and endonuclease G. Cyt c and AIF are involved in the activation of the caspases, which are cysteine proteases with aspartate specificity. Events that regulate the activation of initiator caspases are key determinants of cell death and define two groups of upstream caspases. One of these, exemplified by caspase-9, is activated by conformational change resulting from dimerization. The other, exemplified by caspase-8, is activated by autoproteolysis. Executioner caspases, exemplified by caspase-3, are activated by proteolytic cleavage of precursor zymogens. They are responsible for the morphological and biochemical changes associated with apoptosis, including cleavage of proteins involved in DNA repair and replication, such as poly ADP-ribose) polymerase (PARP). Smac/Diablo binds inhibitors of activated caspases, and AIF and endonuclease G mediate caspase-independent cell death pathways (Friedlander 2003). Three

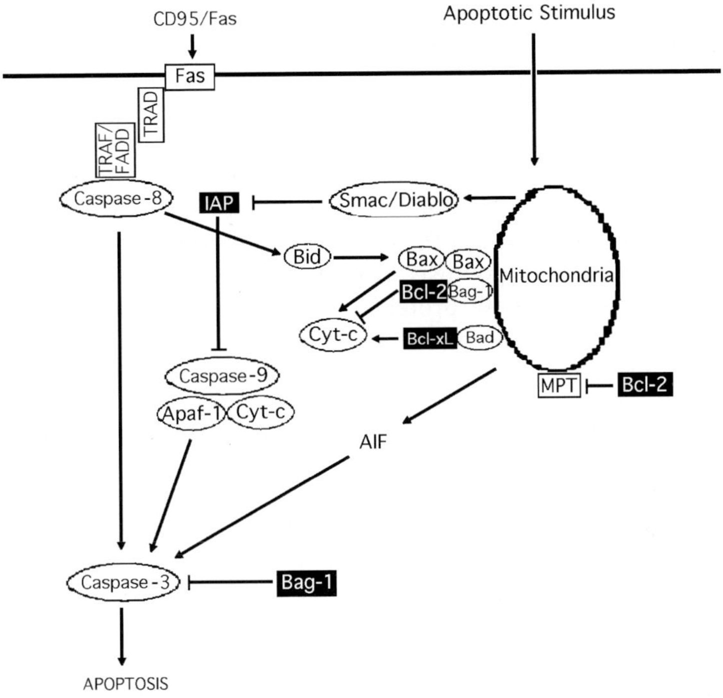

Fig. 1 Apoptotic pathways. Schematic representation focuses on the extrinsic and intrinsic pathways. Apoptosis specific mitochondrial changes include mitochondrial permeability transition (*MPT*) and cytochrome *c* (Cyt c) release. Dimerization or heterodimerization of Bcl-2 family members (Bcl-2, Bax, Bad, Bag-1), cleavage of Bid and mitochondrial release of Smac/Diablo and AIF are shown. Released Cyt c forms a complex with Apaf-1 and caspase-9, leading to activation of the executioner caspases, exemplified by caspase-3. Activation of death receptors by the respective ligand (e.g., Fas) leads to the recruitment of adaptor proteins (such as TRAD and TRAF), activation of caspase-8, and subsequent activation of the proteolytic cascade and apoptosis

apoptotic pathways have been identified (Fig. 1). The intracellular pathway is activated by loss of neurotrophic factors [e.g., nerve growth factor (NGF)], excitotoxic injury, or virus infection. It initiates with Cyt c release and its complexation with pro-caspase-9 and Apaf-1, leading to caspase-9 activation. The extracellular apoptotic pathway is initiated by binding of the ligands Fas or tumor necrosis factor (TNF)-α to their respective receptors. These receptors contain cytoplasmic domains that anchor adaptor proteins which are involved in the recruitment and activation of caspase-8 or -10. Activated initiator caspases, in turn, activate the executioner caspases (Hill et al. 2003). The third apoptotic pathway originates in the endoplasmic reticulum, and it involves activation of caspases-12 and -9 (Morishima et al. 2002).

Three protein families regulate apoptosis. The Bcl-2 proteins, which also function in neurological disorders, consist of antiapoptotic (e.g., Bcl-2) and proapoptotic (e.g., Bad) members that are differentially mobilized by various stimuli. Pro- and antiapoptotic proteins heterodimerize, and the balance between the two classes determines, at least in part, the susceptibility to apoptosis (Chittenden 1998). The proapoptotic function of some of these proteins is inactivated by phosphorylation, whereas that of other members is activated by cleavage or mitochondrial translocation (Lambeng et al. 2003). One antiapoptotic family member, Bag-1, interacts and cooperates with Bcl-2 or functions independently (Schulz et al. 1997). The second class of cell death inhibitors, the inhibitor of apoptosis proteins (IAPs), can bind and inhibit activated caspases-3 and -7, and one member, XIAP, can also associate with Cyt c and inhibit caspase-9. However, Smac/Diablo, which is also released from mitochondria on apoptotic stress, can neutralize the function of the IAPs. Smac/Diablo may also potentiate the caspase cascade by displacing XIAP from mature caspase-9 (Hill et al. 2003), underscoring the strict cross-regulation of the apoptotic pathways by various "check and balance" mechanisms (Fig. 1). The role of IAPs in neuronal apoptosis is still unknown.

Heat shock proteins (Hsp) are also involved in apoptosis regulation. Most Hsp have antiapoptotic activity and are overexpressed in tumor tissues, where they contribute to tumor progression and resistance to chemotherapy. Hsp70 proteins prevent Cyt c release from mitochondria, inhibit the JNK and p38MAPK signaling cascades, and function downstream of caspase-3. They also bind and sequester activated caspases, APAF and AIF. Hsp90 is required for the maintenance of the c-Raf-1 function, and Hsp27 sequesters Cyt c (Beere 2001). However, the antiapoptotic activity of the Hsp in neurological disorders is controversial (Latchman 2004). The first Hsp with proapoptotic activity (H11) was cloned in our laboratory. It acquires antiapoptotic activity after single-site mutation (Smith et al. 2000b; Gober et al. 2003).

3
Apoptosis Is a Predominant Cell Death in Neurodegeneration

Neuronal survival is stimulated by trophic factors that function through specific receptors. The MEK/ERK pathway serves as a conduit for signaling from various receptors, including receptors for neurotrophic factors (e.g., NGF), glutamate [e.g., kainic acid (KA), *N*-methyl-D-aspartate (NMDA)], or other neurotransmitters [e.g., nicotinic acetylcholine (nAChR) and dopamine D1/D5 (DAR)], leading to CREB activation. The dopamine receptor also activates CREB via adenylate cyclase (AC) and the cAMP/PKA pathway that channels into ERK through the Rap-1/B-Raf module (Fig. 2). The signaling pathway that is activated as a result of survival signals is determined, at least in part, by the specific phosphorylation site on the receptor (Chao 2003). The neurotransmitter glutamate functions at excitatory synapses of the brain. Different glutamate

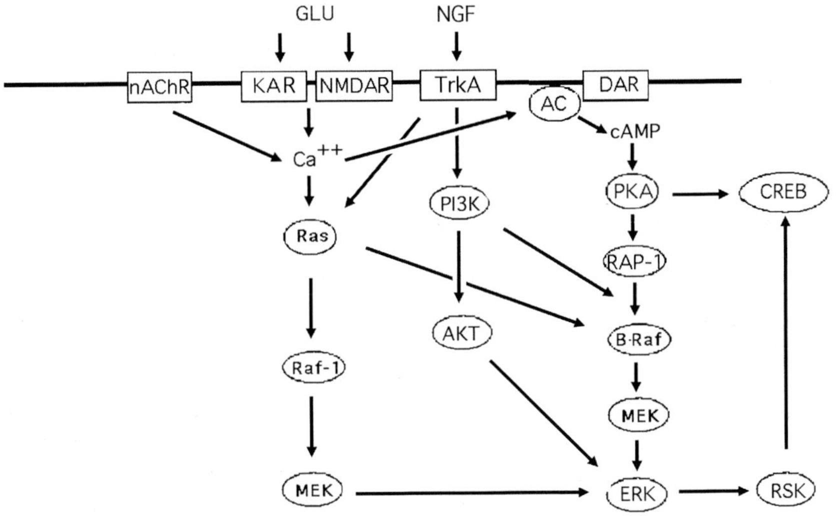

Fig. 2 Survival signaling pathways in neurons. Neuronal survival is stimulated by trophic factors that function through specific receptors. The MEK/ERK pathway serves as a conduit for signaling from the various receptors. Exemplified are the TrkA receptor for the neurotrophic factor NGF, the glutamate receptors [e.g. kainic acid (*KAR*) and *N*-methyl-D-aspartate (*NMDAR*)], and receptors for other neurotransmitters [e.g., nicotinic acetylcholine (*nAChR*) and dopamine D1/D5 (*DAR*)], leading to CREB activation. The dopamine receptor also activates CREB via adenylate cyclase (*AC*) and the cAMP/PKA pathway that channels into ERK through the Rap-1/B-Raf module

receptors in the postsynaptic membrane transduce the signal released from the presynaptic terminal into electric and biochemical events in the postsynaptic neurons. Different patterns of activation of NMDA-type glutamate receptors can trigger long-term potentiation (LTP) or long-term depression (LTD) of synaptic strength. LTP and LTD induction, as characterized in the CA1 and CA3 regions of the hippocampus, occur in postsynaptic neurons and require Ca^{2+} influx through the NMDA receptor. These long-lasting forms of synaptic plasticity are pathways for encoding memories in the brain. The molecular details of plasticity related to these stimuli include activation of signaling pathways mediated by Ras (required for LTP), Rap-1 (required for LTD), and phosphoinositide 3-kinase (PI3-K) modules (Sheng and Kim 2002). The convergence of distinct pathways on final effectors may cause signal amplification and could provide a fail-safe mechanism in case one pathway becomes nonfunctional (Fig. 2).

Apoptosis often results from interference with the strict regulation of these pathways. Stimuli implicated in apoptosis causation include oxidative stress, genetic defects, accumulated burden of endogenous or exogenous factors, loss of neurotrophic support, excessive release of neurotransmitters known as excitotoxins, or virus infection. When inappropriate in timing or extent, apoptosis can trigger or account for progression of neurodegeneration in acute and chronic diseases. Excessive amounts of glutamate, or glutamate function for prolonged intervals, trigger neuronal cell death by overstimulating cognate receptors (Arundine and Tymianski 2004). Excitotoxicity is involved in the pathogenesis of ischemic brain injury, epilepsy, and neurodegenerative diseases. Cerebral ischemia was associated with activation of caspases-1, -3, -8, -9, and -11 and the release of Cyt c (Kang et al. 2003). Ischemic release of glutamate, IL-1β, TNF-α, or reactive oxygen species (ROS) can affect surrounding neurons, a bystander cell death that is associated with activation of caspase-1 (Friedlander 2003). Bcl-2, Bag-1, and Hsp proteins appear to be targets of the caspase-dependent pathways in ischemia. However, Hsp27-mediated protection was minimal, with loss of CA3 hippocampal neurons reduced from 38% in the wild-type animals to 17% in the transgenics (Latchman 2004). Frequently, a mix of both apoptotic and necrotic cell death follows ischemic injury, with the core of the lesion characterized by necrotic cell death while apoptosis occurs in the penumbra, where the degree of hypoxia is less severe.

Apoptosis is the predominant form of cell death in chronic neurodegenerative diseases. In some cases it is associated with genetic mutations that activate the caspase cascade. One example is the Ts16 mouse, which is considered to be a model of Down syndrome (trisomy 21), a genetic defect that is believed to confer vulnerability to neurodegeneration (Perkins et al. 2002b). Another example of such a disorder is amyotrophic lateral sclerosis (ALS), which is characterized by the progressive loss of motor neurons in the brain, brain stem, and spinal cord and is associated with mutations in the gene encoding the free radical -scavenging enzyme Cu, Zn superoxide dismutase 1 (SOD1) in 3%–4% of ALS cases. The proapoptotic activity of the mutant SOD1 is presumably due to an unstable conformation that leads to increased levels of intracellular free radicals and was associated with activation of caspases-1 −3 and -9, Cyt c release, and proapoptotic changes in the Bcl-2 protein family (Pasinelli et al. 2000). Virus-induced neuronal cell death results from a multiplicity of virus-host cell interactions with distinct pathogenetic outcomes and can be necrotic, apoptotic, or both.

4
Caspase-Independent and Stress-Activated Neuronal Cell Death

Caspase-3 activation was implicated in neuronal functions unrelated to cell death (Rohn et al. 2004), and altered cross-talk between the caspases and other signaling pathways was associated with the development or progression of neurodegenerative disorders. For example, significant glial response and IL-1β production were documented during early stages of ALS, whereas induction of TNF-α and pro-caspase-8 activation were detected in spinal cords, late during disease progression (Guegan et al. 2002). Transcriptional upregulation of the caspases, rather than their activation, was reported in both chronic (e.g., ALS) and acute (e.g., stroke) neurological disorders, and caspase-independent apoptosis was implicated in neuronal cell death after stroke (Plesnila et al. 2004). Nonapoptotic programmed cell death was also recently described, including the Ras-activated "autophagic" cascade (Chi et al. 1999) and the cytoplasmic or "trophotoxic" cascade that is associated with the activation of a trophic and mitogenic factor receptor (Castro-Obregon et al. 2004).

Stress-activated kinase cascades implicated in neuronal apoptosis include the c-Jun N-terminal kinase (JNK) and p38 MAPK. JNK induced apoptosis is generally transcription dependent, involving c-Jun as a target. In PC12 cells, JNK3 can trigger apoptosis or increase NGF-induced neurite outgrowth (differentiation) concomitant with target switch from ATF-2 in the apoptotic context to c-Jun in the differentiation context (Waetzig and Herdegen 2003). Transcription-independent JNK- or p38MAPK-induced apoptosis was also described, for example, involving JNK-mediated phosphorylation of the proapoptotic protein Bad on Ser^{128}, inhibiting the interaction of Ser^{136} phosphorylated Bad with the sequestering protein 14-3-3 (Bhakar et al. 2003). The cell type, stimulus, and differentiation state can also affect the way in which cells die and their vulnerability to death-inducing signal. In NGF-differentiated PC12 cells, apoptosis caused by loss of trophic growth support (NGF removal) was associated with a significant decrease in ERK phosphorylation, whereas both JNK and p38MAPK were associated with apoptosis in PC12 cells differentiated with NGF together with cAMP (Lambeng et al. 2003). Similarly, early activation of p38MAPK was shown to protect neurons from TNF-α cytotoxicity, whereas late activation of JNK and p38MAPK coincided with apoptosis (Roulston et al. 1998). Overall, available data underscore the functional versatility of the apoptosis-related molecules in physiological and pathological conditions, stress the complexity and specificity of their cross-talk, and caution against facile conclusions relating to therapeutic interventions.

5 Signaling Pathways Involved in Neuronal Survival

The extracellular signal-related kinase (ERK) and PI3-K/Akt pathways are generally involved in neuronal cell proliferation, differentiation, development, cell cycle, and transmission of survival and mitogenic signals (Lowes et al. 2002). These pathways are linked to G protein-linked cell surface receptors and receptor kinases, including neurotransmitters. They operate through sequential phosphorylation events to activate (phosphorylate) transcription factors and regulate downstream signaling proteins. Thus, in response to survival stimuli, the membrane-bound G protein Ras adopts an active, GTP-bound state and it, in turn, coordinates the activation of a multitude of downstream effectors. The ERK survival pathway begins with the activation of c-Raf-1 kinase. It is followed by the activation of MAP kinase kinase (MEK) and ERK and culminates in the activation of transcription factors. The ERK survival pathway overrides the effects of apoptotic signals, apparently by upregulating antiapoptotic Bcl-2 proteins through transcription-dependent and -independent mechanisms. The former involves activation of transcription factors, such as the calcium/cAMP response element-binding protein (CREB) that binds to the cAMP response element (CRE) and activates gene transcription in response to a wide variety of extracellular signals, including growth factors, hormones ,and neurotransmitters. Transcriptional activation of CREB is controlled through phosphorylation at Ser^{133} by the ERK1/2-mediated activation of the pp90 ribosome S6 kinase (Rsk). Other transcription factors that are targets of the ERK pathway include c-fos, which is also regulated by Rsk and was implicated in cell growth, differentiation, and development, and Elk-1, which induces gene transcription in response to serum and growth factors (Nebreda and Gavin 1999; Kaplan and Miller 2000). Transcription-independent survival mechanisms mediated by activated ERK involve Rsk-2 (one of the three Rsk isotypes), which phosphorylates the proapoptotic protein Bad, thereby causing its inactivation and favoring ERK-mediated cell survival (Fig. 2). In addition to mediating neuronal survival, activated ERK also increases axonal growth and enhances axonal regeneration after axotomy (Atwal et al. 2000) and plays an important role in synaptic plasticity, learning, and memory (Adams and Sweatt 2002).

Cell surface receptors can also activate PI3-K, which in turn activates Akt, involving phosphorylated phosphatidylinositides (PI-3,4-P_2 and PI-3,4,5-P_3) generated in the cell membrane (Kandel and Hay 1999). ERK/Rsk-2 and PI3-K/Akt converge at the level of Bad inactivation to promote cell survival. Ras activation by the NGF receptor TrkA is independent of PI3-K, but PI3-K is required for the activation of Rap-1, and both pathways converge on MEK (York et al. 2000) (Fig. 2).

6 ERK and the Integration of Distinct Signals

How do pathways common to many systems integrate signals from a wide spectrum of activities? The answer may be in the duration and intensity of signaling through ERK, which appear to dictate its subcellular compartmentalization and/or trafficking. This, in turn, dictates whether ERK-expressing cells enter apoptosis, survival, or differentiation. The dual-specificity phosphatases, MKPs, might be the link between the kinetics of ERK activation and its subcellular localization (Colluci-D'Amato et al. 2003). Thus ERK stimulation by NGF is both rapid and sustained. Sustained activation depends on signaling through the Rap1/B-Raf module, whereas the Ras/c-Raf-1 module accounts for the immediate response (Vaudry et al. 2002). Translational control by ERK signaling was also implicated in long-term synaptic plasticity and memory (Kelleher et al. 2004). However, chronic ERK activation was implicated in neurodegeneration (Colluci-D'Amato et al. 2003) and ERK was associated with caspase-independent neuronal injury after ER stress (Arai et al. 2004). Moreover, enzymatic activation may not be sufficient for the successful propagation of a signal. Recent studies described scaffoldlike molecules that tether components of a specific cascade into oligomeric protein complexes and increase their local concentration or exclude illegitimate cross-interaction. One such molecule, MEK partner 1 (MP1), enhances the activation of ERK-1, but not ERK-2, suggesting that it helps to discriminate between the two ERK isoforms (Schaeffer et al. 1999). This may reflect the different functions of the two isoforms. Another scaffold protein, Raf kinase inhibitor protein (RKIP), inhibits the ERK pathway (Chong et al. 2003).

The overall conclusion is that under various cellular milieus, ERK may activate different transcription factors or promote cell survival or growth arrest by a transcription-independent mechanism. The neuron type (CNS vs. PNS, cortical, striatal, or hippocampal, sensory or motor) and its state of differentiation potentially contribute to the outcome of ERK activation. Presumably, apoptosis, differentiation, survival, and proliferation result from the balance of various signaling modules and their targets. In turn, these depend on the combination of neurotrophins and receptors as well as other first messengers available in the specific cellular milieu that is under investigation. Because viruses can hijack all of these pathways, the outcome of infection is virus- and cell type specific.

7 HSV-1, but Not HSV-2, Triggers Apoptosis in CNS Neurons

Both HSV serotypes can trigger or block apoptosis in a cell type-specific fashion. Most efforts have focused on identifying viral gene products involved in apoptosis prevention. However, relatively little is known about the proapoptotic viral genes and their mechanism of action. In the human epithelial cell line HEp-2, apoptosis was triggered by HSV-1 mutants containing a deletion in the viral immediate-early (IE) regulatory genes ICP4 or ICP27, or by virus infection with wild-type virus in the presence of cycloheximide, which only allows expression of the IE genes. These findings were interpreted to indicate that apoptosis is mediated by IE genes (Koyama and Adachi 1997; Sanfilippo et al. 2004). However, in HeLa and BHK cells, HSV-1 evaded caspase activation, maintained Bcl-2 RNA, and protected from the cisplatin-induced decrease in Bcl-2 RNA, apparently involving the antiapoptotic activity of ICP27 and ICP4

(Leopardi and Roizman 1996; Aubert and Blaho 2001). In HSV-1-infected primary human embryonic lung (HEL) fibroblasts, AIF was translocated to the nucleus, but apoptosis did not ensue. Cyt c release and PARP cleavage were seen in HEL fibroblasts infected with an ICP4-deleted mutant, but the cells were resistant to DNA degradation (Zhou and Roizman 2000). HSV-2 induced apoptosis in a small fraction of Hep-2 cells, but apoptosis was not observed in Vero or HeLa cells. Both HSV-1 and HSV-2 triggered apoptosis in dendritic cells that are critical for the stimulation of naïve T cells (Jones et al. 2003), a likely mechanism of immune evasion.

HSV-1 antiapoptotic genes include Us3, which codes for a serine-threonine protein kinase (PK) and was implicated in protection from apoptosis induced by thermal or osmotic stress, UV irradiation, and Fas-induced cell death in various cell lines. Underscoring the contribution of cellular factors, a Us3-deleted mutant functioned in a caspase-independent manner in the neuronal cell line SK-N-SH and a caspase-dependent manner in Hep-2 cells (Hagglund et al. 2002). Us3 was shown to block caspases that cleave Bad at sites predicted to render it more proapoptotic (Benetti et al. 2003) and to function at undefined sites downstream of Bax (Ogg et al. 2004). However, the HSV-1 mutant deleted in Us3 lost virtually all neurovirulence after ic inoculation (Cartier et al. 2003), suggesting that Us3 is involved in HSV-1 neurovirulence. $\gamma_1$34.5 is another antiapoptotic HSV-1 gene. The protein binds/activates protein phosphatase 1α, which dephosphorylates the translation initiation factor eIF2. This is associated with protection from apoptosis induced by the activated double-strand RNA-dependent kinase (PKR) (Tan and Katze 2000). HSV-1 mutants deleted in $\gamma_1$34.5 were impaired in latency establishment and CNS invasion (Bolovan et al. 1994). US5 that codes for glycoprotein J was implicated in protection from Fas-induced death in HSV-1 infected cells (Jerome et al. 1999), and apoptosis was also blocked in SK-NS-H cells by the HSV-1 glycoprotein gD, a function that requires protein glycosylation (Zhou and Roizman 2000). Us11 was also shown to enhance survival in heat-shocked HeLa cells, apparently by binding the homeodomain-interacting protein kinase 2 (HIPK2) and altering its intracellular localization, and it blocked PKR -mediated apoptosis (Giraud et al. 2004). Finally, nuclear translocation of NF-κB was implicated in apoptosis inhibition in HSV-1-infected Hep-2 cells (Goodkin et al. 2003), but this was disputed by others (Taddeo et al. 2003).

Underscoring the cell specificity of the apoptotic response, not one of the HSV-1 antiapoptotic proteins prevented virus-induced apoptosis in CNS neurons. Indeed, HSV-1 triggered apoptosis in primary hippocam-

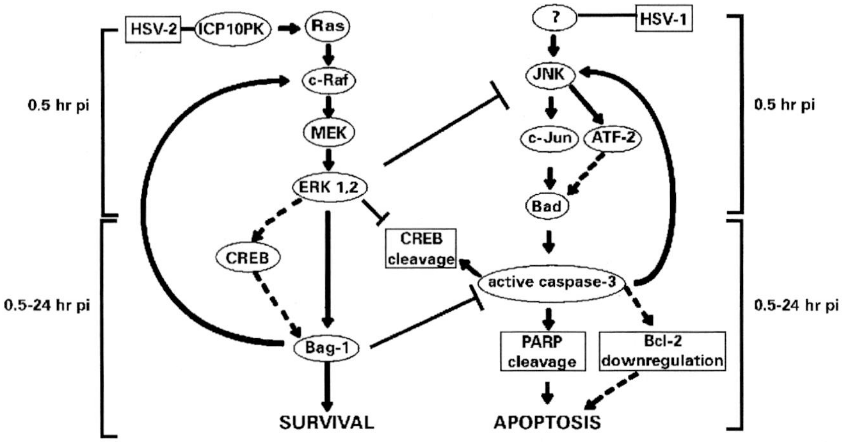

Fig. 3 Schematic representation of apoptosis modulation in CNS neurons infected with HSV-2 or HSV-1. *Dotted lines* represent potentially involved factors (from Perkins et al. 2003a)

pal cultures, involving activation of the JNK/c-Jun pathway (Perkins et al. 2003a,b). HSV-1 infection increased the expression of the three JNK isotypes (JNK1/2/3) and caused intense phosphorylation of JNK1/2. It also induced phosphorylation (activation) of the JNK target c-Jun on Ser^{63} and Ser^{73}, residues that are phosphorylated by P-JNK (Kaplan and Miller 2000). HSV-1-induced apoptosis of hippocampal neurons was inhibited by the pharmacological JNK inhibitor SP600125, indicating that apoptosis is mediated by the activation of the JNK stress pathway. Caspase-3 activation and cleavage of its target PARP were also involved in the HSV-1-induced apoptotic cascade in hippocampal neurons, as was the destabilization of the antiapoptotic protein Bcl-2, and increased expression of the proapoptotic protein Bad (Fig. 3). HSV-1 infection also inhibited phosphorylation of the transcription factor ATF-2, which is activated by JNK and dimerizes with c-Jun to cooperatively induce neuronal apoptosis (Eilers et al. 2001). By contrast, HSV-2 did not trigger apoptosis in hippocampal neurons. Studies of mutants deleted in the N- or C-terminal domains of the large subunit of ribonucleotide reductase (also known as ICP10) indicated that this is due to the antiapoptotic activity of the N-terminal PK activity (Perkins et al. 2002a, 2003a). Dorsal root ganglia neurons infected with HSV-2 also did not undergo apoptosis, but apoptosis was seen in infected glial cells (Ozaki et al. 1997). Collectively the data indicate that apoptosis is the default outcome of CNS infection with HSV-1, but not HSV-2.

8
Apoptosis Is a Component of HSV-1-Induced Encephalitis

In adults and older children (>6 months), HSV-1 causes a nonepidemic, sporadic, acute focal encephalitis (HSE) that accounts for 10%–20% of viral encephalitis cases among adults and older children in the United States. Its prevalence is approximately 1 in 250,000–500,000 individuals annually. Mortality from HSE in the presence of antiviral therapy is 50%–60%, and survivors are left with neurological sequelae involving impairments in memory, cognition, and personality (Sauerbrei et al. 2000). Conditions that determine the ability of HSV-1 to reach and infect the CNS are still unclear. In adults, transport is believed to occur primarily by a neuronal route, via the olfactory nerves or tracts, or along the branches of the trigeminal nerve innervating the basal meninges (Craig and Nahmias 1973). The olfactory route is more widely accepted because the limbic system includes the central olfactory pathways (Esiri 1982), and virus spread to the brain along the olfactory pathways was demonstrated in mice given intranasal infections (Tomlinson and Esiri 1983). In adults and older children, HSV-2 infection of the CNS is commonly restricted to a self-limiting, nonfatal meningitis that is predominantly seen in patients with primary genital disease (Sauerbrei et al. 2000). The outcome of HSV-2 meningitis is excellent even without antiviral therapy, except in the immunocompromised, in which the infection is more severe. The different transport pathway (intraneuronal for HSV-1 and hematogenous for HSV-2) was implicated in the different outcomes of infection (Craig and Nahmias 1973), but this was not confirmed. Notwithstanding, pathogenesis is likely to be determined by virus-mediated modulation of cellular signaling/apoptotic pathways as evidenced by the finding that HSV-1, but not HSV-2, triggers apoptosis in the CNS.

Cytopathologic manifestations of HSE include lytic (necrotic) effects resulting from virus replication and inflammation. However, the severity of histopathologic changes and neurological symptoms does not correlate with the viral burden in the brain, suggesting that mechanisms other than virus-induced necrotic cell death are also involved in disease pathogenesis. We found evidence of apoptosis (TUNEL+ neurons) in most (75%) of the studied HSE brains. Because apoptotic cells are rapidly cleared in vivo (Johnson et al. 1997), the percentage of apoptotic neurons represents a "snapshot" in time rather than being indicative of the real extent of apoptosis in HSE brains. Significantly, however the TUNEL+ brains also stained with antibodies to activated JNK, caspase-3, and cleaved PARP, and there was a strong correlation between these apoptosis-related findings and the presence of HSV-1 infection at the site of apoptosis. The data suggest that apoptosis is HSV-1 induced and implicate JNK activation in HSE pathogenesis. Consistent with these conclusions, apoptosis-related bystander death was reported in HSV-1-infected mouse brain stems (Shaw et al. 2002), and apoptosis was implicated in neurovirulence as evidenced for an HSV-1 mutant that replicated well in the CNS and PNS but was attenuated for neurovirulence (Pelosi et al. 1998). Presumably, the failure of HSV-2 to cause HSE in adults is due to the antiapoptotic activity of HSV-2-specific proteins.

HSV-1 has also been associated with Alzheimer disease (AD), a chronic neurodegenerative disorder characterized by intracellular accumulation of neurofibrillary tangles, abnormal deposition of the β-amyloid (Aβ) protein into focal plaques, and extensive neuronal death also related to apoptosis. The hypothesis is that people who harbor latent HSV-1 in the brain suffer from mild encephalitis (and apoptosis) related to periodic virus reactivation and this problem is augmented by the age-related decline in the immune system. Individuals homozygous or heterozygous for the type ϵ4 allele of the apolipoprotein E gene (ApoE4) may have a higher frequency of virus reactivation (Dobson et al. 2003). In support of this hypothesis, viral DNA was detected in the brain of many elderly people and HSV-1 glycoprotein B sequences were shown to be homologous to Aβ, interact with the ApoE gene, and cause neuronal death (Cribbs et al. 2000). However, other studies failed to detect the HSV-1 genome in the brain or an association between its presence and ApoE4 (Marques and Straus 2001).

Other members of the Herpesviridae family can also occasionally affect the CNS, but apoptosis does not appear to be involved. Epstein-Barr virus (EBV) causes infectious mononucleosis, which is accompanied by a mild form of encephalitis in about 1 in 100 cases. Other neurological complications, including seizures, Guillain-Barre syndrome, Bell palsy, transverse myelitis, meningitis, and cranial nerve palsies may also occur in about 36% of cases (Domachowske et al. 1996). Cytomegalovirus (CMV) encephalitis is a complication seen in immunocompromised patients, where it has a high mortality rate (about 50%). At autopsy, up to 12% of HIV patients had evidence of CMV in the brain (Arribas et al. 1996). Complications associated with reactivation of Varicella Zoster virus (VZV) include encephalitis, transverse myelitis, cerebral palsy, and postherpetic neuralgia. Adults account for 55% of varicella-related deaths, with encephalitis having a 10%–30% mortality rate. VZV induces apoptosis in vitro in fibroblast and immune cells, but not in sensory neurons (Hood et al. 2003), and reactivation of the HHV-6A variant causes neurological complications (including encephalitis) in AIDS and transplant patients (Kong et al. 2003). However, all these neurological disorders appear to be independent of apoptosis, suggesting that neuronal apoptosis is a default property specific for HSV-1.

9 The Role of Apoptosis in HSV Latency

What is the role of apoptosis in the HSV life cycle? On infection of skin and mucosal membranes (primary infection), HSV is transported retrogradely to trigeminal and cervical (HSV-1) or sacral (HSV-2) sensory ganglia, where the virus establishes latency. Latency is characterized by viral DNA persistence in sensory neurons in a largely untranscribed state, followed by periodic episodes of virus reactivation. Neurons are nonpermissive for virus replication at the time of latency establishment/maintenance, but during latency reactivation they become permissive and virus replication ensues. How does HSV override the "default" apoptotic program in sensory neurons in order to establish and reactivate from latency? The answer is still unclear. One possibility is that viral proteins are not directly responsible for latency establishment and/or reactivation, which depends on the availability of cellular factors required for virus replication. One such factor could be C1 (HCF, host cell factor) that is essential for VP16-mediated transactivation of the viral IE genes and virus replication. Indeed, C1 is sequestered in the cytoplasm of latently infected sensory neurons but translocates to the nucleus under conditions associated with virus reactivation (Kristie et al. 1995). However, the general consensus is that viral genes are involved in HSV latency, although their identity is still controversial. The latency-associated transcripts (LATs) were implicated in HSV-1 latency. The primary LAT transcript (minor LAT) is approximately 8.3 kb long. In latently infected ganglia, transcription of the HSV genome is restricted to LAT, primarily the stable 2.0-kb intron that is spliced from the primary transcript. Because the 2.0-kb LAT overlaps the 3′ end of ICP0 mRNA and ICP0 is a major transactivator of HSV gene expression, it has been suggested that LAT facilitates latency establishment/maintenance by reducing viral gene expression through antisense suppression of ICP0 (Thompson and Sawtell 1997). An alternative hypothesis was that ICP0 expression is suppressed by the ORF P protein (potentially encoded by the minor LAT), by inhibiting the splicing of ICP0 transcripts. However, both spliced and intron-containing ICP0 transcripts were found in latently infected ganglia, and their levels were not increased by LAT or ORF P (Coen 2002).

Independent studies concluded that LAT is not absolutely required for latency establishment and/or reactivation, but it is required for efficient latency reactivation (Leib et al. 1989; Trousdale et al. 1991). One group has recently reported the expression of a LAT protein encoded by an ORF within the 2.0-kb intron, which acts somewhat like ICP0 in that it enhances lytic genes and promotes interaction with cellular transcription factors (Thomas et al. 2002). However, most studies attributed the role of LAT in latency reactivation to its antiapoptotic activity (Perng et al. 2002; Thompson and Sawtell 2000; Ahmed et al. 2002). Presumably, this ensures a large pool of latently infected cells that contributes to efficient virus reactivation. In the HSV-1 rabbit eye model, antiapoptotic activity was localized to LAT nucleotides 1–76 and 447–1499 and shown to involve inhibition of caspase-9 activation (Jin et al. 2003). It is thought that LAT counteracts apoptosis induced by corticosteroids that are upregulated by reactivation-inducing stress stimuli, and it may also stabilize the translational complex, thereby leading to preferential expression of cellular proteins that aid in cell survival (Ahmed et al. 2002). Still, the antiapoptotic function of LAT in latency reactivation is controversial. Although LAT enhances reactivation in the rabbit eye model, it does not seem to be required in small-animal models (Jones 2003). Studies of the LAT antiapoptotic activity were not done in neurons, although apoptosis is cell type specific, and there is some concern that the LAT antiapoptotic activity is unique to certain HSV-1 strains. More importantly, apoptosis (caspase-3 activation)-induced HSV-1 reactivation (Hunsperger and Wilcox 2003) and latency reactivation were associated with LAT downregulation by stress-induced cAMP early repressors (ICER) (Colgin et al. 2001). Therefore, if LAT is involved in latency reactivation, it may be by a mechanism other than apoptosis, likely immune evasion. Indeed, STAT1 (signal transducer and activator of transcription 1) binds the LAT promoter, suggesting that cytokines may initiate or contribute to LAT-mediated virus reactivation.

The failure of the HSV-2 LAT to (a) substitute for its HSV-1 counterpart in promoting latency reactivation (Hill et al. 2003) and (b) modulate latency reactivation or its establishment (Wang et al. 2001) suggests that LAT is not involved in HSV-2 reactivation. A protein that appears to be involved in HSV-2 latency is the PK domain of the large subunit of HSV-2 ribonucleotide reductase (ICP10). The ICP10 promoter is the only viral promoter with AP-1 *cis*-response elements (Aurelian 1998), and basal ICP10 expression is controlled by AP-1 transcription factors that are upregulated by latency-reactivating stimuli (Zhu and Aurelian 1997). An HSV-2 mutant deleted in the PK domain of ICP10 was significantly impaired in latency establishment/ reactivation (Wachsman et al. 2001), and virus reactivation (by ganglia cocultivation) was inhibited by an antisense oligonucleotide specific for ICP10, but not by other unrelated oligonucleotides (Aurelian and Smith 2000).

10
HSV-2 Antiapoptotic Genes and Neuronal Survival

HSV-2 does not trigger apoptosis in hippocampal or cortical neurons, likely involving the contribution of viral antiapoptotic genes. The Us3 protein was shown to protect from apoptosis induced by osmotic shock, but not in neurons, and only early in infection (Asano et al. 1999). The HSV-2 UL14 protein was also implicated in protection of Hep-2 cells from apoptosis induced by osmotic shock, apparently involving its Hsp-like properties (Yamauchi et al. 2003). Their contribution to protection from apoptosis in neurons is still unclear. Our recent microarray studies (unpublished) indicated that HSV-2 infection upregulates cellular genes that modulate neuronal functions. They include, among others, the p75 neurotrophin receptor, CLK1, which induces neuronal cell differentiation in a manner akin to that employed by NGF (Myers et al. 1994), the corticotropin-releasing factor receptor 1 (CRFR1), which is critical for maintaining neuronal cell homeostasis (Bale et al. 2002), GAD-65, an isoform of the brain enzyme glutamate decarboxylase involved in synaptic transmission (Buss et al. 2001), and the P2X purinoceptor that is involved in glutamate release at primary sensory synapses and modulates long-term potentiation in the hippocampus (Pankratov et al. 2002). These genes were not upregulated by HSV-1, underscoring the different functions of the two HSV serotypes in the CNS. However, their contribution to apoptosis inhibition is unknown.

The HSV-2 ICP10PK gene is the only viral antiapoptotic gene with an established mechanism of action in the CNS. ICP10PK is a serine-threonine PK located at the N-terminus of the large subunit (R1) of viral RR. The ICP10PK catalytic core (at position 176–259) is preceded by a single transmembrane (TM) helical segment (position 85–105) followed by a basic amino acid that is responsible for TM anchorage within the cell membrane. Protein sequences upstream of the TM are localized to the cell surface; those downstream of the TM, including the core catalytic domain, are cytoplasmic (Aurelian 1998). Targets of the ICP10PK kinase activity include calmodulin, the ubiquitous calcium sensor protein that is involved in almost all intracellular events, and Ras-GAP, a negative regulator of Ras that is inactivated by phosphorylation on serine-threonine (Smith et al. 2000). In HSV-2 infected cells, ICP10PK binds Ras-GAP at N-SH2 and PH modules, respectively involving ICP10PK phosphotyrosine (pTyr) residues pT^{117} and pT^{141} and a WD40-like sequence at position 106–173. ICP10PK also binds the Grb_2-Sos complex involving the ICP10PK proline-rich motif at position 396 and the Grb_2 C-terminal SH3 motif. As a result of its interaction with Grb_2-Sos and the phosphorylation (inactivation) of Ras-GAP, ICP10PK activates Ras (converts it to the GTP-loaded form), thereby initiating activation of the MEK/ERK cascade. The outcome is increased expression and metabolic stabilization of the c-fos transcription factor (Nelson et al. 1996; Smith et al. 2000a). ICP10PK-mediated activation of the Ras/MEK/ERK pathway is required for expression of the IE genes ICP4, ICP27 and ICP22, and timely onset of virus growth (Smith et al. 1998. 2000a). ICP10PK is located in the virion tegument (Smith and Aurelian 1997), and its synthesis in infected cells is regulated with IE kinetics (Zhu and Aurelian 1997). Its antiapoptotic activity in neurons does not require virus replication.

Studies of HSV-2 mutants respectively deleted in the PK (ICP10ΔPK) or RR (ICP10ΔRR) domains of ICP10 indicated that ICP10PK inhibits virus-induced apoptosis in primary hippocampal and cortical cultures. The antiapoptotic activity of ICP10PK was c-Raf-1 dependent, as evidenced by its loss in cells transfected with a dominant-negative c-Raf-1 mutant or a pharmacological inhibitor of c-Raf kinase (Perkins et al. 2003a). This is in contradistinction to the survival of uninfected hippocampal cultures, which was PI3-K dependent (Perkins et al. 2002a), suggesting that survival pathways are switched by infection with HSV-2. ICP10PK-mediated protection from virus-induced apoptosis was MEK/ERK dependent and involved induction of Bag-1 expression, as evidenced by loss of protection and inhibition of Bag-1 upregulation with the MEK-specific pharmacological inhibitor U0126. Protection was also associated with Bcl-2 stabilization, presumably through interaction with Bag-1 and CREB activation (Fig. 3). Significantly, apoptosis induced by HSV-1 or ICP10ΔPK was blocked by ectopic delivery of ICP10PK or Bag-1, indicating that ICP10PK overrides virus-induced apoptosis.

Moreover, neuronal protection by transfected ICP10PK or Bag-1 was approximately two- to threefold higher than the percentage of transfected cells, indicative of a bystander effect in which neurons rescued from apoptosis by transfection produce trophic factors and/or form synapses that stimulate the survival and adaptive responses of surrounding neurons (Perkins et al. 2003a).

Recent studies indicate that ICP10PK also upregulates Rap-1 and adenylate cyclase and activates the Rap-1/B-Raf module in virus-infected hippocampal neurons (Smith et al., manuscript in preparation). Rap-1 is a small GTPase that has an effector domain which is virtually identical to that of Ras. In vitro, the active (GTP loaded) form of Rap-1 binds most Ras effectors, a finding that was interpreted to indicate that Rap-1 is an antagonist of Ras signaling. However, this hypothesis is not supported by recent findings which indicate that Ras and Rap-1 can be activated by the same survival signal (Zwartkruis and Bos 1999). The exact cross-talk between the activated Ras/Raf-1 and Rap-1/B-Raf modules in HSV-2-infected hippocampal neurons is still unclear. Ras activation may be required to sensitize the B-Raf/MEK/ERK pathway to GTP loaded Rap-1. Rap-1 may also activate phosphorylated Raf-1, or Rap-1 may require, in addition to GTP loading, a posttranslational modification (such as phosphorylation) to stimulate B-Raf/MEK activation (Bouschet et al. 2003). According to this interpretation, ICP10PK activates Ras, leading to Raf-1 recruitment to the plasma membrane and its subsequent phosphorylation. In turn, the phosphorylated Raf-1 interacts with the intracellular GTP-loaded Rap-1 and allows subsequent ERK activation. Rap-1 phosphorylation would also abolish its suppressive activity on Raf-1, and hence, favor Raf-1 interaction with Ras. This interpretation visualizes the Ras/Raf-1 and Rap-1/B-Raf modules as converging on the MEK/ERK cascade and is consistent with the finding that both the Ras- and Rap-1-dependent signaling pathways are blocked by a dominant-negative c-Raf-1 mutant (Perkins et al. 2003a). ICP10PK also activates the PI3-K/Akt pathway in stably transfected PC12 cells, although this pathway is not activated in virus-infected hippocampal neurons (unpublished). Collectively the data suggest that ICP10PK can activate a wide spectrum of survival pathways, likely in a neuronal and stimulus-specific fashion (Fig. 4).

In addition to its inhibitory effect on HSV-1 induced apoptosis, ICP10PK also blocks apoptosis caused by trophic factor deprivation, excitotoxic injury in intrastriatally injected mice, or genetic defects, such as in the Ts16 (Perkins et al. 2002b; Golembewski et al. 2003). ICP10PK also protects from apoptosis induced by mutant SOD1, as determined

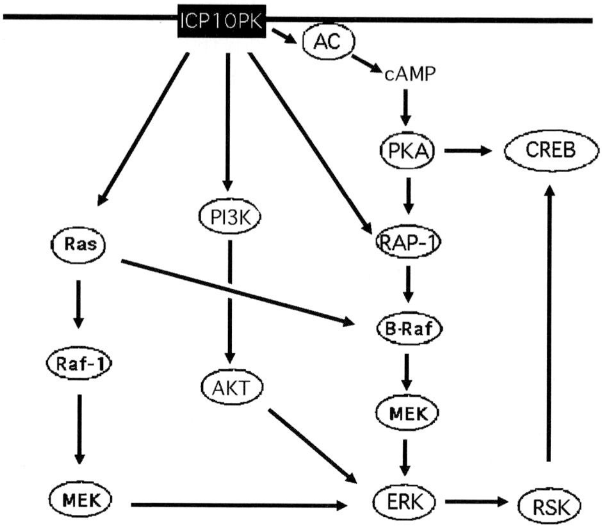

Fig. 4 ICP10PK functions as a receptor that activates various survival modules including Ras/c-Raf-1, Rap-1/B-Raf, and PI3-K/Akt, providing a promising potential antiapoptotic therapeutic for CNS disorders

from the study of mouse neuroblastoma cells that constitutively express wild-type or mutated SOD1. Specifically, the cells were grown in serum-free medium to differentiate to a neuronal phenotype, transfected with the ICP10PK expression vector, and 24 h later treated with 100 μM xanthine and 10 U/ml xanthine oxidase (X/XO) to generate superoxide anion and hydrogen peroxide. Under such circumstances, the survival of the wild-type cell line is significantly higher than that of the cells with the mutant SOD1, indicating that the mutant SOD1 confers increased vulnerability to oxidative stress (Pasinelli et al. 2000). The survival of the X/XO-treated mutant cells that expressed ICP10PK was significantly (p<0.01 by ANOVA) higher than that of the nontransfected cells or cells similarly transfected with a PK-negative mutant of ICP10PK (unpublished). The pathways involved in ICP10PK-mediated protection in these other paradigms of neuronal cell death are still unknown, and are presently under investigation. Nonetheless, it is tempting to conclude that ICP10PK was coopted from a cellular gene (Smith et al. 2000) and evolutionarily conserved to ensure neuronal cell survival, particularly in face of other proapoptotic viral genes. The questions that arise are why does the HSV-1 homolog of ICP10PK lack similar functions, and by extension, why did HSV-1 retain this coopted cellular gene.

11
HSV Vectors for CNS Disease

Present interest in the use of HSV vectors for gene therapy of CNS diseases was stimulated by the ability of HSV to infect postmitotic neurons and its large foreign gene insertion capacity (up to 50 kb). To overcome toxic effects related to neurovirulence, two types of vectors were developed, both based on HSV-1. The first type, known as amplicons, are plasmids engineered to contain an HSV-1 origin of replication and an HSV-1 "packaging" site along with a bacterial origin of replication. The resulting hybrid plasmid is propagated in bacteria and subsequently cotransfected with a defective HSV-1 helper recombinant, creating a mixed population of viral particles that contain either the defective HSV-1 genome or concatemers of plasmid packaged into the HSV-1 capsid. Progress was recently achieved by novel purification procedures that resulted in significant reduction in helper virus. Still, the drawbacks of this system include low vector titers and the high probability of recombination leading to the potential production of replication-competent virus.

The second type of HSV-1-based vectors are mutants deleted in multiple genes to reduce/eliminate their neurovirulence. They can be categorized as replication-competent and replication-defective. The replication-competent vectors maintain a limited ability to replicate, a feature that is desirable in order to achieve a more widespread dissemination in vivo but is associated with cell damage (Huang et al. 1992). These vectors are therefore suited for gene therapy of brain tumors, where they are used as oncolytic vectors. Glioblastoma, the most common malignant primary brain tumor, is an attractive target for anticancer gene therapy because the tumors are highly localized and distant metastases occur only rarely. An HSV-1 mutant (with deletions in the ICP6 and $\gamma_1$34.5 loci) was avirulent on intracerebral inoculation, killed tumor cells, decreased tumor growth, and prolonged survival of tumor-bearing nude mice (Mineta et al. 1995). A broad range of therapeutic genes can be transduced into tumor cells by viral vectors. They include cytokines that stimulate the immune response directed against tumor (e.g., IL-2, IL-12) (Parker et al. 2000) or enhance the effect of gamma irradiation (e.g., TNF-α) (Moriuchi et al. 1998) as well as antiangiogenic factors (Im et al. 1999). Another approach (named "suicide gene therapy") is based on the ability of HSV thymidine kinase (TK) enzyme to preferentially activate the prodrug ganciclovir (GCV), an inhibitor of DNA replication. GCV kills both tumor cells containing the TK gene (delivered with an HSV-1, retrovirus, or adenovirus vector) and surrounding tumor cells.

This approach also involves a “bystander effect,” whereby the number of cells killed significantly exceeds the number of cells expressing viral TK (Freeman et al. 1996). However, these vectors are not suitable for gene therapy of chronic or acute neurological diseases.

HSV-1 vectors have been used to express foreign genes that protect neurons from an ischemic insult by limiting neuronal death (e.g., Bcl-2) (Antonavich et al. 1999) or restoring energy metabolism (e.g., glucose transporter) (Dumas et al. 1999). HSV-1 vectors have also been used to deliver growth factors [e.g., glial cell-derived neurotrophic factor (GDNF)] to dopaminergic neurons in experimental models of Parkinson disease, thereby halting neurodegeneration (Tomac et al. 1995). Transfer of neurotrophic factor genes promoted survival of responsive neuronal populations and had a beneficial role in modulation of local enzymatic and neurotransmitter levels. Overall, however, these vectors suffer from toxicity problems, and recently the focus of attention has shifted toward maintaining neuronal function in addition to promoting survival (Dumas et al. 1999; McLaughlin et al. 2000).

Given the apoptotic activity of HSV-1, at least in the CNS, the use of the HSV-1-based vectors for gene therapy of neurodegenerative diseases needs re-evaluation. HSV-2, which does not cause encephalitis or apoptosis in the CNS, may be a better candidate for the development of vectors for gene therapy of neurological disorders. In this context, ICP10 PK is a particularly promising candidate, because it activates the ERK pathway that was implicated in LTP (determines cognitive functions) (English and Sweatt 1996). This interpretation is supported by our recent findings that a growth-defective HSV-2 mutant that retains the anti-apoptotic gene ICP10PK protects from NMDA-induced excitotoxic injury in mice (Golembewski et al. 2003).

12
Conclusions and Perspective

Encephalitis as a neurological complication of HSV-1 infection is a devastating disease, particularly in the absence of antiviral therapy. CNS pathogenicity relates to multiple mechanisms including lytic effects due to virus replication, inflammatory responses due, at least in part, to cytokine production by microglial cells, and neuronal cell apoptosis, which is a determining aspect of virus-induced CNS pathogenesis. Apoptotic cascades are affected by signaling pathways, and none of these function in isolation. The range of connections that allow integration between the various signaling and apoptotic pathways, and their modulation by virus infection, are still poorly understood. The ERK, PI3-K/Akt, and cAMP/PKA pathways are ubiquitous and pivotal to cellular growth, survival, and function. The JNK and p38MAPK stress-induced pathways and caspase cascades are generally associated with cell death. However, ERK-associated cell death and JNK-mediated cell survival have also been reported. Modular design of proteins and their expression patterns determine the signaling systems that are unique to each cell type. In trying to reconcile studies done in different cell types it is crucial, therefore, to consider the isoforms of the involved players and their intracellular localization. As these are often coupled to feedback and feedforward loops, all considerations must be done in a temporally logical context. Thus changes in individual components of signaling cascades may be transient, affecting a specific outcome (e.g., survival), or long-term, affecting different outcomes (e.g., apoptosis). HSV infection can hijack any one of these pathways, altering this sensitive balance. Indeed, the ability of HSV to trigger and prevent apoptosis has been unequivocally determined in various cultured cells, and the available data suggest that apoptosis modulation is independent of virus replication. In CNS neurons, HSV-1 triggers apoptosis, involving JNK/c-Jun and caspase activation and culminating in the induction of the proapoptotic protein Bad. These pathways are also associated with virus-induced encephalitis in adult humans. By contrast, HSV-2 does not trigger apoptosis in CNS neurons nor cause encephalitis. This is likely due to the antiapoptotic activity of the ICP10PK protein, which involves activation of the ERK survival pathway culminating in the upregulation of Bag-1 and the stabilization of Bcl-2, both of which are antiapoptotic.

HSV-1-induced activation of the hypothalamic-pituitary-adrenocortical (HPA) axis and production of brain-derived IL-1 and prostaglandin 2 were implicated in clinical and behavioral manifestations of encephalitis, independent of virus replication (Ben-Hur et al. 2003). Indeed, it is becoming increasingly evident that the outcome of virus infection in the brain is complicated by the contribution of glial cells (astrocytes, oligodendrocytes, and microglia). Under physiological conditions, microglia are quiescent. They begin to proliferate in response to pathological stress, phagocytose damaged cells, and when hyperactivated, such as by virus infection, may cause neuronal degeneration. IL-6, TGF-β, and TNF-α produced by microglial cells are involved in excitotoxic neuronal damage (Acarin et al. 2000). However, microglia can also produce neurotrophic factors, such as brain-derived neurotrophic factor (BDNF), NGF, and GDNF that are neuroprotective (Suzuki et al. 2001). It is still not entirely clear whether microglia protect or harm neurons or whether TNF-α is beneficial or toxic. Some studies suggest that TNF-α enhances injury (Barone et al. 1997), whereas others indicate that it induces antiapoptotic factors and is protective. Still other studies suggest that TNF-α has a deleterious effect during the acute response that occurs in the traumatized brain but plays a key role in the long-term behavioral recovery and tissue repair (Scherbel et al. 1999). After an ischemic brain insult, the injured cells produce ATP, which activates microglia to produce TNF-α, thereby protecting neurons from glutamate toxicity. Both ERK and JNK are involved in TNF-α expression, and p38MAPK is involved in the nucleo-cytoplasmic transport of TNF-α mRNA (Suzuki et al. 2004). Astrocytes reduce neuronal cell death following a variety of cellular stresses, such as excitotoxicity and oxidative stress. They protect neurons from apoptosis caused by serum deprivation, at least in part by the release of TGF-β, which activates the JNK/c-Jun pathway in neurons via the TGF-β type II receptors. In this system, pathway activation leads to increased transcription of neuroprotective genes (Dhandapani and Brann 2003), presumably because the proapoptotic function of JNK/c-Jun is counteracted by neurotrophic factors in the CNS milieu. Significantly, microglia do not support HSV replication, but they respond to infection with the production of considerable amounts of cytokines, including TNF-α and IL-6 (Lokensgard et al. 2001). Recent studies also suggest that microglia that express the Toll-like receptor TLR2 secrete inflammatory cytokines in the brain, and they are associated with HSV-1-induced encephalitis (Kurt-Jones et al. 2004). The response of astrocytes and microglia to HSV-2 infection is still unclear.

Understanding the cross-talk processes that interlink signaling systems in defined cell types is crucial to our understanding of cell function and disease and ultimately to the development of appropriate therapies. The therapeutic approach dictated by the complex multifactorial nature of neurodegenerative disorders includes the targeting of both apoptotic cell death and the decline in cognitive processes. The use of HSV vectors in gene therapy may provide new avenues for treatment of neurological disorders that involve apoptosis. However, the development of effective HSV-1 vectors for the CNS is hampered by their cytotoxicity and apoptotic activity. Data reviewed here suggest that HSV-2-based vectors may have an additional advantage compared to HSV-1, in that the ICP10 PK gene confers broad antiapoptotic activity in neurons. This includes inhibition of apoptosis triggered by virus infection, neurotrophic factor deprivation (a known etiologic factor of neurodegenerative disorders) genetic defects (including ALS), and excitotoxicity (models of stroke and epilepsy). In addition, ICP10 PK activates the ERK/CREB LTP pathway, providing an improvement in cognitive functions, a benefit sought after, but never achieved, by the current therapeutic strategies of neurodegenerative disorders. However, additional studies are needed to elucidate the contribution of microglia and astrocytes and determine the optimal conditions for ERK activation in order to avoid potentially negative outcomes.

Acknowledgements The studies done in our laboratory were supported by Public Health Service Grants NS-45169 from the National Institute of Neurological Disorders and Stroke and AR-42647 from the National Institute of Arthritis and Musculoskeletal Diseases.

References

Acarin L, Gonzalez B, Castellano B (2000) Neuronal, astroglial and microglial cytokine expression after an excitotoxic lesion in the immature rat brain. Eur J Neurosci 12:3505–3520

Adams JP, Sweatt JD (2002) Molecular psychology: roles for the ERK MAP kinase cascade in memory. Annu Rev Pharmacol Toxicol 42:135–163

Ahmed M, Lock M, Miller CG, Fraser NW (2002) Regions of the herpes simplex virus type 1 latency-associated transcript that protect cells from apoptosis in vitro and protect neuronal cells in vivo. J Virol 76:717–729

Antonavich FJ, Federoff HJ, Davis JN (1999) Bcl-2 transduction, using a herpes simplex virus amplicon, protects hippocampal neurons from transient global ischemia. Exp Neurol 156:130–137

Arai K, Lee SR, van Leyen K, Kurose H, Lo EH (2004) Involvement of ERK MAP kinase in endoplasmic reticulum stress in SH-SY5Y human neuroblastoma cells. J Neurochem 89:232–239

Arribas JR, Storch GA, Clifford DB, Tselis AC (1996) Cytomegalovirus encephalitis. Ann Intern Med 125:577–587

Arundine M, Tymianski M (2004) Molecular mechanisms of glutamate-dependent neurodegeneration in ischemia and traumatic brain injury. Cell Mol Life Sci 61:657–668

Asano S, Honda T, Goshima F, Watanabe D, Miyake Y, Sugiura Y, Nishiyama Y (1999) US3 protein kinase of herpes simplex virus type 2 plays a role in protecting corneal epithelial cells from apoptosis in infected mice. J Gen Virol 80:51–56

Atwal JK, Massie B, Miller FD, Kaplan DR (2000) The TrkB-Shc site signals neuronal survival and local axon growth via MEK and P13-kinase. Neuron 27:265–277

Aubert M, Blaho JA (2001) Modulation of apoptosis during herpes simplex virus infection in human cells. Microbes Infect 3:859–866

Aurelian L (1998) Herpes simplex virus type 2: unique biological properties include neoplastic potential mediated by the PK domain of the large subunit of ribonucleotide reductase. Front Biosci 3:d237–d249

Aurelian L, Smith CC (2000) Herpes simplex virus type 2 growth and latency reactivation by co-cultivation are inhibited with antisense oligonucleotides complementary to the translation initiation site of the large subunit of ribonucleotide reductase (RR1). Antisense Nucleic Acid Drug Dev 10:77–85

Bale T.L, Picetti R, Contarino A, Koob GF, Vale WW, Lee KF (2002) Mice deficient for both corticotropin-releasing factor receptor 1 (CRFR1) and CRFR2 have an impaired stress response and display sexually dichotomous anxiety-like behavior. J Neurosci 22:193–199

Barone FC, Arvin B, White RF, Miller A, Webb CL, Willette RN, Lysko PG, Feuerstein GZ (1997) Tumor necrosis factor-alpha. A mediator of focal ischemic brain injury. Stroke 28:1233–1244

Beere HM (2001) Stressed to death: regulation of apoptotic signaling pathways by the heat shock proteins. Sci STKE 93:1–6

Benetti L, Munger J, Roizman B (2003) The herpes simplex virus 1 US3 protein kinase blocks caspase-dependent double cleavage and activation of the proapoptotic protein BAD. J Virol 77:6567–6573

Ben-Hur T, Cialic R, Weidenfeld J (2003) Virus and host factors that mediate the clinical and behavioral signs of experimental herpetic encephalitis. A short autoreview. Acta Microbiol Immunol Hung 50:443–451

Bhakar AL, Howell JL, Paul CE, Salehi AH, Becker EB, Said F, Bonni A, Barker PA (2003) Apoptosis induced by p75NTR overexpression requires Jun kinase-dependent phosphorylation of Bad. J Neurosci 23:11378–11381

Bolovan CA, Sawtell NM, Thompson R L (1994). ICP34.5 mutants of herpes simplex virus type 1 strain 17syn+ are attenuated for neurovirulence in mice and for replication in confluent primary mouse embryo cell cultures. J Virol 68:48–55

Bouschet T, Perez V, Fernandez C, Bockaert J, Eychene A, Journot L (2003) Stimulation of the ERK pathway by GTP-loaded Rap1 requires the concomitant activation of Ras, protein kinase C, and protein kinase A in neuronal cells. J Biol Chem 278:4778–4785

Buss K, Drewke C, Lohmann S, Piwonska A, Leistner E (2001) Properties and interaction of heterologously expressed glutamate decarboxylase isoenzymes GAD(65 kDa) and GAD(67 kDa) from human brain with ginkgotoxin and its 5′-phosphate. J Med Chem 44:3166–3174

Cartier A, Komai T, Masucci MG (2003) The Us3 protein kinase of herpes simplex virus 1 blocks apoptosis and induces phosphorylation of the Bcl-2 family member Bad. Exp Cell Res 291:242–250

Castro-Obregon S, Rao RV, del Rio G, Chen SF, Poksay KS, Rabizadeh S, Vesce S, Zhang XK, Swanson RA, Bredesen DE (2004) Alternative, nonapoptotic programmed cell death: mediation by arrestin 2, ERK2, and Nur77. J Biol Chem 279:17543–17553

Chao MV (2003) Neurotrophins and their receptors: a convergence point for many signalling pathways. Nat Rev Neurosci 4:299–309

Chi S, Kitanaka C, Noguchi K, Mochizuki T, Nagashima Y, Shirouzu M, Fujita H, Yoshida M, Chen W, Asai A, Himeno M, Yokoyama S, Kuchino Y (1999) Oncogenic Ras triggers cell suicide through the activation of a caspase-independent cell death program in human cancer cells. Oncogene 18:2281–2290

Chittenden T (1998) Mammalian Bcl-2 family genes. In Wilson J.W., Booth C., Potten C.S., eds., Apoptosis genes, Kluwer Acad. Publishers pp 37–83
Chong H, Vikis HG, Guan KL (2003) Mechanisms of regulating the Raf kinase family. Cell Signal 15:463–469
Coen DM (2002) Neither LAT nor open reading frame P mutations increase expression of spliced or intron-containing ICP0 transcripts in mouse ganglia latently infected with herpes simplex virus. J Virol 76:4764–4772
Colgin MA, Smith RL, Wilcox CL (2001) Inducible cyclic AMP early repressor produces reactivation of latent Herpes simplex virus type 1 in neurons in vitro. J Virol 75:2912–2920
Colucci-D'Amato L, Perrone-Capano C, di Porzio U (2003) Chronic activation of ERK and neurodegenerative diseases. Bioessays 25:1085–1095
Craig CP, Nahmias AJ (1973) Different patterns of neurologic involvement with herpes simplex virus types 1 and 2: isolation of herpes simplex virus type 2 from the buffy coat of two adults with meningitis. J Infect Dis 127:365–372
Cribbs DH, Azizeh BY, Cotman CW, LaFerla FM (2000) Fibril formation and neurotoxicity by a herpes simplex virus glycoprotein B fragment with homology to the Alzheimer's Aβ peptide. Biochemistry 39:5988–5994
Dhandapani KM, Brann DW (2003) Transforming growth factor-beta: a neuroprotective factor in cerebral ischemia. Cell Biochem Biophys 39:13–22
Dobson CB, Wozniak MA, Itzhaki RF (2003) Do infectious agents play a role in dementia? Trends Microbiol 11:312–317
Domachowske JB, Cunningham CK, Cummings DL, Crosley CJ, Hannan WP, Weiner LB (1996) Acute manifestations and neurologic sequelae of Epstein-Barr virus encephalitis in children. Pediatr Infect Dis J 15:871–875
Dumas T, McLaughlin J, Ho D, Meier T, Sapolsky R (1999) Delivery of herpes simplex virus amplicon-based vectors to the dentate gyrus does not alter hippocampal synaptic transmission in vivo. Gene Ther 6:1679–1684
Eilers A, Whitfield J, Shah B, Spadoni C, Desmond H, Ham J (2001) Direct inhibition of c-Jun N- terminal kinase in sympathetic neurons prevents c-jun promoter activation and NGF withdrawal-induced death. J. Neurochem 76:1439–1454
English JD, Sweatt JD (1996) Activation of p42 Mitogen activated protein kinase in hippocampal long term potentiation. J Biol Chem 271:24329–24332
Esiri MM (1982) Herpes simplex encephalitis. J Neurol Sci 54:209–226
Fisman DN, Lipsitch M, Hook EW 3rd, Goldie SJ (2002) Projection of the future dimensions and costs of the genital herpes simplex type 2 epidemic in the United States. Sex Transm Dis 29:608–622
Freeman SM, Whartenby KA, Freeman JL, Abboud CN, Marrogi AJ (1996) In situ use of suicide genes for cancer therapy. Semin Oncol 23:31–45
Friedlander RM (2003) Apoptosis and caspases in neurodegenerative diseases. N Engl J Med 348:1365–1375
Giraud S, Diaz-Latoud C, Hacot S, Textoris J, Bourette RP, Diaz JJ (2004) US11 of herpes simplex virus type 1 interacts with HIPK2 and antagonizes HIPK2-induced cell growth arrest. J Virol 78:2984–2993
Gober M, Smith CC, Ueda K, Toretsky J, Aurelian L (2003) Forced expression of the H11 heat shock protein can be regulated by DNA methylation and trigger apoptosis in human cells. J Biol Chem 278:37600–37609

Golembewski EK, Wales SQ, Aurelian L, Yarowsky PJ (2003) ICP10 PK as a therapy for acute excitotoxic injury in vivo and its mechanism of anti-apoptotic activity. Program No. 741.10 Abstract Viewer/Itinerary Planner. Washington, DC: Society for Neuroscience

Goodkin ML, Ting AT, Blaho JA (2003) NF-kappaB is required for apoptosis prevention during herpes simplex virus type 1 infection. J Virol 77:7261–7280

Guegan C, Vila M, Teismann P, Chen C, Onteniente B, Li M, Friedlander RM, Przedborski S, Teissman P (2002) Instrumental activation of bid by caspase-1 in a transgenic mouse model of ALS. Mol Cell Neurosci 20:553–562

Hagglund R, Munger J, Poon AP, Roizman B (2002) U(S)3 protein kinase of herpes simplex virus blocks caspase 3 activation induced by the products of U(S)1.5 and U(L)13 genes and modulates expression of transduced U(S)1.5 open reading frame in a cell type-specific manner. J Virol 76:743–754

Hill J, Patel A, Bhattacharjee P, Krause P (2003) An HSV-1 chimeric containing HSV-2 latency associated transcript (LAT) sequences has significantly reduced adrenergic reactivation in the rabbit eye model. Curr Eye Res 26:219–224

Hill MM, Adrain C, Martin SJ (2003) Portrait of a killer: the mitochondrial apoptosome emerges from the shadows. Mol Interv 3:19–26

Hood C, Cunningham AL, Slobedman B, Boadle RA, Abendroth A (2003) Varicella-zoster virus-infected sensory neurons are resistant to apoptosis, yet human foreskin fibroblasts are susceptible: evidence for a cell-type-specific apoptotic response. J Virol 77:12852–12864

Huang Q, Vonsattel JP, Schaffer PA, Martuza RL, Breakfield XO, DiFiglia M (1992) Introduction of a foreign gene (*Escherichia coli* lacZ) into rat neostriatal neurons using herpes simplex virus mutants: a light and electron microscopy study. Exp Neurol 115:303–316

Hunsperger EA, Wilcox CL (2003) Caspase-3-dependent reactivation of latent herpes simplex virus type 1 in sensory neuronal cultures. J Neurovirol 9:390–398

Im SA, Gomez-Manzano C, Fueyo J, Liu TJ, Ke LD, Kim JS, Lee HY, Steck PA, Kyritsis AP, Yu WK (1999) Antiangiogenesis treatment for gliomas: transfer of antisense-vascular endothelial growth factor inhibits tumor growth in vivo. Cancer Res 59:895–900

Jerome KR, Fox R, Chen Z, Sears AE, Lee H, Corey L (1999) Herpes simplex virus inhibits apoptosis through the action of two genes, US5 and US3. J Virol 73:8950–8957

Jin L, Peng W, Perng GC, Brick DJ, Nesburn AB, Jones C, Wechsler SL (2003) Identification of herpes simplex virus type 1 latency-associated transcript sequences that both inhibit apoptosis and enhance the spontaneous reactivation phenotype. J Virol 77:6556–6561

Johnson Webb S, Harrison DJ, Wyllie AH (1997) Apoptosis: an overview of the process and its relevance in disease. Adv Pharmacol 41:1–31

Jones C (2003) Herpes simplex virus type 1 and bovine herpesvirus 1 latency. Clin Microbiol Rev 16:79–95

Jones CA, Fernandez M, Herc K, Bosnjak L, Miranda-Saksena M, Boadle RA, Cunningham A (2003) Herpes simplex virus type 2 induces rapid cell death and functional impairment of murine dendritic cells in vitro. J Virol 77:11139–11149

Kandel ES, Hay N (1999) The regulation and activities of the multifunctional serine/threonine kinase Akt/PKB. Exp. Cell Res 253:210–229

Kang SJ, Sanchez I, Jing N, Yuan J (2003) Dissociation between neurodegeneration and caspase-11-mediated activation of caspase-1 and caspase-3 in a mouse model of amyotrophic lateral sclerosis. J Neurosci 23:5455–5460

Kaplan DR, Miller FD (2000) Neurotrophin signal transduction in the nervous system. Curr Opin Neurobiol 10:381–391

Kelleher RJ 3rd, Govindarajan A, Jung HY, Kang H, Tonegawa S (2004) Translational control by MAPK signaling in long-term synaptic plasticity and memory. Cell 116:467–479

Kong H, Baerbig Q, Duncan L, Shepel N, Mayne M (2003) Human herpesvirus type 6 indirectly enhances oligodendrocyte cell death. J Neurovirol 9:539–550

Koyama AH, Adachi A (1997) Induction of apoptosis by herpes simplex virus type 1. J Gen Virol 78:2909–2912

Kristie TM, Pomerantz JL, Twomey TC, Parent SA, Sharp PA (1995) The cellular C1 factor of the herpes simplex virus enhancer complex is a family of polypeptides. J Biol Chem 270:4387–4394

Kurt-Jones EA, Chan M, Zhou S, Wang J, Reed G, Bronson R, Arnold MM, Knipe DM, Finberg RW (2004) Herpes simplex virus 1 interaction with Toll-like receptor 2 contributes to lethal encephalitis. Proc Natl Acad Sci USA 101:1315–1320

Lambeng N, Willaime-Morawek S, Mariani J, Ruberg M, Brugg B (2003) Activation of mitogen-activated protein kinase pathways during the death of PC12 cells is dependent on the state of differentiation. Brain Res Mol Brain Res 111:52–60

Latchman DS (2004) Protective effect of heat shock proteins in the nervous system. Curr Neurovasc Res 1:21–27

Leib DA, Bogard CL, Kosz-Vnenchak M, Hicks KA, Coen DM, Knipe DM, Schaffer PA (1989) A deletion mutant of the latency-associated transcript of herpes simplex virus type 1 reactivates from the latent state with reduced frequency. J Virol 63:2893–2900

Leopardi R, Roizman B (1996) The herpes simplex virus major regulatory protein ICP4 blocks apoptosis induced by the virus or by hyperthermia. Proc Natl Acad Sci USA 93:9583–9587

Lokensgard JR, Hu S, Sheng W, vanOijen M, Cox D, Cheeran MC, Peterson PK (2001) Robust expression of TNF-alpha, IL-1beta, RANTES, and IP-10 by human microglial cells during nonproductive infection with herpes simplex virus. J Neurovirol 7:208–219

Lowes VL, Ip NY, Wong YH (2002) Integration of signals from receptor tyrosine kinases and G protein- coupled receptors. Neurosignals 11:5–19

Marques AR, Straus SE (2001) Lack of association between HSV-1 DNA in the brain, Alzheimer disease and apolipoprotein E4. J Neurovirol 7:82–83

McLaughlin J, Roozendaal B, Dumas T, Gupta A, Ajilore O, Hsieh J, Ho D, Lawrence M, McGaugh JL, Sapolsky R (2000) Sparing of neuronal function post-seizure with gene therapy. Proc Natl Acad Sci USA 97:12804–12809

Mineta T, Rabkin SD, Yazaki T, Hunter WD, Martuza RL (1995) Attenuated multimutated herpes simplex virus-1 for the treatment of malignant gliomas. Nat Med 1:938–943

Morishima N, Nakanishi K, Takenouchi H, Shibata T, Yasuhiko Y (2002) An endoplasmic reticulum stress-specific caspase cascade in apoptosis. Cytochrome *c*-independent activation of caspase-9 by caspase-12. J Biol Chem 277:34287–34294

Moriuchi S, Oligino T, Krisky D, Marconi P, Fink D, Cohen J, Glorioso JC (1998) Enhanced tumor cell killing in the presence of ganciclovir by herpes simplex virus type 1 vector-directed co-expression of human tumor necrosis factor-alpha and herpes simplex virus thymidine kinase. Cancer Res 58:5731–5737

Myers MP, Murphy MB, Landreth G (1994) The dual-specificity CLK kinase induces neuronal differentiation of PC12 cells. Mol Cell Biol 14:6954–6961

Nebreda AR, Gavin A-C (1999) Cell survival demands some Rsk. Science 286:1309–1310

Nelson J, Zhu J, Smith CC, Kulka M, Aurelian L (1996) ATP and SH3 binding sites in the protein kinase of the large of herpes simplex virus ribonucleotide reductase (ICP10). J Biol Chem 271:17021–17027

Ogg PD, McDonell PJ, Ryckman BJ, Knudson CM, Roller RJ (2004) The HSV-1 Us3 protein kinase is sufficient to block apoptosis induced by overexpression of a variety of Bcl-2 family members. Virology 319:212–224

Ozaki N, Sugiura Y, Yamamoto M, Yokoya S, Wanaka A, Nishiyama Y (1997). Apoptosis induced in the spinal cord and dorsal root ganglion by infection of herpes simplex virus type 2 in the mouse. Neurosci Lett 228:99–102

Pankratov YV, Lalo UV, Krishtal OA (2002) Role for P2X receptors in long-term potentiation. J Neurosci 22:8363–8369

Parker JN, Gillespie GY, Love CE, Randall S, Whitley RJ, Markert JM (2000) Engineered herpes simplex virus expressing IL-12 in the treatment of experimental murine brain tumors. Proc Natl Acad Sci USA 97:2208–2213

Pasinelli P, Houseweart MK, Brown RH Jr, Cleveland DW (2000) Caspase-1 and -3 are sequentially activated in motor neuron death in Cu,Zn superoxide dismutase-mediated familial amyotrophic lateral sclerosis. Proc Natl Acad Sci USA 97:13901–13906

Pelosi E, Rozenberg F, Coen DM, Tyler KL (1998) A herpes simplex virus DNA polymerase mutation that specifically attenuates neurovirulence in mice. Virology 252:364–372

Perkins D, Pereira EFR, Gober M, Yarowsky PJ, Aurelian L (2002a) The herpes simplex virus type 2 R1 PK (ICP10 PK) blocks apoptosis in hippocampal neurons involving activation of the MEK/MAPK survival pathway. J Virol 76:1435–1449

Perkins D, Yu YX, Bambrick LL, Yarwosky PJ, Aurelian L (2002b) Expression of herpes simplex virus type 2 protein ICP10 PK rescues neurons from apoptosis due to serum deprivation or genetic defects. Exp Neurol 174:118–122

Perkins D, Pereira EFR, Aurelian L (2003a). The HSV-2 R1 PK (ICP10 PK) functions as a dominant regulator of apoptosis in hippocampal neurons by activating the ERK survival pathway and upregulating the anti-apoptotic protein Bag-1. J Virol 77:1292–1305

Perkins D, Gyure KA, Pereira EFR, Aurelian L (2003b) Herpes simplex virus type 1 induced encephalitis has an apoptotic component associated with activation of c-Jun N-terminal kinase. J. Neurovirol 9:101–111

Perng G-C, Maguen B, Jin L, Mott KR, Osorio N, Slanina SM, Yukht A, Ghiasi H, Nesburn AB, Inman M, Henderson G, Jones C, Wechsler SL (2002) A gene capa-

ble of blocking apoptosis can substitute for the herpes simplex virus type 1 latency-associated transcript gene and restore wild-type reactivation levels. J Virol 76:1224–1235
Plesnila N, Zhu C, Culmsee C, Groger M, Moskowitz MA, Blomgren K (2004) Nuclear translocation of apoptosis-inducing factor after focal cerebral ischemia. J Cereb Blood Flow Metab 24:458–466
Rohn TT, Cusack SM, Kessinger SR, Oxford JT (2004) Caspase activation independent of cell death is required for proper cell dispersal and correct morphology in PC12 cells. Exp Cell Res 295:215–225
Roulston A, Reinhard C, Amiri P, Williams LT (1998) Early activation of c-Jun N-terminal kinase and p38 kinase regulate cell survival in response to tumor necrosis factor alpha. J Biol Chem 273:10232–10239
Sanfilippo CM, Chirimuuta FN, Blaho JA (2004) Herpes simplex virus type 1 immediate-early gene expression is required for the induction of apoptosis in human epithelial HEp-2 cells. J Virol 78:224–239
Sauerbrei A, Eichhorn U, Hottenrott G, Wutzler P (2000). Virological diagnosis of herpes simplex encephalitis. J Clin Virol 17:31–36
Schaeffer HJ, Catling AD, Eblen ST, Collier LS, Krauss A, Weber MJ (1998) MP1: a MEK binding partner that enhances enzymatic activation of the MAP kinase cascade. Science 281:1668–1671
Scherbel U, Raghupathi R, Nakamura M, Saatman KE, Trojanowski JQ, Neugebauer E, Marino MW, McIntosh TK (1999) Differential acute and chronic responses of tumor necrosis factor-deficient mice to experimental brain injury. Proc Natl Acad Sci USA 96:8721–8726
Schulz JB, Bremen D, Reed JC, Lommatzsch J, Takayama S, Wullner U, Loschmann P-A, Klockgether T, Weller M (1997) Cooperative interception of neuronal apoptosis by Bcl-2 and Bag-1 expression: prevention of caspase activation and reduced production of reactive oxygen species. J Neurochem 69:2075–2086
Shaw MM, Gurr WK, Thackray AM, Watts PA, Littler E, Field HJ (2002) Temporal pattern of herpes simplex virus type 1 infection and cell death in the mouse brain stem: influence of guanosine nucleoside analogues. J Virol Methods 102:93–102
Sheng M, Kim MJ (2002) Postsynaptic signaling and plasticity mechanisms. Science 298:776–780
Smith CC, Aurelian L (1997) The large subunit of herpes simplex virus type 2 ribonucleotide reductase (ICP10) is associated with the virion tegument and has PK activity. Virology 234:235–242
Smith CC, Peng T, Kulka M, Aurelian L (1998) The PK domain of the large subunit of herpes simplex virus type 2 ribonucleotide reductase (ICP10) is required for immediate early gene expression and virus growth. J Virol 72:9131–9141
Smith CC, Nelson J, Aurelian L, Gober M, Goswami BB (2000a) Ras-GAP binding/phosphorylation by HSV-2 RR1PK (ICP10) and activation of the Ras/MEK/MAPK mitogenic pathway are required for timely onset of virus growth. J Virol 74:10417–10429
Smith CC, Yu YX, Kulka M, Aurelian L (2000b) A novel human gene similar to the PK coding domain of the large subunit of herpes simplex virus type 2 ribonucle-

otide reductase (ICP10) codes for a serine-threonine PK and is expressed in melanoma cells. J Biol Chem 275:25690–25699
Suzuki T, Hide I, Ido K, Kohsaka S, Inoue K, Nakata Y (2004) Production and release of neuroprotective tumor necrosis factor by P2X7 receptor-activated microglia. J Neurosci 24:1–7
Suzuki H, Imai F, Kanno T, Sawada M (2001) Preservation of neurotrophin expression in microglia that migrate into the gerbil's brain across the blood-brain barrier. Neurosci Lett 312:95–98
Taddeo B, Luo TR, Zhang W, Roizman B (2003) Activation of NF-kappaB in cells productively infected with HSV-1 depends on activated protein kinase R and plays no apparent role in blocking apoptosis. Proc Natl Acad Sci USA 100:12408–12413
Tan S-L, Katze MG (2000) HSV.com: maneuvering the internetworks of viral neuropathogenesis and evasion of the host defense. Proc Natl Acad Sci USA 97:5684–5686
Thomas SK, Lilley CE, Latchman DS, Coffin RS (2002) A protein encoded by the herpes simplex virus (HSV) type 1 2-kilobase latency-associated transcript is phosphorylated, localized to the nucleus, and overcomes the repression of expression from exogenous promoters when inserted into the quiescent HSV genome. J Virol 76:4056–4067
Thompson RL, Sawtell NM (1997) The herpes simplex virus type 1 latency-associated transcript gene regulates the establishment of latency. J Virol 71:5432–5440
Thompson RL, Sawtell NM (2000) HSV latency-associated transcript and neuronal apoptosis. Science 289:1651
Tomac A, Lindqvist E, Lin LFH, Ogre SO, Young D, Hoffer BJ, Olson L (1995) Protection and repair of the nigrostriatal dopaminergic system by GDNF in vivo. Nature 373:335–339
Tomlinson AH, Esiri MM (1983) Herpes simplex encephalitis. J Neurol Sci 60:473–484
Trousdale MD, Steiner I, Spivack JG, Deshmane SL, Brown SM, MacLean AR, Subak-Sharpe JH, Fraser NW (1991) In vivo and in vitro reactivation impairment of a herpes simplex virus type 1 latency-associated transcript variant in a rabbit eye model. J Virol 65:6989–6993
Vaudry D, Stork PJ, Lazarovici P, Eiden LE (2002) Signaling pathways for PC12 cell differentiation: making the right connections. Science 296:1648–1649
Wachsman M, Kulka M, Smith CC, Aurelian L (2001) A growth and latency defective herpes simplex virus type 2 mutant (ICP10ΔPK) has prophylactic and therapeutic protective activity in guinea pigs. Vaccine 19:1879–1890
Waetzig V, Herdegen T (2003) A single c-Jun N-terminal kinase isoform (JNK3-p54) is an effector in both neuronal differentiation and cell death. J Biol Chem 278:567–572
Wang K, Pesnicak L, Guancial E, Krause PR, Straus SE (2001) The 2.2-kilobase latency-associated transcript of herpes simplex virus type 2 does not modulate viral replication, reactivation, or establishment of latency in transgenic mice. J Virol 75:8166–8172
Yamauchi Y, Daikoku T, Goshima F, Nishiyama Y (2003) Herpes simplex virus UL14 protein blocks apoptosis. Microbiol Immunol 47:685–689

York RD, Molliver DC, Grewal SS, Stenberg PE, McCleskey EW, Stork PJ (2000) Role of phosphoinositide 3-kinase and endocytosis in nerve growth factor-induced extracellular signal-regulated kinase activation via Ras and Rap1. Mol Cell Biol 20:8069–8083

Zhou G, Roizman B (2000) Wild-type herpes simplex virus 1 blocks programmed cell death and release of cytochrome *c* but not the translocation of mitochondrial apoptosis-inducing factor to the nuclei of human embryonic lung fibroblasts. J Virol 74:9048–9053

Zhu J, Aurelian L (1997) AP-1 *cis*-response elements are involved in basal expression and Vmw110 transactivation of the large subunit of herpes simplex virus types 2 ribonucleotide reductase (ICP10). Virology 231:301–312

Zwartkruis FJ, Bos JL (1999) Ras and Rap1: two highly related small GTPases with distinct function. Exp Cell Res 253:157–165

CTMI (2005) 289:113–130

The Role of Apoptosis in Defense Against Baculovirus Infection in Insects

R. J. Clem

Division of Biology, Kansas State University, 232 Ackert Hall, Manhattan, KS 66506, USA
rclem@ksu.edu

Abstract The baculoviruses make up a large, diverse family of DNA viruses that have evolved a number of fascinating mechanisms to manipulate their insect hosts. One of these is the ability to regulate apoptosis during infection by expressing proteins that can inhibit caspase activation and/or activity, including the caspase inhibitor P35 and its relatives, and the inhibitor of apoptosis (IAP) proteins. Experimental manipulations of the expression of these antiapoptotic genes, either by genetic deletions or by RNAi, have shed light on the effectiveness of apoptosis in combating baculovirus infection. The results of these experiments indicate that apoptosis can be an extremely powerful response to baculovirus infection, reducing viral replication, infectivity, and the ability of the virus to spread within the insect host even if a successful infection is established. Apoptosis is especially effective when it is combined with other innate antiviral defenses, which are largely unexplored in insects to date.

1
The Baculoviruses

1.1
Baculovirus Characteristics

Members of the virus family *Baculoviridae* are distinguished by having enveloped, rod-shaped nucleocapsids containing large, circular, double-stranded DNA genomes (for reviews of baculoviruses, see Miller 1997 and Friesen and Miller 2001). All known baculoviruses infect only arthropods. and most have highly restricted host ranges, being able to productively infect only a few closely related species. The vast majority of baculoviruses discovered to date infect lepidopteran (moth or butterfly) insect species of agricultural or silvicultural importance. Most of the known baculoviruses produce two morphologically distinct types of virions during infection, the budded virus (BV) form and the occluded virus (OV) form. These two forms of virus have distinct roles in the infection cycle in the host insect, with OV being responsible for horizontal infections between hosts and BV being responsible for spreading infection within a host.

BV particles are produced first during infection and exit the cell by budding through the plasma membrane, acquiring an envelope in the process. BV particles are infectious in most cell types and are responsible for spreading infection throughout the tissues of the infected insect; they are also the form of virus that is used to infect cultured cells during the use of baculoviruses as vectors for expression of foreign genes. BV particles enter cells by endocytosis and subsequent fusion of the viral envelope with the endosomal membrane (Fig. 1A), and they are capable of entering a wide variety of cells, not only from their insect host but from other insects and even from mammalian species. Indeed, there is interest in using baculoviruses as human gene therapy vectors, because none of the viral genes is expressed in mammalian cells, but they can efficiently express foreign genes if an appropriate promoter is engineered into the viral genome (Huser and Hofmann 2003).

OV, on the other hand, are produced later in infection when nucleocapsids remain in the nucleus of the infected insect cell, acquire an envelope through an unknown process, and become enmeshed in large, paracrystalline protein occlusions. The enveloped virions within OV, known as occlusion-derived virus (ODV), are highly specialized for infection of midgut epithelial cells, and, reflecting their ability to infect this particular cell type, the proteins in the envelope of ODV are distinct

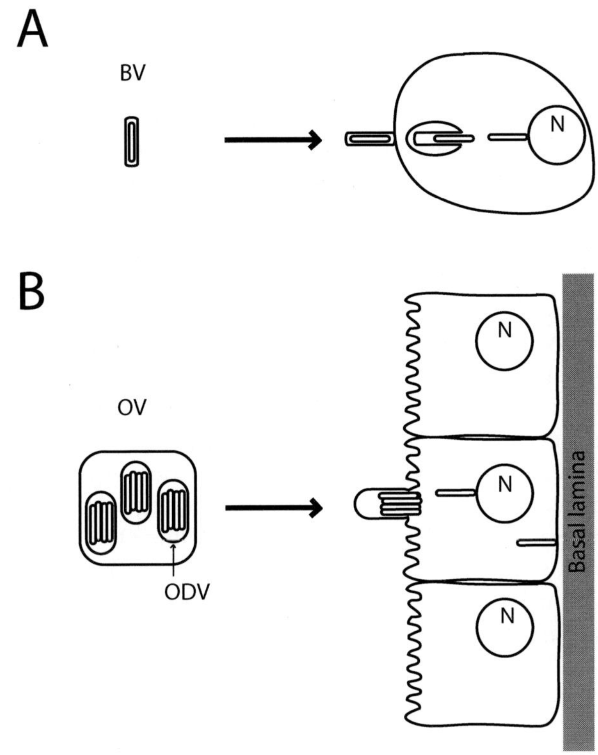

Fig. 1 Infection by different baculovirus morphotypes. **A** Budded virus (*BV*) consists of enveloped nucleocapsids that are produced by budding from the plasma membrane of infected cells. BV attach to many different cell types and enter by endocytosis, where fusion of the viral and endosomal membranes results in release of the nucleocapsid into the cytoplasm. The nucleocapsid travels to the nucleus (*N*) and delivers the viral genetic material. **B** Occluded virus (*OV*) accumulates later in the nuclei of infected cells and consists of enveloped occlusion-derived virions (*ODV*) that are embedded in a paracrystalline matrix. In some baculoviruses, such as AcMNPV, ODV consist of clusters of nucleocapsids that are enveloped within a single membrane. On entering the midgut of the lepidopteran host, the OV matrix dissolves, releasing ODV that attach and enter midgut epithelial cells by direct fusion between the viral and plasma membranes. In the case of multiply enveloped viruses such as AcMNPV, some nucleocapsids travel to the nucleus while others travel through the epithelial cell and bud from the basal surface as BV. The BV that is produced must then traverse the basal lamina in order to successfully enter the host

from those found in BV. OV are highly resistant to environmental extremes and, if protected from light, can persist for long periods of time until they are ingested as a contaminant of the insect's food. Once they enter the highly alkaline environment of the lumen of the lepidopteron midgut, OV dissolve, releasing ODV that infect the epithelial cells lining the midgut by a process believed to involve direct fusion of the ODV envelope and the midgut cell plasma membrane (Fig. 1B).

Currently two genera are recognized within the family *Baculoviridae*, *Nucleopolyhedrovirus* and *Granulovirus* (referred to as NPV and GV, respectively). The NPV have larger OV that contain many ODV particles, whereas the OV form of GV (called granules) are smaller and contain a single ODV particle. The complete genome sequences for nearly 20 different baculovirus species from both genera have been determined, and partial sequences exist for many others, allowing for phylogenetic analysis of a number of conserved viral genes (Herniou et al. 2003). The distinction between the two genera is also strongly supported by these types of analyses. Within the OV of some NPV the ODV contain multiple nucleocapsids within each envelope (MNPV), whereas others contain only a single nucleocapsid per envelope (SNPV). The type species of the NPV genus, *Autographa californica* nucleopolyhedrovirus (AcMNPV), is an MNPV.

1.2 Baculovirus Infection at the Cellular Level

Once the nucleocapsid enters the cytoplasm and the viral DNA is delivered to the nucleus, the regulation of baculovirus gene expression is similar to that of other DNA viruses, with genes being expressed in a temporal cascade fashion in early, late, and very late phases. An immediate-early class of genes is also sometimes distinguished. Early viral genes are expressed from promoters that resemble typical eukaryotic promoters, and early expression is accomplished by host RNA polymerase II (Friesen 1997). Among the early genes are included a number of late expression factor (*lef*) genes that are necessary for the replication of the viral genome and expression of the late and very late classes of viral genes (Rapp et al. 1998). A key viral transactivator is the IE-1 protein, which is produced throughout the infection (Friesen 1997). At the end of the early phase, viral DNA replication begins by a process thought to be similar to the rolling circle mechanism (Lu et al. 1997). Once viral DNA replication is initiated, expression of the late class of genes begins. Interestingly, the expression of the late and very late classes is carried out by a virally encoded RNA polymerase that is made up of at least four subunits (Guarino et al. 1998). The use of a viral RNA polymerase is unique among the nuclear replicating DNA viruses. The late and very late promoters recognized by this polymerase are relatively simple in structure and very different from host eukaryotic RNA pol II promoters. During the late phase, BV is produced in large amounts. At some point, the transcription of late promoters decreases and the viral RNA polymerase begins to recognize and transcribe primarily very late promoters, which are few in number and control the expression of proteins that are necessary for the production of OV. The very late promoters are transcribed at extremely high levels and thus are commonly used for driving the expression of foreign genes in baculovirus expression systems. OV are formed in the nucleus of the infected cells, and eventually the infected cells lyse in a manner that appears to be necrotic cell death.

2 Baculovirus Pathogenesis

2.1 Baculovirus Infection at the Organismal Level

Caterpillars can be experimentally infected by injection of BV into the hemocoel, or interior body cavity. The open circulatory system of insects allows the injected BV to be dispersed rapidly throughout the body of the caterpillar, and most of the tissues of the insect become infected. The natural route of infection, however, is by feeding on material contaminated with OV. Once ODV have successfully fused with midgut epithelial cells, the infection must penetrate the basal lamina underlying the midgut epithelium in order to spread into the rest of the insect. The midgut epithelial cells from many lepidopteran species undergo a process known as sloughing, or loss of cells from the epithelium into the gut lumen. There is evidence suggesting that this sloughing process can be increased in response to infection and thus might represent an effort by the insect to rid itself of infected cells before the virus can penetrate the basal lamina. In cells infected with enveloped ODV from MNPV, which contain multiple nucleocapsids, it has been shown that a fraction of the nucleocapsids enter the nucleus while the rest are transported to the basal surface and directly bud from the cell, thus speeding up the production of BV progeny by several hours (Volkman 1997). Thus the MNPV strategy is thought to be an evolutionary adaptation to the sloughing response of the insect, allowing the successful penetration of the virus into the hemocoel before the infected cell is lost. The nature of midgut cell sloughing has not been investigated in any detail, and it is tempting to speculate that it may be via an apoptotic mechanism.

The mechanism used by the virus to penetrate the basal lamina underlying the midgut epithelium is somewhat controversial and may depend on the virus-host combination. There is evidence that, in at least some cases, the virus is able to infect tracheolar cells that penetrate the basal lamina and supply oxygen to the midgut cells, and thus pass through the basal lamina into the hemocoel (Engelhard et al. 1994). There may also be cases where the virus is able to gain access to the hemocoel through other mechanisms such as small openings or gaps in the basal lamina (Federici 1997). In any case, once past the basal lamina, the virus is thought to spread primarily by circulating through the hemocoel, in many cases infecting hemocytes (blood cells), which are thought to help amplify and spread the infection. However, in at least one case (AcMNPV infection of *Spodoptera frugiperda*) there is evidence suggesting that hemocytes do not play an important role in this process (Clarke and Clem 2002). Organs within the insect are also lined with epithelia that have basal laminae, and again there is evidence that the viruses use tracheal epithelial cells as a conduit into the interior of these organs. In a typical infection, most of the organs of the insect become heavily infected and produce large amounts of both BV and OV, and it is common for a single infected insect to yield more than a billion OV particles. The virus encodes enzymes that help in the degradation of the insect tissues, including a chitinase and a cathepsin-like protease. Expression of these enzymes is necessary for melting or liquefaction of the insect cadaver, which helps in dispersing the OV in the environment.

3 Baculoviral Antiapoptotic Genes

3.1 The *p35* Gene

The first baculovirus antiapoptotic gene was discovered by analysis of a spontaneous AcMNPV mutant called the annihilator. Whereas AcMNPV infection of cultured SF-21 cells, derived from *Spodoptera frugiperda* (fall armyworm), normally results in production of BV and OV and eventual necrotic cell death, infection with the annihilator mutant resulted in apoptosis, as well as decreased production of BV and a complete lack of OV (Clem et al. 1991; Hershberger et al. 1992; Clem and Miller 1993). Complementation analysis revealed that the mutation in the annihilator was due to a deletion in a gene called *p35* (Clem et al. 1991), which had been sequenced previously (Friesen and Miller 1987).

The P35 protein has subsequently been shown to be the most widely acting caspase-inhibiting protein known and has been used extensively to study apoptosis in many different systems. P35 is able to directly bind and inhibit a wide variety of caspases. Its mechanism of action has been well studied and involves cleavage of the P35 protein by the caspase, followed by a conformational change that results in a covalent bond being formed between the caspase and P35 (Xu et al. 2001; Eddins et al. 2002).

P35 appears to be a more effective inhibitor of effector-type caspases than initiator caspases, because it has low activity against human caspase-9 (Vier et al. 2000), the apical *Drosophila* caspase Dronc (Meier et al. 2000), or the presumed equivalent of Dronc in *S. frugiperda*, which has been termed Sf-caspase-X (LaCount et al. 2000). Although homologs of P35 have been described in several other baculoviruses, to date genes with homology to *p35* have not been identified in the genomes of other organisms. One of these other baculovirus proteins with homology to P35, the P49 protein from *Spodoptera litura* NPV (SlNPV), is able to inhibit the initiator caspases Dronc (Jabbour et al. 2002) and Sf-caspase-X (Zoog et al. 2002).

3.2 Baculovirus IAP Proteins

IAP proteins are found in cellular genomes ranging from yeast to humans, and for information on their functions in regulating apoptosis and other cellular functions the reader is referred to recent reviews on IAPs (Salvesen and Duckett 2002; Liston et al. 2003). The focus of this section will be on the baculovirus IAPs. The first *iap* genes were discovered by virtue of their ability to complement the annihilator mutant and were derived from the baculoviruses *Cydia pomonella* GV (Cp-*iap*) and *Orgyia pseudotsugata* NPV (Op-*iap*) (Crook et al. 1993; Birnbaum et al. 1994). At least one *iap* gene is present in almost all of the baculovirus genomes that have been sequenced to date, whereas *p35* genes are only present in a subset of baculoviruses. Genes with homology to baculovirus *iap* have been identified in the genomes of eukaryotes ranging from yeast to humans. In yeast and nematodes, *iap* genes appear to be involved in regulating cell division rather than apoptosis, but in higher animals, including insects and mammals, *iap* genes are important in regulating apoptosis. Interestingly, to date all of the viruses encoding *iap* genes are able to infect arthropods, including baculoviruses, entomopoxviruses, iridoviruses, and African swine fever virus (which infects ticks as an obligate part of its replication cycle), suggesting that IAP proteins are especially important in regulating apoptosis in insects and other arthropods.

In addition to Op-IAP and Cp-IAP, several other baculovirus *iap* genes have been shown to have antiapoptotic activity (Maguire et al. 2000; Ikeda et al. 2004). However, many of the baculovirus *iap* genes that have been tested do not seem to be able to inhibit apoptosis (Clem and Miller 1994; Bideshi et al. 1999; Maguire et al. 2000). The function of these genes is unclear, and it may be that they have anti-apoptotic activity only in certain situations; alternatively, they may function in other cellular or viral processes by virtue of their ability to function as E3 ubiquitin ligases, as has been shown for a number of viral and cellular IAPs (Yang et al. 2000; Imai et al. 2003; Green et al. 2004).

3.2.1 The Op-IAP Protein

The best-studied baculovirus IAP is Op-IAP. Expression of Op-IAP very effectively inhibits apoptosis stimulated by a variety of signals in SF-21 cells, including AcMNPV annihilator mutant infection, UV light, actinomycin D, and expression of Hid, Reaper, and Grim, three proapoptotic proteins from *Drosophila* (Birnbaum et al. 1994; Manji et al. 1997; Vucic et al. 1997, 1998). Suppression of Op-IAP expression by RNAi has been used to demonstrate that this protein is also required to prevent apoptosis during infection of lepidopteran cells by its cognate virus, OpMNPV (Means et al. 2003). However, the mechanism by which Op-IAP inhibits apoptosis is still unclear. Op-IAP contains two predicted BIR domains and a C-terminal RING domain. The roles of these domains in inhibiting apoptosis in the cell line SF-21 have been carefully examined (Vucic et al. 1998; Wright and Clem 2002). The function of BIR1 is unknown, but its removal only slightly reduces Op-IAP's ability to protect against overexpression of Hid. BIR2, on the other hand, is critical for the antiapoptotic function of Op-IAP. Mutations in BIR2 eliminate binding to HID, Reaper, and Grim and also completely abrogate antiapoptotic activity. BIR2 binds the N-terminus of Hid with a mechanism very similar to that of Smac binding to XIAP, and the sequences flanking BIR2 are also important for antiapoptotic function, although not for Hid binding, which is dependent only on the BIR2 core sequence (Wright and Clem 2002). The RING domain is also important for Op-IAP function, as removal of the RING completely eliminates the ability of Op-IAP to inhibit apoptosis due to *p35*-mutant virus infection or actinomycin D, and RING mutants have only a weak ability to protect Sf21 cells from HID overexpression (Vucic et al. 1998; Wright and Clem 2002). The RING domain of Op-IAP has E3 ubiquitin ligase activity and is capable of promoting ubiquitination of itself and of Hid (Green et al. 2004). Op-IAP functions upstream of P35 in Sf21 cells, because expression of Op-IAP blocks processing of Sf-caspase-1, an effector caspase found in these cells, whereas P35 does not (Manji et al. 1997; Seshagiri and Miller 1997). Additional recent evidence indicates that Op-IAP functions upstream of caspase activation altogether in Sf21 cells (Zoog et al. 2002).

Op-IAP is also able to protect mammalian cells against apoptosis induced by a variety of stimuli, including overexpression of caspases or FADD, infection with Sindbis virus, or treatment with tumor necrosis factor (TNF) or anti-fas antibody (Duckett et al. 1996; Hawkins et al. 1996, 1998). Thus the mechanism(s) used by Op-IAP to inhibit apoptosis must be conserved in insects and mammals. Although it was shown that Op-IAP expression inhibited apoptosis in mammalian cells induced by overexpression of caspase-1 or -2 (Hawkins et al. 1998), there are no direct data supporting the hypothesis that Op-IAP can directly interact with and inhibit caspases. In fact, it has been reported that Op-IAP is not able to inhibit human caspase-9 (Huang et al. 2000), and unpublished results indicate that Op-IAP does not inhibit the *Drosophila* caspases Dronc, Drice, or Dcp-1 (C.W. Wright, J.C. Means, T. Penabaz, R.J. Clem, submitted). Interestingly, although Op-IAP can inhibit apoptosis very efficiently in SF-21 cells and mammalian cells, it does not appear to have antiapoptotic activity when expressed in *Drosophila* S2 cells (C.W. Wright, J.C. Means, T. Penabaz, R.J. Clem, submitted) or in the fly eye (B. Hay, personal communication).

4 Evidence for Apoptosis Being an Antiviral Defense

4.1 Antiviral Immunity in Insects

In vertebrate animals, two arms of the immune system exist: innate and acquired immunity. Innate immunity is rapid and relatively nonspecific and does not result in immunologic memory, whereas acquired immunity involves antibodies and T cells that can recognize specific foreign agents and invoke long-term memory. Insects do not appear to possess acquired immunity, but they do possess powerful innate immune systems. Insect innate immunity can be further divided into cellular and humoral mechanisms (Gillespie et al. 1997). The cellular immune response is mediated by the hemocytes, or blood cells, of insects. Hemocytes are capable of phagocytosing small foreign objects, including microbes such as viruses, bacteria, and fungi, or encapsulating larger objects. Encapsulation involves attachment of multiple layers of hemocytes around a foreign agent, cutting off oxygen and nutrient supplies. Hemocytes can also activate the phenyloxidase system, resulting in melanization and death of invading microbes. The humoral immune response involves the production of a variety of antimicrobial lectins and peptides that are capable of binding to and/or causing membrane damage and death in bacteria and fungi. The synthesis of these antibacterial and antifungal peptides is regulated by genetic pathways that have the subject of intense investigation in *Drosophila* (reviewed in Hoffmann 2003).

Although the molecular basis for antibacterial and antifungal immunity in insects is becoming clearer, the mechanisms responsible for antiviral immunity in insects are still poorly understood. Some of the most interesting work to date has come from studies using AcMNPV to infect the lepidopteran insects *Helicoverpa zea* and *Manduca sexta* (Washburn et al. 1996, 2000). In these species, hemocytes are able to encapsulate cells and tissues infected with AcMNPV. Suppression of the cellular immune response with chemicals or by parasitization led to increased virus replication and spread, leading the authors to suggest that recognition and encapsulation of infected cells could be at least partially responsible for limiting infection in this species. The hemocytes from *H. zea* also appear to be highly resistant to infection and may play a role in clearing virus from the hemolymph (Trudeau et al. 2001). At this time, it is not known how widespread this type of immunity is among other insects.

4.2 Infection with AcMNPV *p35*-Mutant Viruses

Evidence that apoptosis can be involved in insect defense against viruses originally came from work that was performed in the laboratory of the late Dr. Lois Miller. The first indication that apoptosis could be detrimental to virus replication came from studies using cultured cells. Infection of SF-21 cells with the annihilator mutant or other strains of

AcMNPV with mutations in the *p35* gene resulted in yields of BV that were decreased at least 100-fold compared to control viruses containing *p35*, as well as a complete lack of OV (Hershberger et al. 1992; Clem and Miller 1993). In contrast to the SF-21 cell line, *p35*-mutant viruses did not induce apoptosis in a different cell line, TN-368, derived from the cabbage looper *Trichoplusia ni*, but instead replicated normally, producing normal yields of both BV and OV (Hershberger et al. 1992; Clem and Miller 1993).

When *p35*-mutant viruses were used to infect *S. frugiperda* larvae by either hemocoelic injection of BV or feeding of OV, they were found to have drastically reduced infectivity compared to wild-type virus. Injection of BV into *S. frugiperda* larvae resulted in an increase in LD_{50} (the dose required for 50% lethality) of the *p35*-mutant virus of more than 1,000-fold compared to control virus containing *p35* (Clem and Miller 1993), whereas feeding of OV to *S. frugiperda* larvae revealed a 25-fold increase in the concentration of OV required for lethality (Clem et al. 1994). Importantly, in *T. ni* larvae, the infectivity of the *p35*-mutant viruses was equivalent to that of the control viruses by both methods of inoculation, further suggesting that the reduced infectivity in *S. frugiperda* larvae was due to an apoptotic response.

If the decreased infectivity in *S. frugiperda* larvae was due to an apoptotic response, then blocking apoptosis with a different antiapoptotic gene should restore full infectivity. This hypothesis was tested by replacing the *p35* gene with Cp-*iap*, which appears to block apoptosis at a different step in the apoptotic pathway. When a virus lacking *p35* but expressing the Cp-*iap* gene was injected into *S. frugiperda* larvae, the virus had infectivity equivalent to wild-type AcMNPV (Clem et al. 1994). This result strongly suggests that apoptosis caused the defect in infectivity of the *p35*-mutant virus, and that apoptosis can serve as an effective antiviral defense mechanism.

These early experiments have been followed up more recently by studies using viruses that express enhanced green fluorescent protein (eGFP) from a constitutive host promoter to be able to follow the progression of infection within the insect. Similar to previous results with untagged viruses, a drastic negative effect on infectivity was seen when eGFP-expressing viruses that either contained or lacked the *p35* gene were used to infect *S. frugiperda* larvae by injection of BV into the hemocoel (Clarke and Clem 2003). Even when very high doses of *p35*-deleted BV (more than 2,000 times the LD_{50} for wild type) were injected into *S. frugiperda*, the majority of the insects showed no signs of eGFP expression either externally or internally, indicating that the insects were able to inactivate and clear large amounts of virus. In the case of those larvae that were successfully infected, most expressed eGFP transiently in the fat body and epithelial tissues for the first 2 days after infection. External eGFP expression was observed only in very localized regions that tended to be on the ventral surface of the caterpillar, and the infection did not spread. These larvae died after a prolonged time, but their death did not seem to be directly due to virus infection. Only in a small proportion of larvae (around 10% of the total number injected) did the infection progress throughout the tissues and appear somewhat normal. However, even in these larvae, the infection spread slower than normal, and titers of virus in the hemolymph were significantly reduced compared to wild-type infections. *S. frugiperda* larvae infected with *p35*-mutant AcMNPV also do not undergo the melting or liquefaction that normally occurs after the death of baculovirus-infected caterpillars (Clem and Miller 1993; Clarke and Clem 2003).

When infected tissues from *S. frugiperda* larvae (as determined by eGFP fluorescence) were examined for signs of apoptosis by TUNEL assay, larvae infected with wild-type virus were negative for TUNEL staining until late times after infection, when weak TUNEL staining was observed (Clarke and Clem 2003). These wild type-infected cells did not appear apoptotic morphologically but instead displayed typical baculovirus cytopathic effects including nuclear swelling, indicating that the TUNEL staining seen at late times was artifactual. Larvae infected with *p35*-mutant virus, however, displayed strong TUNEL staining throughout the infection process, and TUNEL-positive cells had highly fragmented nuclei, indicative of apoptosis. Tissues infected with the *p35*-mutant virus also were highly fragmented in nature, presumably due to large numbers of apoptotic cells.

These results indicate that apoptosis does indeed occur during *p35* mutant infection in *S. frugiperda*, and support the hypothesis that apoptosis can be an effective defense against baculovirus infection. However, during characterization of the infection of *S. frugiperda* with wild-type AcMNPV, it was noted that the larvae were somewhat resistant to infection by the hemocoelic injection route compared to *T. ni* larvae (Table 1) (Clarke and Clem 2002). Whereas the LD_{50} for *T. ni* larvae was found to be less than 1 plaque forming unit (PFU)/larva, that for *S. frugiperda* was calculated at 20 PFU/larva. This difference begs the question of how effective apoptosis would be without this background resistance seen against wild-type AcMNPV in *S. frugiperda*. To resolve this question, similar experiments were carried out in a related species, *Spodoptera exigua* (Table 1). The LD_{50} for *S. exigua* was found to be 4 PFU/larva,

Table 1 Infectivity differences between wild-type AcMNPV and *p35*-mutant AcMNPV in different insect species

Species	LD_{50}[a] Wild-type AcMNPV	LD_{50} *p35*-mutant AcMNPV	Fold difference
T. ni	<1[b]	<1	~1
S. frugiperda	20	$>1\times10^5$	>5,000
S. exigua	4	265	66

[a] Lethal dose required for 50% lethality.
[b] Plaque-forming units (PFU) per larva.

which is intermediate compared to *T. ni* and *S. frugiperda*. Thus *S. exigua* is not as resistant to wild-type AcMNPV as *S. frugiperda*. The LD_{50} for *p35*-mutant AcMNPV in *S. exigua* larvae was 265 PFU/larva, a 66-fold increase compared to wild-type AcMNPV. When *S. exigua* larvae that had been infected with *p35*-mutant virus were examined by TUNEL assay, widespread apoptosis was observed (T.E. Clarke, L. Heaton and R.J. Clem, submitted). Thus apoptosis also appears to be an effective response against AcMNPV infection in *S. exigua*. However, the difference between the LD_{50} of wild type and the *p35* mutant in *S. frugiperda* is much greater than in *S. exigua* (Table 1). Thus it appears that a second mechanism of resistance (such as that operating in *S. frugiperda*) can act synergistically with apoptosis to form a far more effective barrier to infection. However, even in *S. exigua*, where a 66-fold difference exists between the LD_{50} of wild type and the *p35* mutant, apoptosis is quite effective at limiting baculovirus infectivity.

Interestingly, another group has reported that the LD_{50} for AcMNPV in *S. frugiperda* is similar to that of *T. ni*, less than 1 PFU/larva (Haas-Stapleton et al. 2003). The difference in LD_{50} results between the two laboratories is puzzling, because a difference has only been reported with *S. frugiperda*, and not with *T. ni*, for which both laboratories report similar results. Variations in the strains of virus or *S. frugiperda* larvae used have been ruled out, and at this point it appears that the difference lies in the method used to inject the insects. Regardless of the reason, the results indicate that some type of uncharacterized antiviral resistance mechanism can be triggered in *S. frugiperda*.

5 Conclusions

The work done so far in characterizing the effects of apoptosis on AcMNPV infection in *S. frugiperda* and *S. exigua* has provided some of the best evidence to date that apoptosis can be a highly effective response by animal hosts against virus infection. Because many different families of viruses contain antiapoptotic genes, it appears that apoptosis is an ancient response to virus infection, and probably all viruses have had to evolve ways to deal with this cellular defense mechanism. All baculoviruses sequenced to date contain at least one antiapoptotic gene, namely homologs of *p35* or *iap* or both, indicating that apoptosis can deal a heavy blow to the ability of baculoviruses to successfully infect insects.

It appears that an apoptotic response by initially infected cells effectively "raises the bar" in terms of the amount of virus inoculum that is required to successfully establish an infection. However, the effectiveness of apoptosis can be greatly enhanced by other mechanisms of resistance. In *S. frugiperda*, mutation of *p35* results in an increase in LD_{50} of greater than 1,000-fold, whereas in *S. exigua*, which are more susceptible to wild-type AcMNPV infection, the increase is only around 66-fold. In those *S. frugiperda* larvae that are successfully infected, the spread of AcMNPV *p35* mutant infection is also severely limited, much more so than in *S. exigua*. Thus apoptosis appears to be only one mechanism that is used by insects to resist baculovirus infection, and its usefulness is greatly enhanced if more than one mechanism can be brought to bear on the invading virus. The nature of these other resistance mechanisms is largely unexplored but may involve phagocytosis of virally infected cells. So far, all of the in vivo work that has been done on the role of apoptosis in defense against baculovirus infection has been done by injection of BV into larvae. The next phase of research will involve determining whether an apoptotic response is important in the midgut epithelium, which is the initial site of invasion in a natural baculovirus infection.

Acknowledgements The author wishes to thank all of the individuals who have contributed to the research in his laboratory, past and present. Research in Dr. Clem's laboratory has been supported by the Cooperative State Research Education and Extension Service of the U.S. Department of Agriculture, by grant RR-017686 from the National Center for Research Resources of the National Institutes of Health, and by the Kansas Agricultural Experiment Station. This is contribution number 04-374-J from the Kansas Agricultural Experiment Station.

References

Bideshi DK, Anwar AT, Federici BA (1999) A baculovirus anti-apoptosis gene homolog of the *Trichoplusia ni* granulovirus. Virus Genes 19:95–101

Birnbaum MJ, Clem RJ, Miller LK (1994) An apoptosis-inhibiting gene from a nuclear polyhedrosis virus encoding a peptide with cys/his sequence motifs. J Virol 68:2521–2528

Clarke TE, Clem RJ (2002) Lack of involvement of haemocytes in the establishment and spread of infection in *Spodoptera frugiperda* larvae infected with the baculovirus *Autographa californica* M nucleopolyhedrovirus by intrahemocoelic injection. J Gen Virol 83:1565–1572

Clarke TE, Clem RJ (2003) In vivo induction of apoptosis correlating with reduced infectivity during baculovirus infection. J Virol 77:2227–2232

Clem RJ, Fechheimer M, Miller LK (1991) Prevention of apoptosis by a baculovirus gene during infection of insect cells. Science 254:1388–1390

Clem RJ, Miller LK (1993) Apoptosis reduces both the in vitro replication and the in vivo infectivity of a baculovirus. J. Virol. 67:3730–3738

Clem RJ, Miller LK (1994) Control of programmed cell death by the baculovirus genes *p35* and *iap*. Mol Cell Biol 14:5212–5222

Clem RJ, Robson M, Miller LK (1994) Influence of infection route on the infectivity of baculovirus mutants lacking the apoptosis-inhibiting gene *p35* and the adjacent gene *p94*. J Virol 68:6759–6762

Crook NE, Clem RJ, Miller LK (1993) An apoptosis-inhibiting baculovirus gene with a zinc finger-like motif. J Virol 67:2168–2174

Duckett CS, Nava VE, Gedrich RW, Clem RJ, Van Dongen JL, Gilfillan MC, Shiels H, Hardwick JM, Thompson CB (1996) A conserved family of cellular genes related to the baculovirus *iap* gene and encoding apoptosis inhibitors. EMBO J 15:2685–2694

Eddins MJ, Lemongello D, Friesen PD, Fisher AJ (2002) Crystallization and low-resolution structure of an effector-caspase/P35 complex: similarities and differences to an initiator-caspase/P35 complex. Acta Cryst D58:299–302

Engelhard EK, Kam-Morgan LNW, Washburn JO, Volkman LE (1994) The insect tracheal system: a conduit for the systemic spread of *Autographa californica* M nuclear polyhedrosis virus. Proc. Natl Acad Sci USA 91:3224–3227

Federici BA (1997) Baculovirus pathogenesis. In: Miller LK (ed) The Baculoviruses. Plenum Press, New York, pp 33–59

Friesen PD (1997) Regulation of baculovirus early gene expression. In: Miller LK (ed) The Baculoviruses. Plenum Press, New York. pp 141–170

Friesen PD, Miller LK (1987) Divergent transcription of early 35 and 94-kilodalton protein genes encoded by the *Hind*-III K genome fragment of the baculovirus *Autographa californica* nuclear polyhedrosis virus. J Virol 61:2264–2272

Friesen PD, Miller LK (2001) Insect viruses. In: Knipe DM and Howley PM (eds) Fields' Virology. Lippincott Williams and Wilkins, Philadelphia. 1:599–628.

Gillespie JP, Kanost MR, Trenczek T (1997) Biological mediators of insect immunity. Annu Rev Entomol 42:611–643

Green MC, Monser KP, Clem RJ (2004) Ubiquitin protein ligase activity of the anti-apoptotic baculovirus protein Op-IAP3. Virus Res 105:89–96

Guarino LA, Xu B, Jin J, Dong W (1998) A virus-encoded RNA polymerase purified from baculovirus-infected cells. J Virol 72:7985–7991

Haas-Stapleton E, Washburn JO, Volkman LE (2003) Pathogenesis of *Autographa californica* M nucleopolyhedrovirus in fifth instar *Spodoptera frugiperda.* J Gen Virol 84:2033–2040

Hawkins C, Uren A, Hacker G, Medcalf R, Vaux D (1996) Inhibition of interleukin 1 beta-converting enzyme-mediated apoptosis of mammalian cells by baculovirus IAP. Proc Natl Acad Sci USA 93:13786–13790

Hawkins CJ, Ekert PG, Uren AG, Holmgreen SP, Vaux DL (1998) Anti-apoptotic potential of insect cellular and viral IAPs in mammalian cells. Cell Death Differ 5:569–576

Herniou E, Olszewski J, Cory J and O'Reilly DR (2003) The genome sequence and evolution of baculoviruses. Annu. Rev. Entomol. 48:211–234

Hershberger PA, Dickson JA, Friesen PD (1992) Site-specific mutagenesis of the 35-kilodalton protein gene encoded by *Autographa californica* nuclear polyhedrosis virus: cell line-specific effects on virus replication. J Virol 66:5525–5533

Hoffmann JA (2003) The immune response of *Drosophila.* Nature 426:33–38

Huang Q, Deveraux QL, Maeda S, Salvesen GS, Stennicke HR, Hammock BD, Reed JC (2000) Evolutionary conservation of apoptosis mechanisms: lepidopteran and baculoviral inhibitor of apoptosis proteins are inhibitors of mammalian caspase-9. Proc Natl Acad Sci USA 97:1427–1432

Huser A, Hofmann C (2003) Baculovirus vectors: novel mammalian cell gene-delivery vehicles and their applications. Am J Pharmacogenomics 3:53–63

Ikeda M, Yanagimoto K, Kobayashi M (2004) Identification and functional analysis of *Hyphantria cunea* nucleopolyhedrovirus *iap* genes. Virology 321:359–371

Imai N, Matsuda N, Tanaka K, Nakano A, Matsumoto S, Kang W (2003) Ubiquitin ligase activities of *Bombyx mori* nucleopolyhedrovirus RING finger proteins. J Virol 77:923–930

Jabbour AM, Ekert PG, Coulson EJ, Knight MJ, Ashley DM, Hawkins CJ (2002) The p35 relative, p49, inhibits mammalian and *Drosophila* caspases including DRONC and protects against apoptosis. Cell Death Differ 9:1311–1320

LaCount DJ, Hanson SF, Schneider CL, Friesen PD (2000) Caspase inhibitor P35 and inhibitor of apoptosis Op-IAP block in vivo proteolytic activation of an effector caspase at different steps. J Biol Chem 275:15657–15664

Liston P, Fong WG and Korneluk RG (2003) The inhibitors of apoptosis: there is more to life than Bcl2. Oncogene 22:8568–8580

Lu A, Krell PJ, Vlak JM, Rohrmann GF (1997) Baculovirus DNA replication. In: Miller LK (ed) The Baculoviruses. Plenum Press, New York, pp 171–191

Maguire T, Harrison P, Hyink O, Kalmakoff J, Ward VK (2000) The inhibitors of apoptosis of *Epiphyas postvittana* nucleopolyhedrovirus. J Gen Virol 81:2803–2811

Manji GA, Hozak RR, LaCount DJ, Friesen PD (1997) Baculovirus inhibitor of apoptosis functions at or upstream of the apoptotic suppressor P35 to prevent programmed cell death. J Virol 71:4509–4516

Means JC, Muro I, Clem RJ (2003) Silencing of the baculovirus Op-*iap3* gene by RNA interference reveals that it is required for prevention of apoptosis during

Orgyia pseudotsugata M nucleopolyhedrovirus infection of Ld652Y cells. J Virol 77:4481–4488
Meier P, Silke J, Leevers SJ, Evan GI (2000) The *Drosophila* caspase DRONC is regulated by DIAP1. EMBO J 19:598–611
Miller LK, Ed (1997) The baculoviruses. New York, Plenum Press
Rapp JC, Wilson JA, Miller LK (1998) Nineteen baculovirus open reading frames, including LEF-12, support late gene expression. J Virol 72:10197–10206
Salvesen GS, Duckett CS (2002) IAP proteins: blocking the road to death's door. Nat Rev Mol Cell Biol 3:401–410
Seshagiri S, Miller LK (1997) Baculovirus inhibitors of apoptosis (IAPs) block activation of Sf-caspase-1. Proc Natl Acad Sci USA 94:13606–13611
Trudeau D, Washburn JO, Volkman LE (2001) Central role of hemocytes in *Autographa californica* M nucleopolyhedrovirus pathogenesis in *Heliothis virescens* and *Helicoverpa zea*. J Virol 75:996–1003
Vier J, Furmann C, Hacker G (2000) Baculovirus P35 protein does not inhibit caspase-9 in a cell-free system of apoptosis. Biochem Biophys Res Commun 276:855–861
Volkman LE (1997) Nucleopolyhedrovirus interactions with their insect hosts. Adv Virus Res 48:313–348
Vucic D, Kaiser WJ, Harvey AJ, Miller LK (1997) Inhibition of Reaper-induced apoptosis by interaction with inhibitor of apoptosis proteins (IAPs). Proc Natl Acad Sci USA 94:10183–10188
Vucic D, Kaiser WJ, Miller LK (1998) Inhibitor of apoptosis proteins physically interact with and block apoptosis induced by *Drosophila* proteins HID and GRIM. Mol Cell Biol 18:3300–3309
Vucic D, Kaiser WJ, Miller LK (1998) A mutational analysis of the baculovirus inhibitor of apoptosis Op-IAP. J Biol Chem 273:33915–33921
Washburn JO, Haas-Stapleton E, Tan FF, Beckage NE, Volkman LE (2000) Co-infection of *Manduca sexta* larvae with polydnavirus from *Cotesia congregata* increases susceptibility to fatal infection by *Autographa californica* M nucleopolyhedrovirus. J Insect Physiol 46:179–190
Washburn JO, Kirkpatrick BA, Volkman LE (1996) Insect protection against viruses. Nature 383:767
Wright CW, Clem RJ (2002) Sequence requirements for Hid binding and apoptosis regulation in the baculovirus inhibitor of apoptosis Op-IAP: Hid binds Op-IAP in a manner similar to Smac binding of XIAP. J Biol Chem 277:2454–2462
Xu G, Cirilli M, Huang Y, Rich RL, Myszka DG, Wu H (2001) Covalent inhibition revealed by the crystal structure of the caspase-8/p35 complex. Nature 410:494–497
Yang Y, Fang S, Jensen JP, Weissman AM, Ashwell JD (2000) Ubiquitin protein ligase activity of IAPs and their degradation in proteasomes in response to apoptotic stimuli. Science 288:874–877
Zoog SJ, Schiller JJ, Wetter JA, Chejanovsky N, Friesen PD (2002) Baculovirus apoptotic suppressor P49 is a substrate inhibitor of initiator caspases resistant to P35 in vitro. EMBO J 21:5130–5140

CTMI (2005) 289:131–150

The Role of Host Cell Death in *Salmonella* Infections

D. G. Guiney

Department of Medicine, UCSD School of Medicine, 9500 Gilman Dr., La Jolla, CA 92093-0640, USA
dguiney@ucsd.edu

Abstract *Salmonella enterica* is an important enteric pathogen of humans and a variety of domestic and wild animals. Infection is initiated in the intestinal tract, and severe disease produces widespread destruction of the intestinal mucosa. *Salmonella* strains can also disseminate from the intestine and produce serious, sometimes fatal infections with considerable cytopathology in a number of systemic organs. A combination of bacterial genetic and cell biology studies have shown that *Salmonella* uses specific virulence mechanisms to induce host cell death during infection. *Salmonella* produces one set of virulence proteins to promote invasion of the intestine and a different set to mediate systemic disease. Significantly, each set of virulence factors mediates a distinct mechanism of host cell death. The *Salmonella* pathogenicity island-1 (SPI-1) locus encodes a type III protein secretion system (TTSS) that delivers effector proteins required for intestinal invasion and the production of enteritis. The SPI-1 effector SipB activates caspase-1 in macrophages, releasing IL-1β and IL-18 and inducing rapid cell death by a mechanism that has features of both apoptosis and necrosis. Caspase-1 is required for *Salmonella* to infect Peyer's patches and disseminate to systemic tissues in mice. Progressive *Salmonella* infection in mice requires the SPI-2 TTSS and associated effector proteins as well as the SpvB cytotoxin. Apoptosis of macrophages in the liver is found during systemic infection. In cell culture, *Salmonella* strains induce delayed apoptosis dependent on SPI-2 function in macrophages from a variety of sources. This delayed apoptosis also requires activation of TLR4 on macrophages by the bacterial LPS. Downstream activation of kinase pathways leads to balanced pro- and antiapoptotic regulatory factors in the cell. NF-κB and p38 mitogen-activated protein kinase (MAPK) are particularly important for the induction of antiapoptotic factors, whereas the kinase PKR is required for bacterial-induced apoptosis. The *Salmonella* SPI-2 TTSS is essential for altering the balance in favor of apoptosis during intracellular infection, but the effectors involved remain poorly characterized. The SpvB cytotoxin has been shown to play a role in apoptosis in human macrophages by depolymerizing the actin cytoskeleton. A model for the role of bacteria-induced host cell death in *Salmonella* pathogenesis is proposed. In the intestine, the *Salmonella* SPI-1 TTSS and SipB mediate macrophage death by caspase-1 activation, which also releases IL-1β and IL-18, promoting inflammation and subsequent phagocytosis by incoming macrophages and leading to dissemination to systemic tissues. Intracellular secretion of virulence effector proteins by the SPI-2 TTSS facilitates growth of *Salmonella* in these macrophages and the delayed onset of apoptosis in extraintestinal tissues. These infected, apoptotic cells are targeted for engulfment by incoming macrophages, thus perpetuating the cycle of cell-to-cell spread that is the hallmark of systemic *Salmonella* infection.

1 Introduction

Salmonella enterica strains are facultative intracellular pathogens that can produce both localized enteritis and disseminated systemic disease in humans and a variety of other vertebrates (Ohl and Miller 2001). Natural *Salmonella* infections are acquired exclusively through ingestion, with subsequent colonization of the intestine. The organisms are able to attach to the intestinal epithelial cells and induce uptake of the bacteria into specialized membrane-bound vesicles termed *Salmonella*-containing vacuoles (SCV) (Holden 2002; Knodler and Steele-Mortimer 2003). This invasion process requires the function of a type III protein secretion system (TTSS) encoded in the *Salmonella* pathogenicity island-1 (SPI-1) locus (Galan 2001). The SPI-1 TTSS transfers a number of bacterial effector proteins into host epithelial cells, producing physiological and structural changes in the cytoskeleton leading to bacterial uptake. In epithelial cell culture systems, *Salmonella* strains are able to proliferate within the SCV and also traverse the epithelial cell to be released from the basolateral side (Finlay et al. 1988). Studies using polarized epithelial cell monolayers have shown that *Salmonella* infection stimulates a chloride secretory response as well as the induction and release of proinflammatory mediators (Eckmann et al. 1993; Resta-Lenert and Barrett 2002; Mrsny et al. 2004). It is likely that the diarrhea, fever, and abdominal pain characteristic of *Salmonella* enteritis are due to a combination of these secretory and inflammatory processes. Although *Salmonella* infection is known to localize in the Peyer's patch lymphoid tissue

(Jones et al. 1994), the diffuse nature of the mucosal pathology in hosts that develop enteritis suggests that epithelial cells in addition to the specialized M cells of Peyer's patches are targets for *Salmonella* invasion (Santos et al. 2001c). In severe cases, there is diffuse loss of epithelium and the normal mucosal architecture is destroyed, suggesting that *Salmonella* invasion and/or the resulting inflammation causes extensive cell death in the epithelial layer (Santos et al. 2001c). Mucosal macrophages phagocytose *Salmonella*, but a massive influx of neutrophils quickly ensues in animals that develop clinical enteritis. *Salmonella* rapidly disseminate to systemic organs of the reticuloendothelial system by a mechanism dependent on CD18-positive cells (Vazquez-Torres et al. 1999). Although enteritis can be severe, particularly in young hosts, the disease is often brief, and the rapid resolution before the onset of the acquired immune response indicates that elements of innate immunity likely contribute to the termination of intestinal disease. However, the elimination of extraintestinal *Salmonella* depends on a T cell-mediated acquired immune response (Jones and Falkow 1996). In particular, a defect in CD4 T cells, as seen in HIV infection and other states of defective cellular immunity, leads to chronic and severe disseminated salmonellosis (Libby et al. 2002). A second *Salmonella* TTSS, encoded by the pathogenicity island-2 locus (SPI-2), is required for strains to produce systemic disease (Waterman and Holden 2003). Additional virulence genes, including the *Salmonella* intracellular toxin SpvB, promote systemic infection. A large body of experimental evidence indicates that extraintestinal *Salmonella* infection occurs inside tissue macrophages (Fields et al. 1986; Richter-Dahlfors et al. 1997).

The human-adapted pathogen, *Salmonella enterica* serovar *typhi*, is the cause of the clinical syndrome of typhoid fever, a disease seen only in people. Typhoid fever differs in several significant aspects from disease due to nontyphoid *Salmonella* strains. *Typ*hi causes little or no clinical enteritis but instead produces a characteristic prolonged systemic infection despite the development of acquired T cell immunity in normal hosts (Pang et al. 1995). Eventually, the host immune system is able to eliminate the systemic phase of the illness, but often long-term carriage in the biliary system ensues. *Typhi* strains produce a capsule, the Vi antigen, but lack certain genes, notably *spvB*, that are important for systemic virulence of nontyphoid *Salmonella* strains (Roudier et al. 1990).

Pathogens frequently cause cell death in infected tissues. In recent years, increasing attention has been paid to the mechanism of cell death associated with different infectious agents (Weinrauch and Zychlinsky 1999). In the field of pathology, two clearly distinguishable types of cell death have been recognized for many years on the basis of histological features: apoptosis and necrosis. Apoptosis, or programmed cell death, is the principal physiological mechanism for eliminating unnecessary cells during embryogenesis, in the development and regulation of the immune system, and during normal cell turnover in many tissues (Marsden et al. 2002). Necrosis is usually associated with tissue injury from a interruption in the vascular supply, decreased oxygenation, or various noxious environmental factors (Majno and Joris 1995). Many variants of these processes occur, and it is not the purpose of this review to differentiate among the various mixed forms of cell death. Classically, apoptosis is triggered by two separate pathways, extrinsic and intrinsic (Marsden et al. 2002). The extrinsic mechanism involves extracellular ligands binding to cell membrane receptors that possess death domains on the cytoplasmic side. Recruitment of adapter proteins leads to activation of caspase-8 and eventually caspase-3, the primary executioner of programmed cell death. The intrinsic pathway is triggered by a variety of stimuli that alter the balance between regulatory systems that have pro- or antiapoptotic effects. Predictably, apoptosis is a carefully controlled process with complex checks and balances to avoid inappropriate triggering of the pathway. These regulatory systems include the large family of Bcl-2 proteins that provide connections between the apoptosis machinery and physiological processes in the cell. Eventually, proapoptotic proteins alter the function of the mitochondrial membrane, releasing cytochrome *c* into the cytoplasm and initiating formation of the "apoptosome," consisting of cytochrome *c*, the adaptor Apaf-1, and caspase-9. The apoptosome activates the executioners of apoptosis, including caspase-3. It is also possible for initiator caspases to be activated by regulatory proteins, in turn then stimulating cytochrome *c* release and the remainder of the pathway. Activated caspase-3 is a protease with many cellular targets, with one activation pathway resulting in cleavage of chromosomal DNA between nucleosomes, release of nucleosomes into the cytoplasm, and nuclear condensation. Notably, cellular metabolic activity and cell membrane integrity persist until late in the apoptotic process, and cells display surface markers that stimulate their uptake and removal by tissue macrophages. In contrast, cell membrane integrity is lost early in necrosis; the membrane leaks cellular contents, and the cell lyses.

Bacterial pathogens can induce apoptosis or necrosis by a variety of direct and indirect mechanisms (Weinrauch and Zychlinsky 1999). In many pathogenesis studies, a comprehensive analysis of the mechanism of cell death due to bacterial infection has not been undertaken. Instead, the process has been reported as necrosis or apoptosis based on the use of one or a small number of cell biology markers. Occasionally, a mixed process has been described by terms such as "oncosis" (Majno and Joris 1995) and "programmed necrosis" (Hernandez et al. 2003; Assuncao Guimaraes and Linden 2004). Although this review will generally cite the terms used by the authors of a particular study, the specific experimental data for classifying important cell death processes initiated by *Salmonella* will usually be reported.

2
Overview of Cell Death Induced by Gram-Negative Pathogens

A variety of gram-negative pathogens induce cell death in host cells, and recent work has begun to define the bacterial virulence factors and host cell signal transduction pathways involved. Figure 1 summarizes a current model primarily based on the interactions of *Salmonella enterica*, *Shigella flexneri*, and *Yersinia* species with macrophages (Knodler and Finlay 2001; Monack et al. 2001b; Hsu et al. 2004). Two general mechanisms of pathogen-induced cell death have been identified. One pathway is mediated by effector proteins delivered by type III protein secretion systems (TTSSs) leading to activation of caspase-1 and subsequent cell death by a process that shares features of both necrosis and apoptosis (Knodler and Finlay 2001; Monack et al. 2001b). Active caspase-1 also processes the precursors of IL-1β and IL-18 to the active cytokines that are then released from the dying cell. The second mechanism involves activation of TLR4 by bacterial LPS (Hsu et al. 2004). The cytoplasmic domain of TLR4 contains sites for adapter proteins that initiate a complex signal transduction cascade, leading to the induction of both pro- and antiapoptotic factors. Kinase pathways activate the mitogen-activated protein kinase (MAPK) JNK as well as IKK, with subsequent activation of the transcription regulatory factors AP-1 and NF-κB. AP-1 and NF-κB act together to induce synthesis of proinflammatory cytokines, some of which also have proapoptotic activity, such as TNFα. However, TNF does not generally induce apoptosis in macrophages, because of the production of antiapoptotic factors under the control of NF-κB and the MAPK isoform p38. Inhibition of p38 or NF-κB markedly increases macrophage apoptosis in response to LPS (Park et al. 2002; Hsu et al. 2004). In addition, TLR4 ligation by LPS leads to activation of PKR, initially described as a dsRNA-responsive kinase (Hsu et al. 2004). PKR phosphorylates the elongation factor eIF2α, which leads to inhibition of

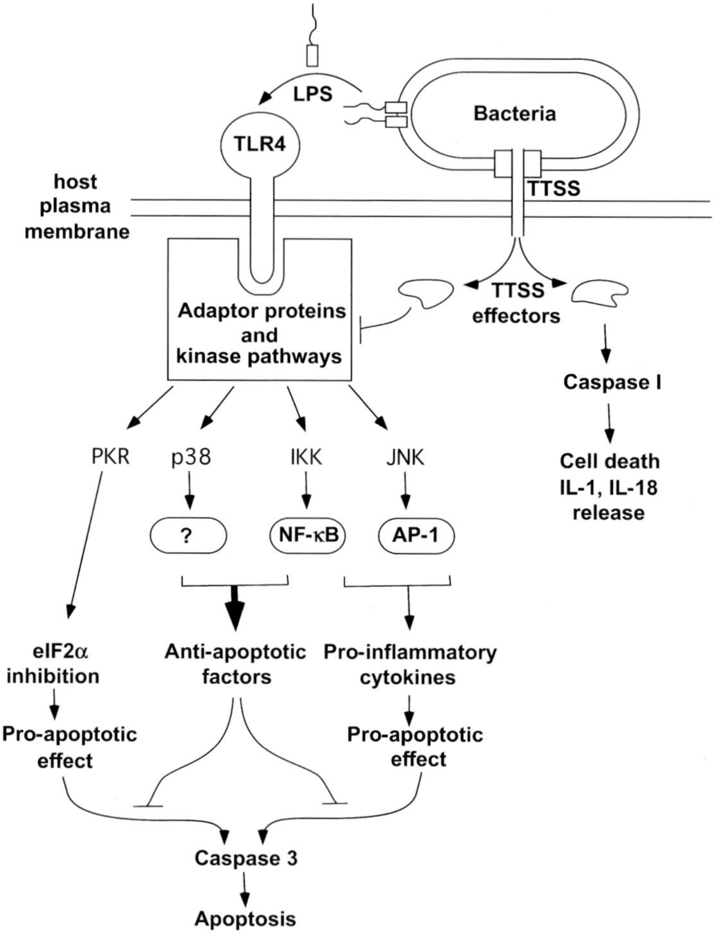

Fig. 1 Pathways leading to the induction of macrophage cell death by gram-negative enteric bacterial pathogens. *Salmonella*, *Shigella*, and *Yersinia* species activate TLR4 and also deliver specific virulence effector proteins through type III secretion systems (TTSS). Both *Salmonella* and *Shigella* produce homologous proteins (SipB and IpaB, respectively) that activate caspase-1, resulting in rapid cell death and release of the proinflammatory cytokines IL-1β and IL-18. The separate TLR4 pathway involves a complicated cascade of adapter proteins and cellular kinases. IKK and JNK activate the transcription factors NF-κB and AP-1, which are key inducers of proinflammatory cytokine genes, including TNFα. The proapoptotic effects of TNFα and other cytokines are countered by antiapoptotic factors that are also under NF-κB control. In addition, the MAPK p38 has an essential role in the induction of antiapoptotic proteins through a poorly understood transcriptional control system. PKR is another critical kinase that is activated downstream from TLR4, leading to phos-

protein synthesis. The resulting proapoptotic effect may be due in part to the decrease in synthesis of antiapoptotic proteins. PKR also appears to have a proapoptotic effect independent of eIF2α phosphorylation as well (Hsu et al. 2004). The balance of these apoptosis regulatory factors determines whether the cell will survive or enter the programmed cell death pathway, with caspase-3 as the major executioner of the characteristic apoptotic changes. Certain pathogens, such as *Salmonella* and *Yersinia*, deliver effector proteins through type III systems to alter important signal transduction pathways and affect the balance between pro- and antiapoptotic factors in the host cell. The *Salmonella* virulence proteins that promote cell death are discussed in detail below.

3
Induction of Epithelial Cell Death by *Salmonella* Infection

Intestinal epithelial cells are a primary host target during the initial phase of *Salmonella* infection (Ohl and Miller 2001). Although some *Salmonella* organisms may bypass epithelial cells and directly enter phagocytic cells, both genetic and histological evidence support the importance of *Salmonella* invasion of epithelial cells in the pathogenesis of enteritis (Santos et al. 2001c; Zhang et al. 2002). The invasion process is mediated by effector proteins secreted by the SPI-1-encoded TTSS. Several studies have detected invasion-dependent apoptosis of intestinal epithelial cells during *Salmonella* infection in tissue culture. Studies by Kim et al. (Kim et al. 1998) and Paesold et al. (Paesold et al. 2002) reported the delayed induction of apoptosis in human colonic epithelial cell lines infected with serotype *typhimurium* and an even greater effect with serotype *dublin*. Epithelial cell apoptosis was characterized by reduced mitochondrial membrane potential, cell surface exposure of phosphatidylserine, caspase-3 activation, cytokeritin cleavage, nuclear con-

phorylation and inhibition of eIF2α, a factor required for protein synthesis. The cellular effect of PKR activation is the promotion of apoptosis. Therefore, after activation of TLR4 by gram-negative bacteria, the cell is balanced between the opposing effects of pro- and antiapoptotic factors. *Salmonella* and *Yersinia* alter the balance in favor of apoptosis by the action of virulence effector proteins secreted by specific type III secretion systems: the *Salmonella* SPI-1 and SPI-2 and the *Yersinia* Yop systems. (Hsu et al. 2004) densation, and DNA fragmentation, all markers of the classic programmed cell death pathway. The induction of apoptosis was shown to require *Salmonella* invasion and intracellular bacterial protein synthesis in the first 6 h of infection. After this time, the cells were committed to the apoptotic pathway but the execution phase was delayed until 18–24 h after invasion. Apoptosis was mediated in part by epithelial production of TNFα and nitric oxide (NO) and was also shown to require the function of the bacterial SPI-2 TTSS and the *spv* virulence genes. The mechanism of apoptosis in *Salmonella*-infected epithelial cells has not been determined. However, *Salmonella* produces several effector proteins that could affect apoptotic pathways. The SPI-1 TTSS mediates the secretion of SptP, AvrA, SigD (SopB), and SspH1 (also secreted by SPI-2) (Galan 2001). SptP has tyrosine phosphatase activity and can downmodulate activation of MAP kinases, particularly Erk (Murli et al. 2001). Along with SspH1, SptP inhibits the expression of NF-κB-dependent genes and could affect the balance of pro- and antiapoptotic factors in the cell (Haraga and Miller 2003). AvrA also inhibits the activation of NF-κB. In the presence of a constitutive *phoP* mutation in serotype *typhimurium*, an *avrA* mutant was found to induce less apoptosis in epithelial cells (Collier-Hyams et al. 2002). SigD (also designated SopB) has been found to activate Akt/protein kinase B, a signal transduction kinase that can exert prosurvival effects. It is postulated that this pathway could be responsible for the delay in apoptosis seen in epithelial cells. The SpvB protein, which requires the SPI-2 TTSS function, produces actin depolymerization during *Salmonella* infection of epithelial cells and is required for apoptosis (Paesold et al. 2002). On the basis of these studies, it is likely that the induction of apoptosis in epithelial cell lines by *Salmonella* is a complex process involving several bacterial proapoptotic effector proteins as well as cellular inflammatory products such as TNF and NO. Although the function of the SPI-2 TTSS system is clearly required for apoptosis in epithelial cells, the role of SPI-1 has been difficult to assess independent of its function in cell invasion.

The importance of epithelial cell apoptosis during *Salmonella* enteritis in vivo remains uncertain. The studies on cell culture systems reviewed above must be interpreted with caution, because the regulation of apoptotic pathways in malignant cell lines is abnormal. However, histological studies in the bovine ileal loop model of *Salmonella* enteritis have detected apoptosis in the mucosa and lymphoid nodules 12 h after infection (Santos et al. 2001b). By this time, inflammation and fluid secretion are well established in this model. The time course for apoptosis in vivo agrees fairly well with the delayed apoptosis seen in the in vitro cell culture systems. Because longer times are difficult to analyze in the bovine intestinal loop system, the extent of apoptosis later in the course of enteritis remains uncertain.

4 Cell Death in Macrophages and Dendritic Cells Mediated by SPI-1 Effectors

Host cell death mediated by effectors of type III secretion systems involved in invasion functions has been reported for both *Shigella* and *Salmonella* strains (Knodler and Finlay 2001; Monack et al. 2001b; Jarvelainen et al. 2003). In both cases, the mechanisms share common features. Homologous effector proteins (IpaB for *Shigella*, SipB for *Salmonella*) activate caspase-1, resulting in the processing of precursor molecules of IL-1β and IL-18 into the mature cytokines, and also triggering cell death. Careful analysis of the mechanism of the *Salmonella* SPI-1-dependent cell death associated with caspase-1 activation has shown this process to be a variant of cell necrosis, termed "programmed necrosis" (Brennan and Cookson 2000; Monack et al. 2001b). This phenomenon has been studied primarily in host cells that contain large amounts of caspase-1, such as macrophages and dendritic cells (Knodler and Finlay 2001; Monack et al. 2001b; van der Velden et al. 2003). When *Salmonella* are grown under conditions that favor expression of SPI-1 and then used to infect macrophages, they produce an early cytotoxic effect within 1–2 h. The infected cells rapidly lose membrane integrity and release cytoplasmic enzymes such as lactic dehydrogenase. Unlike apoptosis, caspase-3 is not activated and PARP, a nuclear protein target of caspase-3, remains uncleaved (Brennan and Cookson 2000). The SPI-1-dependent cytotoxicity can be blocked by glycine, which inhibits ion fluxes through the cytoplasmic membrane in cells undergoing necrosis (Brennan and Cookson 2000). Several studies have shown that caspase-1 in the host cell and SipB secretion by the bacteria are required for this cytotoxic process, and purified SipB has been reported to interact with caspase-1 in vitro (Hersh et al. 1999; Monack et al. 2000). Caspase-2 also appears to contribute to this process (Jesenberger et al. 2000). Early SipB-dependent cell death can be blocked by caspase-1 inhibition and does not occur in macrophages from caspase-1-deficient mice. Furthermore, microinjection of SipB into cells reproduces the cytotoxicity (Hersh et al. 1999). However, the mechanism of cell death following caspase-1 activation has not been determined. SPI-1 effectors also activate the Rho family GTPases Cdc42 and Rac-1, and inhibition of Cdc42 and Rac-1 prenylation inhibits *Salmonella*-induced macrophage death (Forsberg et al. 2003).

The role of caspase-1 in *Salmonella* pathogenesis has been studied with knockout mice (Monack et al. 2000). The LD_{50} for oral infection with serovar *typhimurium* in caspase-$1^{-/-}$ mice is approximately 1,000-fold higher than in control, wild-type mice, but there is no difference in the LD_{50} when the mice are infected by the intraperitoneal route. These results indicate that caspase-1 plays a major role in promoting *Salmonella* infection and/or dissemination from the intestinal tract. Studies on the course of infection showed that the caspase-1-deficient mice had low and inconsistent levels of infection in Peyer's patches as well as systemic organs, Therefore, the resistance of the caspase-1 knockout mice to *Salmonella* appeared to be at the level of the gut-associated lymphoid tissue, the main site of intestinal infection in the murine model.

Several experimental problems have prevented a definitive determination of the role of SipB- and caspase-1-dependent cell death in *Salmonella* pathogenesis. Because SipB is both an effector protein required for cell invasion and a component of the TTSS required for translocation of other SPI-1 effectors, SipB mutants have a pleotrophic effect on virulence. It has not yet been possible to genetically separate the caspase-1 activation property of SipB from its other activities. A similar problem exists for caspase-1, because it functions to activate the proinflammatory cytokines IL-1β and IL-18 as well as trigger the cell death pathway described above. It is not known whether these are separate or interdependent functions. Therefore, the role of caspase-1 in promoting *Salmonella* infection from the intestine could be due to the release of proinflammatory cytokines and/or to the death of macrophages and dendritic cells in the Payer's patches. Further studies will be necessary to address these issues.

Delayed SipB-dependent cytotoxic effects have also been detected in caspase-1-deficient macrophages. By 4–6 h after infection of macrophages from caspase-$1^{-/-}$ mice with *Salmonella*, changes consistent with apoptosis are detected, including release of cytochrome *c* from mitochondria and activation of caspases-2, -3, -6, and -8 (Jesenberger et al. 2000). A separate group of investigators described the localization of SipB to mitochondria in caspase-1-deficient macrophages and the accumulation of structures resembling autophagic vesicles containing mitochondrial and endoplasmic reticulum membrane (Hernandez et al. 2003). This mitochondrial cytopathology was shown by an inhibitor experiment to be independent of all caspase activity. Although these studies demonstrate that SipB can produce cytotoxicity in the absence of caspase-1, the significance of these changes is uncertain. As noted above, *Salmonella* have markedly reduced virulence in caspase-1-deficient mice. In normal macrophages, the caspase-1-mediated cell death occurs much more rapidly than the delayed changes just described, making it unlikely that mitochondrial damage would be the dominant process producing cytopathology.

5 Cell Death in Macrophages Dependent on the SPI-2 TTSS

When macrophages are infected with *Salmonella* grown under conditions that repress expression of the SPI-1 TTSS, they undergo delayed cytopathic changes similar to the apoptosis described above in epithelial cells (Lindgren et al. 1996; Libby et al. 2000; van der Velden et al. 2000; Santos et al. 2001a). However, in macrophages cell entry is mediated by phagocytosis and is markedly increased by opsonization. In this process, findings characteristic of apoptosis, including nucleosome cleavage and DNA fragmentation, occur about 24 h after infection (Libby et al. 2000; Hsu et al. 2004). The use of SPI-1 mutants, including knockouts of SipB, has shown conclusively that this delayed apoptosis is induced by a mechanism that does not involve the SPI-1 TTSS or its effectors (Libby et al. 2000; van der Velden et al. 2000; Knodler and Finlay 2001; Santos et al. 2001a; Browne et al. 2002). In contrast, this late cytopathology is dependent on the SPI-2 TTSS. A requirement for the SpvB intracellular toxin has also been demonstrated in human macrophages, similar to the findings in epithelial cells (Libby et al. 2000; Browne et al. 2002). The mechanism of this SPI-2-dependent apoptosis has not been identified, but a partial requirement for caspase-1 has been reported with macrophages from knockout mice (Monack et al. 2001a). However, it is not known whether the effect of caspase-1 is due to its ability to activate proinflammatory cytokines or to other intracellular activities.

SPI-2-dependent apoptosis is markedly reduced in macrophages deficient in TLR4-mediated signal transduction, but the early SPI-1 mediated cell death is preserved in these macrophages (Hsu et al. 2004). This result supports the model illustrated in Fig. 1. *Salmonella* LPS stimulates TLR4, leading to activation of signal transduction pathways mediated by adapter proteins and cellular kinases. Inhibition of NF-κB or p38 MAP kinase results in dramatic increases in apoptosis, indicating that these pathways are required for the production of antiapoptotic factors in macrophages. The PKR kinase pathway is also required for delayed *Salmonella*-mediated apoptosis, and macrophages from mice deficient in PKR are remarkably resistant to SPI-2-dependent programmed cell death, although they remain sensitive to early, SipB-mediated cytotoxicity. PKR promotes *Salmonella*-induced apoptosis partly through inhibition of protein synthesis mediated by phosphorylation of eIF2α. In addition, PKR and type I interferons activate interferon response factor 3 (IRF3), which also induces proapoptotic factors in macrophages. As predicted, apoptosis is reduced in *Salmonella*-infected macrophages from mice deficient in IRF3 (Hsu et al. 2004). Phagocytosis of *Salmonella* also activates tyrosine kinases and phosphatidylinositol-3 kinase (PI-3K), which lead to activation of Akt, an antiapoptotic factor also induced by invasion of epithelial cells as described in the section above. Inhibition of Akt expression in a human macrophage-like cell line increases apoptosis after phagocytosis of *Salmonella* (Forsberg et al. 2003). Therefore, Akt may be active in delaying SPI-2-dependent apoptosis in both epithelial cells and macrophages.

The signal transduction cascade originating from TLR4 (Fig. 1) leads to the production of a balance of pro- and antiapoptotic factors in macrophages after infection with *Salmonella*. The use of mutants has clearly demonstrated that the function of the *Salmonella* SPI-2 TTSS is required to tip this balance in favor of apoptosis. The SPI-2 system is essential for *Salmonella* proliferation in macrophages, yet the absence of apoptosis seen with SPI-2 mutants is not simply due to decreased survival or lack of growth (van der Velden et al. 2000; Browne et al. 2002). Instead, it is likely that the delayed apoptosis is mediated by the transfer of *Salmonella* effector proteins that alter cellular processes responsible for inducing programmed cell death. The *Salmonella* intracellular toxin SpvB has been linked to apoptosis in human monocyte-derived macrophages (Libby et al. 2000; Browne et al. 2002). The SpvB protein is an ADP-ribosylating toxin that modifies actin and leads to the depolymerization of the actin cytoskeleton in infected cells (Lesnick et al. 2001; Tezcan-Merdol et al. 2001). SpvB is inactive from the extracellular space and requires entry of *Salmonella* into the SCV together with the function of the SPI-2 TTSS to produce actin depolymerization in host cells (Lesnick et al. 2001; Browne et al. 2002). These results suggest that SpvB is translocated by the SPI-2 TTSS from the SCV into the host cell. Cytopathology due to SpvB activity is detectable in some cells by 12 h after *Salmonella* entry into human macrophages and continues to increase until a majority of infected cells are involved. Free nucleosomes and DNA fragmentation in infected cells are found after 24 h. These findings indicate that the SpvB protein is a *Salmonella* virulence factor active in SPI-2-dependent apoptosis. It is possible that other SPI-2-dependent effectors are also involved in this process.

A significant difference between *Salmonella* serovar *typhi* and the nontyphoid serovars associated with systemic disease in people (such as *typhimurium*, *dublin*, and *choleraesuis*) is that *typhi* lacks the *spv* gene locus, including *spvB* (Roudier et al. 1990). On this basis, *typhi* should be less cytotoxic to macrophages during prolonged infection. Experimental evidence comparing *typhimurium* and *typhi* suggests that *typhi* may cause less cytopathology in human macrophages (Schwan et al. 2000). *Typhi* is only found in humans and produces a prolonged disease which may relapse, suggesting that long-term infection of macrophages without the induction of cell death could be an important feature of the pathogenesis of typhoid fever.

Apoptosis has been detected during systemic infection with nontyphoid *Salmonella* serovars in mice (Richter-Dahlfors et al. 1997). Multiple lines of evidence indicate that macrophages are a principal target cell for *Salmonella* in extraintestinal tissue (Ohl and Miller 2001). Confocal microscopy has been used to visualize *Salmonella* in the livers of infected mice (Richter-Dahlfors et al. 1997). Bacteria were found within tissue macrophages, and cell death was demonstrated by TUNEL staining for DNA fragmentation. However, the bacterial genetic requirements for this cytopathology were not determined.

6
Model for the Roles of Host Cell Death in *Salmonella* Pathogenesis

The evidence presented in this review strongly suggests that specific *Salmonella* virulence mechanisms promote host cell death at several stages of the pathogenic process. Figures 2 and 3 present a model for the roles of host cell death during both the intestinal and systemic phases of *Salmonella* infection. The initial interaction between *Salmonella* and intestinal epithelial cells (Fig. 2) results in secretion of SPI-1 effector proteins, leading to cell invasion and production of chemotactic factors on both the luminal (hepoxilin) and basolateral (IL-8, other chemokines, and proinflammatory cytokines) sides. Evidence in the mouse model suggests that $CD18^+$ cells can access the intestinal lumen and engulf *Salmonella*, leading to early dissemination. If the bacteria are expressing SPI-1 genes, these $CD18^+$ cells could be subjected to SipB-mediated cell death as the bacteria are carried to systemic organs. Chemotactic factors

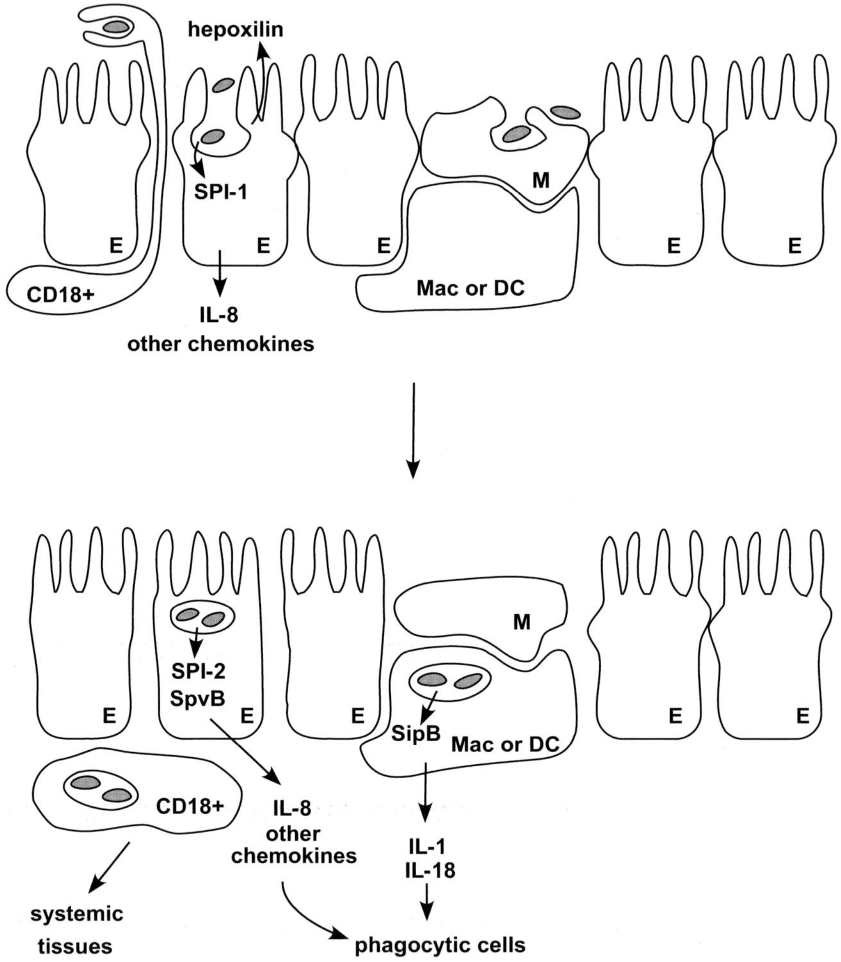

Fig. 2 Early events leading to death of intestinal cells during *Salmonella* infection. The *top panel* shows a model for the initial events in the pathogenesis of *Salmonella* enteritis. *E*, epithelial cells with microvilli; *M*, specialized epithelial cells lacking microvilli that cover Payer's patch tissue; *Mac*, macrophages; *DC*, dendritic cells. Hepoxilin is a recently identified chemotactic factor that is released on the luminal side by epithelial cells infected with *Salmonella* (Mrsny et al. 2004). IL-8 and other chemokines are released on the basolateral side (Eckmann et al. 1993). $CD18^+$ cells have been shown to mediate the rapid uptake of *Salmonella* from the intestine by a poorly characterized process that may involve partial or complete migration between epithelial cells (Vazquez-Torres et al. 1999). This early $CD18^+$ cell-mediated uptake is not dependent on active invasion mediated by *Salmonella*, because it occurs with SPI-1 mutants. However, in the normal situation, *Salmonella* at the epithelial surface

released by the intestinal epithelium may enhance this process as the infection develops. In the calf model of enteritis, most of the local intestinal inflammatory reaction depends on SPI-1 effectors secreted by the bacteria. Intestinal macrophages underlying M cells in Peyer's patches probably take up *Salmonella* that have transcytosed through the overlying M cells. These bacteria may still express the SPI-1 TTSS and secrete SipB, producing caspase-1-dependent macrophage death and release of IL-1β and IL-18. The combination of chemokines and these proinflammatory cytokines stimulates an intense inflammatory infiltrate that spreads through large areas of the epithelium in hosts that develop clinical enteritis. Secretion of SPI-2 effectors and SpvB could lead to epithelial cell death by apoptosis and contribute to the widespread epithelial erosion seen in severe enteritis. The *spv* genes appear to contribute to the severity of enteritis in some strains of serovar *dublin* (Libby et al. 1997). These later stages of the intestinal phase of the disease are depicted in Fig. 3. Phagocytic cells move into the mucosa to engulf apoptotic epithelial cells and the debris of macrophages dying of caspase-1-mediated "programmed necrosis." Both processes of cell death lead to the display of phosphatidylserine on the dying cell surface, serving as a target for a macrophage receptor that stimulates phagocytosis (Fadok et al. 2000). In the process of cellular phagocytosis, *Salmonella* are also taken up by the incoming macrophages. These bacteria will now be expressing SPI-2 and *spv* genes, rather than the SPI-1 locus used for epithelial invasion, and the bacteria will be primed for intracellular pathogenesis. Some of these macrophages probably serve as sites for *Salmonella* proliferation in the mucosa and submucosa, whereas others carry *Salmonella* to systemic tissues. This model accounts for the caspase-1 role in *Salmonella* pathogenesis, because the death of intestinal macrophages with release of IL-1β would strongly promote the influx of circulating inflam-

are likely to express SPI-1 and could produce cytopathology in the CD18$^+$ cells. *Salmonella* invasion through epithelial cells requires the function of the SPI-1 TTSS and secretion of effector proteins. The *lower panel* depicts subsequent events leading to the inflammatory response and early dissemination by CD18$^+$ cells that have ingested bacteria. In epithelial cells, induction of the *Salmonella* SPI-2 TTSS and *spvB* facilitates the intracellular proliferation of bacteria (Paesold et al. 2002). Rapid transcytosis of bacteria through the M cells may result in continued secretion of SipB into underlying macrophages and dendritic cells, resulting in release of IL-1 and IL-18 as proinflammatory mediators

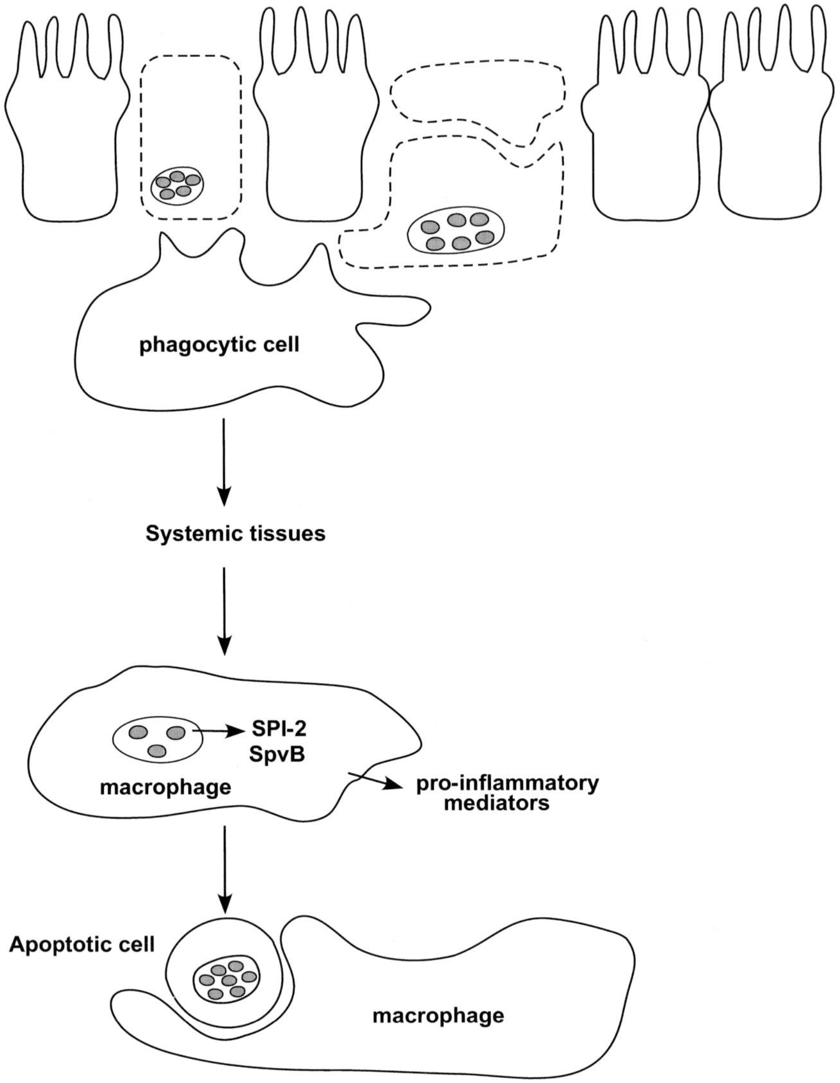

Fig. 3 Consequences of host cell death induced by *Salmonella* virulence factors. Dying cells are depicted by the *dashed lines*. Infected intestinal macrophages and dendritic cells may be the first to die by the SipB- and caspase-1-dependent mechanism. Epithelial cells likely undergo delayed apoptosis mediated by SPI-2 TTSS effectors and SpvB. Dying cells are recognized by receptors on macrophages, which engulf the cell debris containing viable *Salmonella*. The bacteria are then carried by these infected macrophages to systemic tissues. Secretion of the SPI-2 TTSS effectors and the SpvB toxin induces delayed apoptosis of these macrophages, with the formation

matory cells and increase the uptake of *Salmonella* by macrophages in the subepithelial compartment, promoting dissemination. The SipB- and caspase-1-dependent mechanism of macrophage death is also favored at this stage because intestinal macrophages do not express CD14 and are likely to be deficient for TLR4-mediated signal transduction and activation (Smith et al. 2001), a requirement for SPI-2- but not SPI-1-mediated cell death (Fig. 1) (Hsu et al. 2004).

The infected macrophages transport the replicating intracellular *Salmonella* to the mesenteric lymph nodes and then into the circulation (Fig. 3). The action of SPI-2 effectors and SpvB promotes *Salmonella* growth and subsequent apoptosis of these host macrophages, stimulating another round of phagocytosis of infected apoptotic bodies by tissue-based macrophages in systemic organs, particularly the liver and spleen. From these sites, successive rounds of intracellular growth, apoptosis, and subsequent phagocytosis of infected apoptotic bodies perpetuate the infection and ensure cell-to-cell spread. The ineffectiveness of extracellular antibiotics, such as gentamicin, in the treatment of salmonellosis, both in experimental animals and in human disease (Fierer et al. 1990), provides strong evidence in favor of this model of cell-to-cell spread.

References

Assuncao Guimaraes C, Linden R (2004) Programmed cell deaths. Apoptosis and alternative deathstyles. Eur J Biochem 271:1638–1650

Brennan MA, Cookson BT (2000) Salmonella induces macrophage death by caspase-1-dependent necrosis. Mol Microbiol 38:31–40

Browne SH, Lesnick ML, Guiney DG (2002) Genetic requirements for salmonella-induced cytopathology in human monocyte-derived macrophages. Infect Immun 70:7126–7135

Collier-Hyams LS et al. (2002) Cutting edge: Salmonella AvrA effector inhibits the key proinflammatory, anti-apoptotic NF-kappa B pathway. J Immunol 169:2846–2850

of apoptotic cells containing large numbers of bacteria that are phagocytized by additional macrophages responding to proinflammatory factors. Continued cycles of phagocytosis, bacterial growth, and apoptosis mediate cell-to-cell spread of *Salmonella* in systemic tissues

Eckmann L, Kagnoff MF, Fierer J (1993) Epithelial cells secrete the chemokine interleukin-8 in response to bacterial entry. Infect Immun 61:4569–4574

Fadok VA, Bratton DL, Rose DM, Pearson A, Ezekewitz RA, Henson PM (2000) A receptor for phosphatidylserine-specific clearance of apoptotic cells. Nature 405:85–90

Fields PI, Swanson RV, Haidaris CG, Heffron F (1986) Mutants of *Salmonella typhimurium* that cannot survive within the macrophage are avirulent. Proc Natl Acad Sci USA 83:5189–5193

Fierer J, Hatlen L, Lin JP, Estrella D, Mihalko P, Yau-Young A (1990) Successful treatment using gentamicin liposomes of *Salmonella dublin* infections in mice. Antimicrob Agents Chemother 34:343–348

Finlay BB, Gumbiner B, Falkow S (1988) Penetration of *Salmonella* through a polarized Madin-Darby canine kidney epithelial cell monolayer. J Cell Biol 107:221–230

Forsberg M, Druid P, Zheng L, Stendahl O, Sarndahl E (2003) Activation of Rac2 and Cdc42 on Fc and complement receptor ligation in human neutrophils. J Leukoc Biol 74:611–619

Galan JE (2001) Salmonella interactions with host cells: type III secretion at work. Annu Rev Cell Dev Biol 17:53–86

Haraga A, Miller SI (2003) A *Salmonella enterica* serovar typhimurium translocated leucine-rich repeat effector protein inhibits NF-kappa B-dependent gene expression. Infect Immun 71:4052–4058

Hernandez LD, Pypaert M, Flavell RA, Galan JE (2003) A *Salmonella* protein causes macrophage cell death by inducing autophagy. J Cell Biol 163:1123–1131

Hersh D, Monack DM, Smith MR, Ghori N, Falkow S, Zychlinsky A (1999) The *Salmonella* invasin SipB induces macrophage apoptosis by binding to caspase-1. Proc Natl Acad Sci USA 96:2396–2401

Holden DW (2002) Trafficking of the *Salmonella* vacuole in macrophages. Traffic 3:161–169

Hsu LC et al. (2004) The protein kinase PKR is required for macrophage apoptosis after activation of Toll-like receptor 4. Nature 428:341–345

Jarvelainen HA, Galmiche A, Zychlinsky A (2003) Caspase-1 activation by *Salmonella*. Trends Cell Biol 13:204–209

Jesenberger V, Procyk KJ, Yuan J, Reipert S, Baccarini M (2000) Salmonella-induced caspase-2 activation in macrophages: a novel mechanism in pathogen-mediated apoptosis. J Exp Med 192:1035–1046

Jones BD, Falkow S (1996) Salmonellosis: host immune responses and bacterial virulence determinants. Annu Rev Immunol 14:533–561

Jones BD, Ghori N, Falkow S (1994) *Salmonella typhimurium* initiates murine infection by penetrating and destroying the specialized epithelial M cells of the Peyer's patches. J Exp Med 180:15–23

Kim JM, Eckmann L, Savidge TC, Lowe DC, Witthoft T, Kagnoff MF (1998) Apoptosis of human intestinal epithelial cells after bacterial invasion. J Clin Invest 102:1815–1823

Knodler LA, Finlay BB (2001) Salmonella and apoptosis: to live or let die? Microbes Infect 3:1321–1326

Knodler LA, Steele-Mortimer O (2003) Taking possession: biogenesis of the *Salmonella*-containing vacuole. Traffic 4:587–599

Lesnick ML, Reiner NE, Fierer J, Guiney DG (2001) The *Salmonella* spvB virulence gene encodes an enzyme that ADP-ribosylates actin and destabilizes the cytoskeleton of eukaryotic cells. Mol Microbiol 39:1464–1470

Libby SJ et al. (1997) The spv genes on the *Salmonella dublin* virulence plasmid are required for severe enteritis and systemic infection in the natural host. Infect Immun 65:1786–1792

Libby SJ et al. (2002) Characterization of the spv locus in *Salmonella enterica* serovar Arizona. Infect Immun 70:3290–3294

Libby SJ, Lesnick M, Hasegawa P, Weidenhammer E, Guiney DG (2000) The *Salmonella* virulence plasmid spv genes are required for cytopathology in human monocyte-derived macrophages. Cell Microbiol 2:49–58

Lindgren SW, Stojiljkovic I, Heffron F (1996) Macrophage killing is an essential virulence mechanism of *Salmonella typhimurium*. Proc Natl Acad Sci USA 93:4197–4201

Majno G, Joris I (1995) Apoptosis, oncosis, and necrosis. An overview of cell death. Am J Pathol 146:3–15

Marsden VS et al. (2002) Apoptosis initiated by Bcl-2-regulated caspase activation independently of the cytochrome c/Apaf-1/caspase-9 apoptosome. Nature 419:634–637

Monack DM, Detweiler CS, Falkow S (2001a) Salmonella pathogenicity island 2-dependent macrophage death is mediated in part by the host cysteine protease caspase-1. Cell Microbiol 3:825–837

Monack DM, Hersh D, Ghori N, Bouley D, Zychlinsky A, Falkow S (2000) Salmonella exploits caspase-1 to colonize Peyer's patches in a murine typhoid model. J Exp Med 192:249–258

Monack DM, Navarre WW, Falkow S (2001b) Salmonella-induced macrophage death: the role of caspase-1 in death and inflammation. Microbes Infect 3:1201–1212

Mrsny RJ et al. (2004) Identification of hepoxilin A3 in inflammatory events: a required role in neutrophil migration across intestinal epithelia. Proc Natl Acad Sci USA 101:7421–7426

Murli S, Watson RO, Galan JE (2001) Role of tyrosine kinases and the tyrosine phosphatase SptP in the interaction of *Salmonella* with host cells. Cell Microbiol 3:795–810

Ohl ME, Miller SI (2001) Salmonella: a model for bacterial pathogenesis. Annu Rev Med 52:259–274

Paesold G, Guiney DG, Eckmann L, Kagnoff MF (2002) Genes in the *Salmonella* pathogenicity island 2 and the *Salmonella* virulence plasmid are essential for *Salmonella*-induced apoptosis in intestinal epithelial cells. Cell Microbiol 4:771–781

Pang T, Bhutta ZA, Finlay BB, Altwegg M (1995) Typhoid fever and other salmonellosis: a continuing challenge. Trends Microbiol 3:253–255

Park JM, Greten FR, Li ZW, Karin M (2002) Macrophage apoptosis by anthrax lethal factor through p38 MAP kinase inhibition. Science 297:2048–2051

Resta-Lenert S, Barrett KE (2002) Enteroinvasive bacteria alter barrier and transport properties of human intestinal epithelium: role of iNOS and COX-2. Gastroenterology 122:1070–1087

Richter-Dahlfors A, Buchan AM, Finlay BB (1997) Murine salmonellosis studied by confocal microscopy: *Salmonella typhimurium* resides intracellularly inside macrophages and exerts a cytotoxic effect on phagocytes in vivo. J Exp Med 186:569–580

Roudier C, Krause M, Fierer J, Guiney DG (1990) Correlation between the presence of sequences homologous to the vir region of *Salmonella dublin* plasmid pSDL2 and the virulence of twenty-two *Salmonella* serotypes in mice. Infect Immun 58:1180–1185

Santos RL, Tsolis RM, Baumler AJ, Smith R, 3rd, Adams LG (2001a) *Salmonella enterica* serovar typhimurium induces cell death in bovine monocyte-derived macrophages by early sipB-dependent and delayed sipB-independent mechanisms. Infect Immun 69:2293–2301

Santos RL, Tsolis RM, Zhang S, Ficht TA, Baumler AJ, Adams LG (2001b) Salmonella-induced cell death is not required for enteritis in calves. Infect Immun 69:4610–4617

Santos RL, Zhang S, Tsolis RM, Kingsley RA, Adams LG, Baumler AJ (2001c) Animal models of *Salmonella* infections: enteritis versus typhoid fever. Microbes Infect 3:1335–1344

Schwan WR, Huang XZ, Hu L, Kopecko DJ (2000) Differential bacterial survival, replication, and apoptosis-inducing ability of *Salmonella* serovars within human and murine macrophages. Infect Immun 68:1005–1013

Smith PD et al. (2001) Intestinal macrophages lack CD14 and CD89 and consequently are down-regulated for LPS- and IgA-mediated activities. J Immunol 167:2651–2656

Tezcan-Merdol D, Nyman T, Lindberg U, Haag F, Koch-Nolte F, Rhen M (2001) Actin is ADP-ribosylated by the Salmonella enterica virulence-associated protein SpvB. Mol Microbiol 39:606–619

van der Velden AW, Lindgren SW, Worley MJ, Heffron F (2000) *Salmonella* pathogenicity island 1-independent induction of apoptosis in infected macrophages by *Salmonella enterica* serotype typhimurium. Infect Immun 68:5702–5709

van der Velden AW, Velasquez M, Starnbach MN (2003) *Salmonella* rapidly kill dendritic cells via a caspase-1-dependent mechanism. J Immunol 171:6742–6749

Vazquez-Torres A et al. (1999) Extraintestinal dissemination of *Salmonella* by CD18-expressing phagocytes. Nature 401:804–808

Waterman SR, Holden DW (2003) Functions and effectors of the *Salmonella* pathogenicity island 2 type III secretion system. Cell Microbiol 5:501–511

Weinrauch Y, Zychlinsky A (1999) The induction of apoptosis by bacterial pathogens. Annu Rev Microbiol 53:155–187

Zhang S et al. (2002) The *Salmonella enterica* serotype typhimurium effector proteins SipA, SopA, SopB, SopD, and SopE2 act in concert to induce diarrhea in calves. Infect Immun 70:3843–3855

CTMI (2005) 289:151–174

Role of Macrophage Apoptosis in the Pathogenesis of *Yersinia*

Y. Zhang · J. B. Bliska (✉)

Department of Molecular Genetics and Microbiology, Center for Infectious Diseases, SUNY Stony Brook, Stony Brook, NY 11794-5222, USA
jbliska@ms.cc.sunysb.edu

Abstract *Yersinia* species that are pathogenic for humans (*Yersinia pestis*, *Yersinia pseudotuberculosis*, and *Yersinia enterocolitica*) induce apoptosis in macrophages. *Yersinia*-induced apoptosis utilizes the mitochondrial pathway and is executed by activation of caspase cascades. The mechanism of *Yersinia*-induced apoptosis in macrophages has two essential components. One component is the innate immune response of macrophages to the pathogen, which leads to the activation of a survival response and a death response. Recognition of the bacterial cell envelope component lipopolysaccharide by Toll-like receptor 4 (TLR4) constitutes an important part of the innate immune response to the pathogen. The second essential component is YopJ, a protein secreted into *Yersinia*-infected macrophages via a bacterial type III secretion system, which selectively shuts down the survival pathway. In the absence of the survival pathway, the death pathway is executed, and *Yersinia*-infected macrophages undergo apoptosis. In this review, we introduce the basic features of *Yersinia* pathogenesis, summarize our current understanding of *Yersinia*-induced apoptosis, and discuss the role of apoptosis during *Yersinia* infection.

Abbreviations

TLR	Toll-like receptor
LPS	Lipopolysaccharide
M cell	Microfold cell
TTSS	Type III secretion system
MAP	Mitogen-activated protein
MKK	MAP kinase kinase
IκB	Inhibitor κB
NF-κB	Nuclear factor κB
IKK	IκB kinase
TIR	Toll/interleukin-1 receptor
MyD88	Myeloid differentiation factor 88
TIRAP	TIR domain-containing adaptor protein
DD	Death domain
DED	Death effector domain
IRAK	IL-1R-associated kinase
TRAF6	Tumor necrosis factor receptor-associated factor 6
TRIF	TIR domain-containing adaptor inducing IFNβ

1 Overview of *Yersinia* Pathogenesis

1.1 Pathogenic *Yersinia* Species

Three species of bacteria in the genus *Yersinia* are considered important pathogens of humans (Brubaker 1991). *Yersinia enterocolitica* and *Yersinia pseudotuberculosis* cause intestinal infections that are typically self-limiting. *Y. enterocolitica* and *Y. pseudotuberculosis* replicate in Peyer's patches and mesenteric lymph nodes and only rarely cause systemic infections. *Yersinia pestis* is the agent of bubonic, septicemic, and pneumonic plague. *Y. pestis* infections are transmitted by fleas (bubonic, septicemic) or by aerosols (pneumonic) and typically result in acute, fatal disease.

Y. enterocolitica is classified into 5 biogroups and ~60 serotypes (Bottone 1997). *Y. pseudotuberculosis* is divided into 21 serologic groups based on differences in the O-antigen of lipopolysaccharide (LPS) (Skurnik et al. 2000). *Y. pestis,* which lacks O-antigen, is now considered to be a recently evolved subspecies of a *Y. pseudotuberculosis* serogroup O1b strain (Achtman et al. 1999; Skurnik et al. 2000). *Y. pestis* strains have been classified into three biovars (Antiqua, Mediaevalis, and Orientalis) based on minor phenotypic differences (Perry and Fetherston 1997). In this chapter we will use the genus name (*Yersinia*) to refer to the three pathogenic species in a generic sense.

A number of virulence factors have been identified that are common to two or more of the *Yersinia* species (Brubaker 1991; Carniel 1999, 2002; Perry and Fetherston 1997; Revell and Miller 2001). These include virulence factors that promote adherence, invasion, coordinated gene expression, serum resistance, and acquisition of iron (Carniel 1999, 2002; Perry and Fetherston 1997). Another essential virulence factor that is common to the *Yersinia* species is a ~70-kb plasmid. This plasmid, known as pYV in *Y. enterocolitica* and *Y. pseudotuberculosis* and pCD1 in *Y. pestis,* encodes a type III secretion system (TTSS) and a set of secreted proteins known as Yops and LcrV (Plano et al. 2001; Ramamurthi and Schneewind 2002). The Yops and LcrV act to counteract phagocytic mechanisms and cytokine production in macrophages. Two additional plasmids, pMT1 (~101 kb) and pPCP1 (9.6 kb) are uniquely found in *Y. pestis* (Perry and Fetherston 1997). pMT1 and pPCP1 are associated with increased virulence (pMT1 and pPCP1) and vector-borne transmissibility (pMT1) in *Y. pestis* (Perry and Fetherston 1997).

1.2
Pathogenesis of *Y. pestis* Infections

Y. pestis infections that result in bubonic or septicemic plague are initiated when an infected flea takes a blood meal on a human host. During the blood meal the bacteria are introduced directly into the dermis or into the capillary system (Perry and Fetherston 1997). Pneumonic plague infections are initiated when aerosolized bacteria are inhaled into the lungs (Perry and Fetherston 1997). Very little is known about the early events of pathogenesis in plague infections. In the case of infections initiated during a flea blood meal, it is assumed that the bacteria are internalized by blood-derived monocytes, in which the bacteria can survive and replicate (Cavanaugh and Randall 1959). Subsequently, the bacteria are transported or spread from the site of the fleabite to regional lymph nodes (bubonic) or to the bloodstream (primary septicemia). Massive bacterial multiplication in lymph nodes results in the formation of a bubo or swollen node. The bacteria then spread via the bloodstream (secondary septicemia) to major organs such as lungs, spleen, and liver. After extensive bacterial replication in these organs, a tertiary septicemia develops, and the host dies of septic shock and or multiple organ failure (Perry and Fetherston 1997). Bacterial spread to spleen and liver, and replication in these tissues, has been studied in mice challenged intravenously, which mimics the secondary septicemic phase of infection (for example, see Straley and Cibull 1989). Bacterial growth at early stages is associated with the formation of microabscesses that typically contain neutrophils within the central region and monocytic cells at the border. At later stages of infection the microabscesses enlarge into necrotic lesions that are poorly populated with neutrophils or monocytic cells. It is thought that *Y. pestis* replicates as an extracellular pathogen within microabscesses and necrotic lesions. On introduction of *Y. pestis* cells lacking the virulence plasmid (and thus lacking the TTSS) into the bloodstream of mice, the bacteria are taken up into the spleen and liver, but do not replicate, and are contained within granulomas and eliminated (Brubaker 2003).

1.3 Pathogenesis of Intestinal *Yersinia* Infections

Y. enterocolitica and *Y. pseudotuberculosis* are responsible for a broad range of gastrointestinal diseases, from enteritis to mesenteric lymphadenitis (Bottone 1997; El-Maraghi and Mair 1979). Infection is typically initiated after ingestion of contaminated food. The bacteria appear to preferentially invade through the mucosa of the small intestine by passing through microfold (M) cells. M cells are a component of the follicle-associated epithelia that function in antigen uptake (Clark et al. 1998; Marra and Isberg 1997). The invasin protein encoded by *Y. enterocolitica* and *Y. pseudotuberculosis* mediates efficient bacterial entry into M cells (Marra and Isberg 1997; Pepe and Miller 1993). After passing through M cells, the bacteria are transported to the gut associated lymphoid follicles known as Peyer's patches, where they replicate. The bacteria also spread to and replicate within mesenteric lymph nodes, resulting in a characteristic mesenteric lymphadenitis. Only in rare cases does *Y. pseudotuberculosis* or *Y. enterocolitica* cause septicemia in humans. However, *Y. enterocolitica* and *Y. pseudotuberculosis* disseminate to spleen, liver, and lungs from the intestinal tract very efficiently and cause lethal disease in laboratory rodents. In experimental infections of mice, *Y. pseudotuberculosis* and *Y. enterocolitica* replicate as aggregates of extracellular bacteria within microabscesses that are similar in appearance to those formed in response to *Y. pestis* infection (for example, see Hanski et al. 1989; Lian et al. 1987a,b; Simonet et al. 1990). The bacteria appear to effectively resist phagocytosis by neutrophils in abscesses (Hanski et al. 1989; Lian et al. 1987a,b; Simonet et al. 1990). In addition, evidence has been obtained that *Y. pseudotuberculosis* induces apoptosis in macrophages in abscesses (see below) (Monack et al. 1997). Strains of *Y. enterocolitica* or *Y. pseudotuberculosis* that lack the virulence plasmid replicate transiently in murine tissues, but ultimately the plasmid-cured bacteria are contained within granulomas and eliminated from the host.

1.4 Role of the TTSS in Counteracting Protective Immune Responses

The results of studies carried out with mice suggest that activated macrophages are a crucial component of the protective immune response to *Yersinia* (Autenrieth and Heesemann 1993; Brubaker 2003). In the first several days of infection, macrophages activated by natural killer cells seem to play a key role in controlling growth of the pathogen in tissues. Thereafter, if the host survives, a specific T cell response, mediated primarily by $CD4^+$ Th1 cells, allows for granuloma formation and for resolution of the infection. Administration of both IFNγ and TNFα into mice can afford complete protection against lethal doses of *Y. pestis* (Nakajima and Brubaker 1993). Evidence supporting a critical role for TNFα, IFNγ, and a third cytokine, IL-12, in host resistance to *Yersinia* infection has been obtained with neutralizing antibodies (Autenrieth and Heesemann 1993; Autenrieth et al. 1996; Bohn and Autenrieth 1996; Nakajima and Brubaker 1993). TNFα and IL-12 are produced by macrophages as well as by other cells such as dendritic cells. TNFα is a proinflammatory cytokine that has many effects, including the ability to activate macrophages to kill intracellular bacteria (Vazquez-Torres et al. 2001 and references therein). IL-12 plays a key role in activation of NK cells and Th1 cells. IFNγ, produced by activated NK or Th1 cells, is a potent activator of macrophages.

Nakajima and Brubaker have shown that *Y. pestis* can counteract production of IFNα and TNFα in mice during infection, and that the virulence plasmid is required for this activity (Nakajima and Brubaker 1993). It has also been shown that *Y. enterocolitica* counteracts production of TNFα message and protein in Peyer's patches of mice (Beuscher et al. 1995; Burdack et al. 1997). It remains controversial which proteins secreted by the TTSS are directly responsible for counteracting cytokine production. However, it is clear that one function of the TTSS is to counteract production of cytokines to avoid activation of macrophages. A second function of the TTSS is to prevent killing of internalized *Yersinia* by macrophages, either through inhibition of phagocytosis or by induction of apoptosis.

1.5 Molecular Function of the Plasmid-Encoded TTSS

The TTSS encoded on the *Yersinia* virulence plasmid has been extensively studied (Cornelis 2002a; Plano et al. 2001; Ramamurthi and Schneewind 2002). The *Yersinia* TTSS forms a syringelike organelle on the bacterial surface (Cornelis 2002b). Several classes of proteins that are secreted by this system have been identified, including (a) those that are important for regulating the secretion process (e.g., YopN), (b) those that mediate translocation of the effector proteins into the cytosol of the host cell (e.g., YopD and YopB), and (c) the effector proteins themselves (e.g., Yops H, E, and J). LcrV (V antigen) is a unique type of TTSS substrate. LcrV is important for control of secretion, for translocation of effectors, and for counteracting cytokine production (Brubaker 2003). Together with the Ysc secretion machinery, the secreted substrates form an integrated system that allows *Yersinia* to antagonize phagocytosis, prevent cytokine production, and induce apoptosis in macrophages (Cornelis 2002a; Ramamurthi and Schneewind 2002).

The following scenario is consistent with our current understanding of the functioning of the TTSS during *Yersinia*-host cell interactions. Before contact with a host cell the growth of the bacteria at 37°C results in TTSS production and assembly, and a pool of substrates are synthesized and stored within the bacterium. On contact with a target cell such as a macrophage, a number of responses within the bacterium and the host cell are initiated. On the part of the host cell, proteins on the bacterial surface (e.g., invasin) can be recognized by cell receptors (e.g., integrins), and this type of interaction can stimulate bacterial phagocytosis (Isberg and Van Nhieu 1995) or cytokine production (Grassl et al. 2003). Components of the bacterial envelope (e.g., LPS) can be recognized by TLRs to stimulate cytokine production or apoptosis (Haase et al. 2003; Zhang and Bliska 2003). On the part of the bacterium, the TTSS is activated on attachment to the host cell surface. It is thought that YopB, YopD, and LcrV are secreted first because they are required for translocation of the effector Yops across the eukaryotic cell membrane (Cornelis 2002a; Ramamurthi and Schneewind 2002). Interaction of the translocation machinery (e.g., YopB) with the host cell membrane also appears to stimulate a response in the host cell, leading to cytokine production (Viboud et al. 2003) or pore formation (Viboud and Bliska

2001). The pathogen delivers the six effector Yops shortly after contact is made with the host cell in order to counteract host responses. This review will focus on YopJ (also known as YopP in *Y. enterocolitica*) and its role in induction of apoptosis. Information on the other effectors can be found in several recent reviews (Aepfelbacher et al. 1999; Bliska 2000; Cornelis 2002a; Juris et al. 2002).

2
Yersinia-Induced Apoptosis in Macrophages

2.1
Discovery and Characterization of *Yersinia*-Induced Apoptosis in Macrophages

In 1986 Goguen et al. (Goguen et al. 1986) reported that murine macrophages die and release their cytosolic contents between 4 and 6 h after infection with *Y. pestis* or *Y. pseudotuberculosis* (Fig. 1). A *Y. pestis* strain lacking pCD1 did not kill macrophages (Fig. 1), indicating the virulence plasmid, and hence type III secretion, was required for cytotoxicity. In 1997 several groups reported that macrophages infected with *Y. pseudotuberculosis* or *Y. enterocolitica* die of apoptosis (Mills et al. 1997; Monack et al. 1997; Ruckdeschel et al. 1997b). As expected from the earlier work (Goguen et al. 1986) the progress of apoptosis in *Yersinia*-infected macrophages is relatively slow, as compared to the apoptotic events ob-

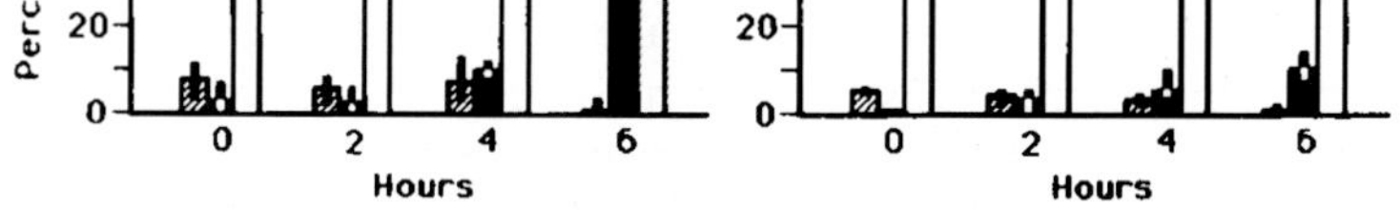

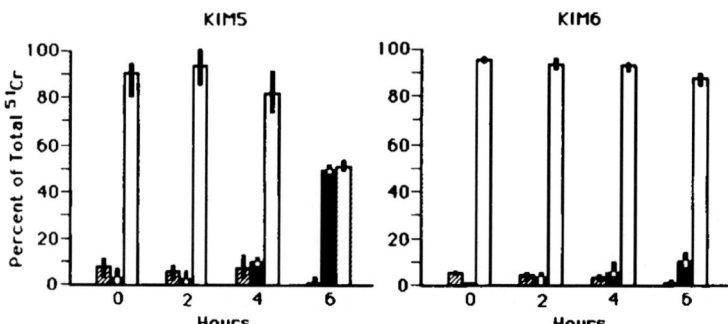

Fig. 1 Release of ^{51}Cr from the P388D1 macrophage cell line infected for the indicated period of time with *Y. pestis* KIM5 (pCD1^{+}) or KIM6 (pCD1^{-}). Percentage of total ^{51}Cr present in detached macrophages (*hatched bars*), free in medium (*black bars*), or in adherent macrophages (*white bars*) is shown. (Adapted from Goguen et al. 1986 with the permission of the publisher)

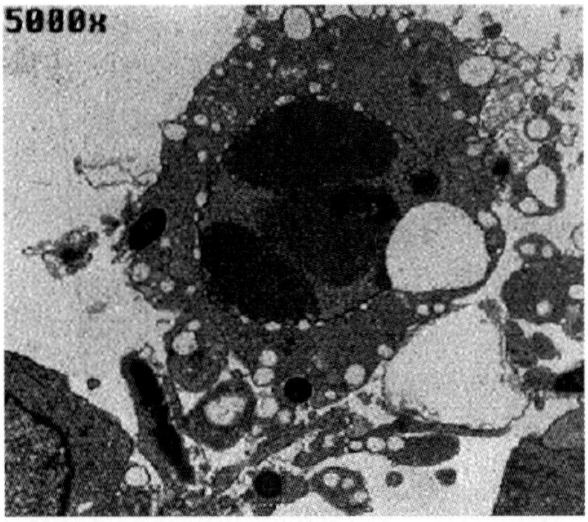

Fig. 2 Transmission electron micrographs of apoptotic RAW264.7 macrophages infected with *Y. pseudotuberculosis*. Macrophages were infected with wild-type *Y. pseudotuberculosis* (*YPIIIpIB1*) for 6 h. Note the chromatin condensation, cytoplasmic vacuolization, and swollen endoplasmic reticulum in macrophages infected with YPIIIpIB1. (Adapted from Monack et al. 1997 with the permission of the publisher)

served on infection of macrophages with *Shigella* or *Salmonella* species (Navarre and Zychlinsky 2000). Collectively, all the hallmarks of apoptosis have been detected in primary macrophages and macrophage-like cell lines infected by *Yersinia*. Transmission electron microscopic images show intense perinuclear chromatin aggregation in infected macrophages (Fig. 2). Scanning electron micrographs reveal extensive membrane blebbing on the surface of dying macrophages (Monack et al. 1997). DNA fragmentation can be detected by TUNEL or gel electrophoresis (Mills et al. 1997; Monack et al. 1997; Ruckdeschel et al. 1997b). Disruption of plasma membrane polarity as indicated by exposure of phosphatidylserine is also observed (Ruckdeschel et al. 1997b). In addition to these specific phenomena of apoptosis, in the late stage of *Yersinia*-induced apoptosis cytosolic contents are released.

Denecker et al. (Denecker et al. 2001) detected cleavage of the Bid protein in *Yersinia*-infected macrophages at 60 min after infection. Bid is a specific substrate of caspase-8, which suggests that caspase-8 is also activated at an early step of the death response in *Yersinia*-infected macrophages (Denecker et al. 2001). tBid, the cleavage product of Bid, triggers cytochrome *c* release from mitochondria, which has also been detected in macrophages beginning at around 80 min after infection with *Yersinia* (Denecker et al. 2001). The release of cytochrome *c* results in assembly of the apoptosome, cleavage and activation of caspase-9, and finally the activation of the so-called executioner caspases-3 and -7. Cleavage of caspases-9, -3, and -7 is detected in macrophages beginning at around 80 min after infection with *Yersinia* (Denecker et al. 2001). These observations are consistent with the idea that *Yersinia*-induced apoptosis is initiated by an extrinsic death signal that is amplified through cytochrome *c* release. This model is also consistent with the demonstration that *Yersinia*-induced apoptosis can be inhibited by broad-spectrum caspase inhibitors (zVAD.fmk or B-D.fmk) or more specific inhibitors of caspase-8 (z-IETD-fmk) or caspase-9 (z-LEHD-fmk) (Denecker et al. 2001; Ruckdeschel et al. 1997b).

2.2
YopJ: A TTSS Effector Required for Apoptosis and Inhibition of Cytokine Production in Macrophages

YopJ is required by all three pathogenic *Yersinia* species to induce macrophage apoptosis (Mills et al. 1997; Monack et al. 1997) (Y. Zhang and J.B. Bliska, unpublished data). YopJ must be delivered into the macrophage to induce apoptosis, and therefore mutations that interfere with type III secretion or translocation will also protect infected macrophages from apoptosis. YopJ was initially identified as a 33-kDa protein secreted by *Y. pestis* (Straley and Bowmer 1986). The 288-residue YopJ protein shares sequence similarity with a family of so-called avirulence proteins found in bacterial pathogens of plants or animals (Orth 2002). The YopJ-related avirulence proteins secreted by plant pathogens are responsible for inducing the hypersensitive response, an apoptosis-like process in plant cells. There appear to be two isoforms of YopJ, one produced by high-virulence strains and another by lower-virulence strains (Ruckdeschel et al. 2001b). A change at residue 143 appears to determine this difference, but it is not clear what aspect of YopJ function is modulated by substitutions at this residue (Ruckdeschel et al. 2001b).

One clue to the mechanism of YopJ-induced apoptosis came from the discovery that YopJ inhibits host signaling pathways that activate mitogen-activated protein (MAP) kinases (Boland and Cornelis 1998; Palmer et al. 1998; Ruckdeschel et al. 1997a) and the transcription factor NF-κB (Ruckdeschel et al. 1998; Schesser et al. 1998). On the basis of these results, it was postulated that YopJ induces apoptosis by blocking expression of antiapoptotic proteins under control of NF-κB (Ruckdeschel et al. 1998). This model was based on earlier demonstrations that antiapoptotic factors under control of NF-κB can prevent cells exposed to stress stimuli from undergoing programmed cell death (Van Antwerp et al. 1996; Wang et al. 1996).

A search for binding partners of YopJ by yeast two-hybrid analysis led to the discovery that YopJ binds to multiple members of the MAP kinase kinase (MKK) family (Orth et al. 1999). Another protein that binds to YopJ is IκB kinase β (IKKβ). The MKKs are upstream activators of the MAP kinases ERK, JNK, and p38, whereas IKKβ is an activator of the NF-κB pathway. IKKβ phosphorylates the inhibitor κBα (IκBα), which leads to its ubiquitination and degradation. Degradation of IκBα exposes the nuclear localization signal in NF-κB and allows it to translocate to the nucleus (Ghosh and Karin 2002). Biochemical studies have shown that YopJ prevents activation of MKKs and IKKβ (Orth et al. 1999). Sequence analysis of YopJ predicts structural similarity between YopJ and a class of cysteine proteases that act on ubiquitin-like (Ubl) proteins (Orth et al. 2000). Indeed, YopJ contains the conserved catalytic His, Glu, and Cys residues that are found in other members of this protease family such as the yeast Ubl-specific protease 1 (Ulp1) and adenovirus-2 protease (AVP). Mutations that disrupt the predicted catalytic site of YopJ abolish its ability to inhibit signaling and to trigger apoptosis (Orth et al. 2000). Orth et al. (Orth et al. 2000) have proposed that YopJ cleaves a ubiquitin or a ubiquitin-like modification from a target protein. Modification of this target protein by ubiquitin is presumably required for activation of MKKs and IKKβ (Orth et al. 2000). However, neither the activity nor the cellular target of YopJ has been demonstrated so far.

2.3
Mechanism of *Yersinia*-Induced Apoptosis

As discussed above, a current model (Ruckdeschel et al. 1998) suggests that YopJ does not activate the apoptosis machinery directly, but rather delivers a "coup de grace" to a macrophage cell that is already traveling down the path to death but is not fully committed to perishing. Macrophages exposed to LPS, a cell envelope component of all gram-negative bacteria, die of apoptosis if protein synthesis is blocked (for example, with a protein synthesis inhibitor such as cyclohexamide) (Karahashi and Amano 1998). Macrophages treated with LPS and MG-132, a nonspecific proteasome inhibitor, undergo apoptosis as well (Ruckdeschel et al. 1998). Proteasome inhibitors prevent proteolysis of IκBα, which prevents release of active NF-κB. Therefore, LPS simulation of TLR4 appears to activate a death response in macrophages. This death response

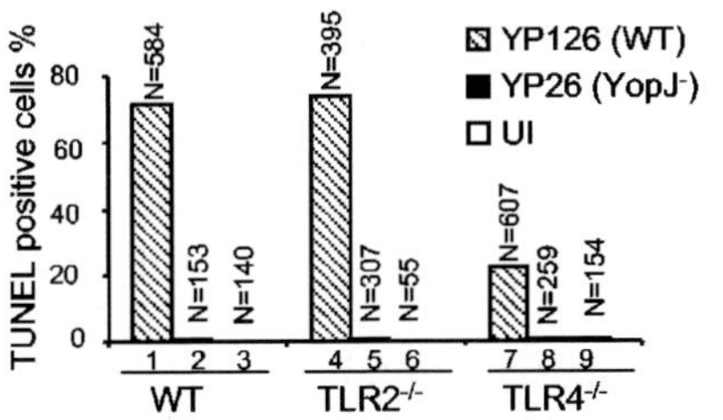

Fig. 3 TLR4-dependent and TLR4-independent apoptosis in macrophages infected with *Yersinia pseudotuberculosis*. Wild-type (*WT*), TLR2$^{-/-}$, or TLR4$^{-/-}$ bone marrow-derived macrophages were left uninfected (*UI*) or infected with wild-type (*YP126*) or YopJ$^-$ (*YP26*) *Y. pseudotuberculosis* for 4 h. Percentages of TUNEL-positive macrophages were determined by fluorescence microscopy. *N*, number of cells counted. (Adapted from Zhang and Bliska 2003 with the permission of the publisher)

can be inhibited by the de novo synthesis of antiapoptosis factors that are under the transcriptional control of NF-κB.

The role of TLR4 in *Yersinia*-induced apoptosis has been investigated with macrophages deficient in TLR receptors (Haase et al. 2003; Zhang and Bliska 2003). Macrophages deficient for expression of TLR4, or which encode nonfunctional TLR4 receptors, are less sensitive (2- to 3-fold) to *Yersinia*-induced apoptosis than their wild-type counterparts (Fig. 3). Therefore, LPS appears to be the major source of apoptotic signaling in *Yersinia*-infected macrophages, and as expected, apoptosis is undetectable in TLR4-deficient macrophages exposed to LPS and MG-132 (Haase et al. 2003). However, TLR4-deficient macrophages are not completely resistant to *Yersinia*-induced apoptosis (Fig. 3) (Haase et al. 2003; Zhang and Bliska 2003), suggesting that there is another source of apoptotic signaling in *Yersinia*-infected macrophages. Another TLR receptor, TLR2, has been shown to promote apoptosis in macrophages in response to stimulation with lipopeptides (Navarre and Zychlinsky 2000). However, TLR2 does not appear to play an important role in *Yersinia*-induced apoptosis, because macrophages deficient in TLR2 are not more resistant to *Yersinia*-induced apoptosis (Haase et al. 2003; Zhang and Bliska 2003). Therefore, apoptotic signaling that is independent of TLR4 and TLR2 must be activated in macrophages infected with *Yersinia*. We have speculated that other components of the bacterial cell, such as flagella, or even a component of the TTSS, could be involved in stimulating apoptosis in macrophages in a TLR4-independent manner (Zhang and Bliska 2003). It is also possible that intracellular pathogen pattern recognition receptors are involved in activating apoptotic signaling during *Yersinia* infection of macrophages. For example, the cytosolic Nod1 and Nod2 proteins can detect peptidoglycan fragments and activate NF-κB (Girardin et al. 2003a,b; Inohara et al. 2003). It is conceivable that peptidoglycan fragments gain access to the cytosolic Nod proteins during *Yersinia* infection because of the formation of pores by the TTSS apparatus (Viboud and Bliska 2001). The possibility that Nod proteins are capable of sensing peptidoglycan and activating an apoptotic response during *Yersinia* infection of macrophages remains to be investigated.

2.4 Mechanism of Apoptotic Signaling via TLR4

Like other members in the TLR family, TLR4 contains extracellular leucine-rich repeats and a cytoplasmic Toll/interleukin-1 receptor (TIR) domain. So far, four TIR-containing adaptor proteins have been identified that function downstream of TLR4 (Fig. 4), and according to gene knockout studies they are roughly divided into two groups, the MyD88-dependent adaptors and the MyD88-independent adaptors (Takeda and Akira 2004). MyD88 (Myeloid differentiation factor 88) contains an N-terminal death domain (DD) and a C-terminal TIR domain. After LPS stimulation, MyD88 is recruited to TLR4 through TIR-TIR domain interaction. IL-1R-associated kinase (IRAK) is then recruited to MyD88 through DD-DD interaction. IRAK is activated by phosphorylation and associates with TRAF6 (tumor necrosis factor receptor-associated factor 6). TRAF6 is an ubiquitin ligase that activates the TAK1 complex, the MKKs and IKKβ, and ultimately NF-κB and MAP kinases (Wang et al. 2001). This pathway is likely to be utilized by all the TLRs. In addition to MyD88, signaling pathways downstream of TLR2 and TLR4 also involve a second TIR domain-containing adaptor, TIR domain-containing adaptor protein (TIRAP), which is also known as MyD88-adaptor-like (Mal). This adaptor contains only the TIR domain, and from gene knockout studies it was concluded that it participates in the MyD88-dependent pathway to activate downstream events (Takeda and Akira 2004). On the other hand, LPS activation of transcription factor IRF-3, and subsequently the production of IFNβ, is considered MyD88 independent. This pathway involves the other two TIR adaptors: TRIF (TIR domain-containing adaptor-inducing IFNβ; also called TIR domain-containing adaptor molecule, TICAM-1) and TRAM (TRIF-related adaptor molecule) (Takeda and Akira 2004). There is evidence suggesting that

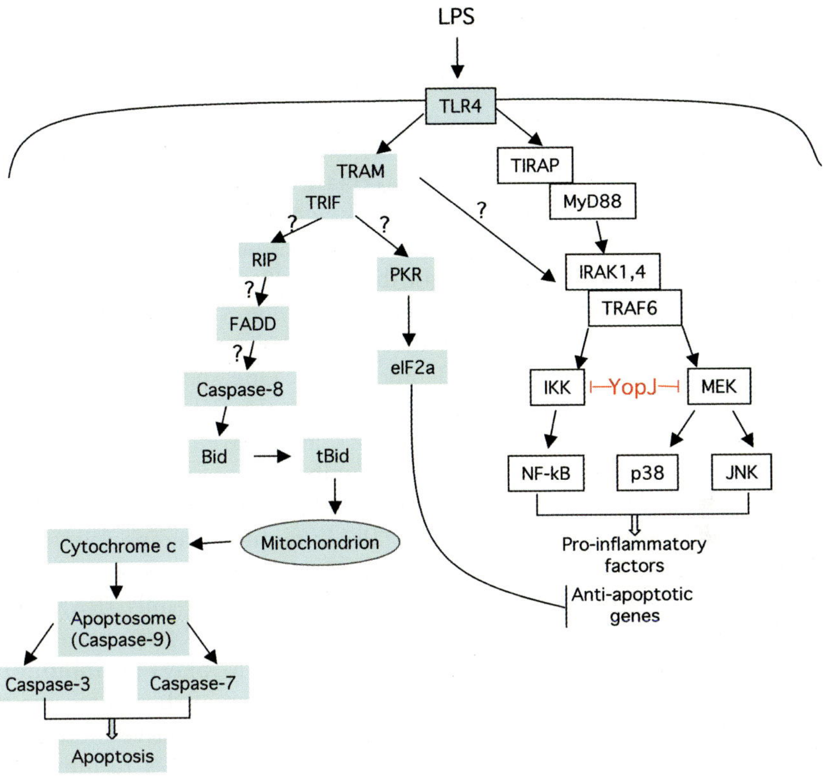

Fig. 4 A schematic diagram of LPS- and TLR4-dependent signal response pathways in macrophages. *Shaded components* are in the proapoptotic pathway, and *nonshaded components* are in the antiapoptotic pathway. *Question marks* indicate tentative pathway steps. Steps at which YopJ interferes with the survival pathway are indicated in *red*

TLR4-induced apoptosis is activated downstream of TRIF, because overexpression of TRIF results in cell death, and LPS does not cause apoptosis in macrophages deficient for TRIF (Han et al. 2004; Hoebe et al. 2003). The diagram in Fig. 4 shows a model for TLR4-dependent signaling pathways that is based on current literature. Below, we discuss the results of several studies that have attempted to elucidate the role of these signaling pathways in *Yersinia*-induced apoptosis.

The roles of MyD88 and TIRAP/Mal in *Yersinia*-induced apoptosis have been studied by introducing dominant-negative versions of these proteins into macrophages by transfection and then incubating the cells with MG-132 and LPS. Overexpression of ΔMyD88 (MyD88 missing its DD) in J774A.1 macrophage-like cells led to partial protection from LPS-induced apoptosis (Haase et al. 2003; Ruckdeschel et al. 2002). Similarly, overexpression of a mutant TIRAP (ΔTIRAP or TIRAP-P125H), which contains a nonfunctional TIR domain, also partially protected macrophages from LPS-induced apoptosis (Haase et al. 2003) These mutant forms of MyD88 and TIRAP have been shown to block NF-κB activation downstream of TLR4 and are considered dominant-negative inhibitors. However, because it is now thought that MyD88 and TIRAP function in the survival pathway downstream of TLR4 (Fig. 4), it is unclear why dominant-negative inhibitors of these proteins would protect macrophages from LPS/YopJ-induced apoptosis. One possibility is that ΔMyD88 and TIRAP-P125H also interfere with the function of TRIF and TRAM in the apoptotic pathway (Fig. 4), and therefore these proteins may not be specific dominant-negative inhibitors.

The role of IRAKs in apoptosis induced by LPS treatment or *Yersinia* infection has also been studied. The IRAK family contains four members: IRAK-1, IRAK-2, IRAK-M, and IRAK-4. These proteins contain an N-terminal DD and a central kinase domain. However, only IRAK1 and IRAK4 are catalytically active. It is thought that IRAK4 is upstream of IRAK1 in LPS-induced signaling and that IRAK-M plays an inhibitory role. According to the ability of truncated versions of these proteins to influence LPS-induced NF-κB activation, dominant-negative mutants of IRAKs have been identified: IRAK1(1–215) contains the N-terminal DD and the region just before the kinase domain, whereas IRAK2(1–96) contains only the DD. Two IRAK4 dominant-negative mutants were identified, one missing the kinase region and one with a point mutation at the catalytic site. Among the these various dominant-negative mutants, IRAK2(1–96) was found to significantly attenuate death in *Yersinia*-infected macrophages (Ruckdeschel et al. 2002). It is possible that IRAK2 functions in the TLR4 apoptotic pathway, or alternatively, the DD of IRAK2(1–96) interfered with the function of other DD-containing proteins in the apoptosis pathway.

IKKβ is pivotal in activating NF-κB (Fig. 4). It has been shown that overexpression of wild-type IKKβ in macrophages results in partial protection from *Yersinia*-induced apoptosis (Ruckdeschel 2001; Zhang and Bliska 2003). Expression of a constitutively active form of IKKβ in macrophages results in complete protection from *Yersinia*-induced apoptosis (Zhang and Bliska 2003). Ablation of IKKβ by gene knockout renders macrophages highly sensitive to the apoptotic activity of LPS (Hsu et al. 2004), whereas overexpression of the NF-κB subunit p65 can protect macrophages from *Yersinia*-induced apoptosis (Ruckdeschel et al. 2001a). It is likely that TRAF6 is important for the activation of IKKβ by LPS, because TRAF6$^{-/-}$ macrophages treated with LPS undergo apoptosis (Hsu et al. 2004). Together, these results are consistent with the idea that activation of NF-κB via TRAF6 and IKKβ results in the production of antiapoptotic proteins that can counteract LPS-induced cell death.

Three categories of apoptosis inhibitors are known to be regulated by NF-κB (Karin and Lin 2002). These categories are represented by factors that block caspase activation, inhibit caspase activity, or prevent mitochondrial depolarization. c-FLIP can interfere with pro-caspase-8 activation (Irmler et al. 1997). This protein contains two death effector domains (DEDs) and a catalytically inactive caspase domain. It can interact with FADD and pro-caspase-8 through homotypic interactions mediated by the DEDs (Irmler et al. 1997). Inhibition of caspase activity can also be mediated by c-IAPs, which directly bind to and inhibit some caspases (Deveraux et al. 1998). The Bcl-2 family members A1 (also called Bfl1) and Bcl-x_L can inhibit apoptosis by preventing mitochondrial depolarization and release of cytochrome *c*. It has been shown that overexpression of A1 can protect macrophages from *Yersinia*-induced apoptosis, but overexpression of A1 also activates NF-κB (Haase et al. 2003). Therefore, whether any of these proteins' functions specifically block the apoptotic pathway activated during *Yersinia* infection remains to be determined.

PKR (interferon-induced, double-stranded RNA-activated protein kinase) has recently been shown to be required for LPS-induced apoptosis (Hsu et al. 2004). Furthermore, LPS-induced activation of PKR requires TRIF (Hsu et al. 2004). Activated PKR phosphorylates eukaryotic translation initiation factor 2α (eIF2α), resulting in inhibition of protein synthesis. PKR is not required for LPS-induced expression of mRNAs encoding apoptosis inhibitors such as c-Flip, c-IAP2, or A1 in macrophages. However, after LPS treatment, PKR activation prevents de novo synthesis of A1 protein, resulting in decreased levels of A1 due to protein turnover. These results are consistent with the idea that PKR induces apoptosis by inhibiting translation of antiapoptosis factors. Interestingly, it has recently been reported that PKR$^{-/-}$ macrophages are resistant to *Yersinia*-induced apoptosis (Hsu et al. 2004). The proposed function of PKR in apoptosis, inhibition of survival factor production, is somewhat analogous to that proposed for YopJ (Fig. 4). PKR acts at the posttranslational level, whereas YopJ acts at the level of gene expression. One would therefore predict that in the presence of YopJ, PKR would be dispensable for *Yersinia*-induced apoptosis. Additional studies will be required to clarify this issue.

Transient transfection experiments indicate that RIP can activate FADD and allow apoptosis to proceed downstream of TRIF (Han et al. 2004). Although a direct connection between TRIF, RIP, and FADD for apoptotic signaling has not been established in macrophages, it is very likely that FADD or another adaptor protein containing a DED and a DD are actively involved in signaling LPS-induced apoptosis (Fig. 4). Studies from Ruckdeschel's group found that expression of dominant-negative FADD, which consists of the DD only, can dramatically decrease apoptosis induced by LPS/MG-132 without activating NF-κB (Haase et al. 2003). The lack of NF-κB activation under these conditions argues that the death domain of FADD specifically perturbs the apoptotic pathway. However, whether FADD itself is directly involved in *Yersinia*-induced apoptosis remains to be determined.

3
Role of Apoptosis During *Yersinia* Infection

Macrophage apoptosis is thought to play an important role in the pathogenesis of several bacterial infections (Monack et al. 2001; Navarre and Zychlinsky 2000). The role of macrophage apoptosis during *Yersinia* infection has not been studied extensively, but there is some evidence to suggest that macrophage apoptosis is important for systemic infection of mice by *Y. pseudotuberculosis* (Monack et al. 1998). In mice challenged intragastrically with wild-type or *yopJ*-mutant *Y. pseudotuberculosis*, a *yopJ* mutant colonized and replicated in Peyer's patches as well as a wild-type strain. However, over time lower numbers of *yopJ* mutants reached the mesenteric lymph nodes or spleen, and growth of the *yopJ* mutant in these tissues was slower than for the wild type. As a result, the 50% lethal dose of the *yopJ* mutant was higher (64-fold) that of the wild-type strain. Because YopJ is important for apoptosis and for counteracting cytokine production, it is difficult to dissect which of these activities are important for pathogenesis. Monack et al. (Monack et al. 1998) obtained evidence with flow cytometry that *Y. pseudotuberculosis* induced apoptosis in Mac1$^+$ cells in mesenteric lymph nodes and spleen and that YopJ was necessary for this cell death. The morphological characteristics of the apoptotic Mac1$^+$ cells were consistent with those of macrophages, rather than those of other Mac1$^+$ cells such as neutrophils or dendritic cells. Histological analysis of abscesses in mesenteric lymph node sections showed the presence of TUNEL-positive cells in the vicinity of bacterial microcolonies. These results are consistent with the idea that macrophages recruited to mesenteric lymph nodes and spleen with the aim of producing cytokines and phagocytosing and killing *Yersinia* are eliminated by apoptosis. However, the finding that a *yopJ*-mutant *Y. pseudotuberculosis* strain colonized Peyer's patches as well as a wild-type strain suggests that apoptosis may be dispensable for pathogenesis in certain tissues. We speculate that this may be due to a decreased sensitivity of macrophages in Peyer's patches to *Yersinia*-induced apoptosis. Ruckdeschel and Richter (Ruckdeschel and Richter 2002) have shown that macrophages preexposed to LPS are resistant to *Yersinia*-induced apoptosis. It is possible that LPS shed from enteric microorganisms can enter into Peyer's patches at sufficiently high levels to result in desensitization of Peyer's patch macrophages to *Yersinia*-induced apoptosis.

Systematic studies of apoptosis during infection by other *Yersinia* species have not been published, but it is likely that apoptosis plays a similar role in the pathogenesis of the other pathogenic *Yersinia* species. Straley and Bowmer (Straley and Bowmer 1986) have studied the virulence of a *Y. pestis yopJ* mutant in the mouse intravenous infection model. As discussed above, the intravenous infection model is thought to mimic a septicemic phase of infection. The LD_{50} of the *Y. pestis yopJ* mutant was only 15-fold higher than that of the wild-type strain (Straley and Bowmer 1986). The relatively small increase in LD_{50} seen in these experiments may indicate that macrophage apoptosis is of greatest importance for *Yersinia* infections initiated from peripheral routes. Alternatively, it is possible that apoptosis is less important for *Y. pestis* pathogenesis than for pathogenesis of the enteric *Yersinia* infections. Additional infection studies will be needed to resolve this issue.

4 Conclusions and Perspectives

A number of bacterial pathogens induce apoptosis in macrophages in a TTSS-dependent manner (Navarre and Zychlinsky 2000). These pathogens include *Salmonella typhimurium* (see the chapter by D.G. Guiney, this volume), *Shigella flexneri* (Navarre and Zychlinsky 2000), and *Pseudomonas aeruginosa* (Hauser and Engel 1999). The *Shigella flexneri* protein IpaB, which is secreted by a TTSS, appears to be sufficient to trigger apoptosis through activation of caspase-1 (Navarre and Zychlinsky 2000). Macrophage apoptosis induced by the pathogenic *Yersinia* species is also a TTSS-dependent process. The YopJ protein has been shown to be required for apoptosis. However, unlike SipB, YopJ is not sufficient for *Yersinia*-induced apoptosis. *Yersinia*-induced apoptosis requires the activation of a death response pathway and the suppression of a survival response by YopJ.

LPS acting through TLR4 appears to be the major death stimulus in cultured macrophages infected by *Yersinia*. Under these conditions, a high concentration of LPS is present because of the relatively high multiplicities of infection used. Furthermore, it is typical in these experiments to add an antibiotic (gentamicin) to the tissue culture medium after several hours of infection to prevent overgrowth of the bacteria. The antibiotic could lead to even greater levels of LPS release from dying bacteria. It is interesting to consider the question of whether concentrations of LPS are sufficiently high in vivo to stimulate a death response via TLR4. It is possible that at early infection times when few numbers of bacteria are present in tissues there may be insufficient LPS levels to stimulate *Yersinia*-induced apoptosis via TLR4. Once the bacteria increase in numbers and are replicating in abscesses, it seems likely that sufficient levels of LPS will be present. It is possible that the TLR4-independent mechanism of apoptosis that has been detected (Zhang and Bliska 2003) could operate under conditions of low multiplicities of infection.

Given that apoptosis plays a critical role in the pathogenesis of other bacteria, it is interesting that *Yersinia yopJ* mutants have relatively subtle virulence phenotypes. One reason for this may be the fact that these bacteria can use other Yops to antagonize phagocytosis and have other means to inhibit cytokine production (e.g., LcrV). Therefore, inducing apoptosis is just one of several mechanisms *Yersinia* use to counteract the functions of macrophages during infection.

It is plausible that macrophages are "hardwired" to commit apoptosis, because after finishing the mission of clearing invaders they must be eliminated to prevent excessive damage to the surrounding tissues. Detection of pathogen-associated molecules by macrophages activates pro- and antiapoptotic pathways, although the survival pathway predominates early during infection. Activation of NF-κB and MAP kinases pathways constitutes an important part of the host defenses against pathogens and promotes macrophage survival through upregulation of the inhibitors of apoptosis. Bacterial manipulation of either pathway may alter the delicate balance between survival and apoptosis and influence the outcome of the infection in favor of the bacteria.

Acknowledgements The authors' work is supported by grants from the National Institutes of Health (AI-43389 and AI-53759) awarded to J.B.B.

References

Achtman M, Zurth K, Morelli G, Torrea G, Guiyoule A, Carniel E (1999) *Yersinia pestis*, the cause of plague, is a recently emerged clone of *Yersinia pseudotuberculosis*. Proc Natl Acad Sci USA 96:14043–14048

Aepfelbacher M, Zumbihl R, Ruckdeschel K, Jacobi CA, Barz C, Heesemann J (1999) The tranquilizing injection of *Yersinia* proteins: a pathogen's strategy to resist host defense. Biol Chem 380:795–802

Autenrieth IB, Heesemann J (1993) In vivo neutralization of tumor necrosis factor alpha and interferon-gamma abrogates resistance to *Yersinia enterocolitica* in mice. Med Microbiol Immunol 181:333–338

Autenrieth IB, Kempf V, Sprinz T, Preger S, Schnell A (1996) Defense mechanisms in Peyer's patches and mesenteric lymph nodes against *Yersinia enterocolitica* involve integrins and cytokines. Infect Immun 64:1357–1368

Beuscher HU, Rodel F, Forsberg A, Rollinghoff M (1995) Bacterial evasion of host immune defense: *Yersinia enterocolitica* encodes a suppressor for tumor necrosis factor alpha expression. Infect Immun 63:1270–1277

Bliska JB (2000) Yop effectors of *Yersinia* spp. and actin rearrangements. Trends Microbiol 8:205–208

Bohn E, Autenrieth IB (1996) IL-12 is essential for resistance against *Yersinia enterocolitica* by triggering IFN-gamma production in NK cells and CD4+ T cells. J Immunol 156:1458–1468

Boland A, Cornelis GR (1998) Role of YopP in suppression of tumor necrosis factor alpha release by macrophages during *Yersinia* infection. Infect Immun 66:1878–1884

Bottone EJ (1997) *Yersinia enterocolitica*: the charisma continues. Clin Microbiol Rev 10:257–276

Brubaker RR (1991) Factors promoting acute and chronic diseases caused by yersiniae. Clin Microbiol Rev 4:309–324

Brubaker RR (2003) Interleukin-10 and inhibition of innate immunity to Yersiniae: roles of Yops and LcrV (V antigen). Infect Immun 71:3673–3681

Burdack S, Schmidt A, Knieschies E, Rollinghoff M, Beuscher HU (1997) Tumor necrosis factor-alpha expression induced by anti-YopB antibodies coincides with protection against *Yersinia enterocolitica* infection in mice. Med Microbiol Immunol (Berl) 185:223–229

Carniel E (1999) The *Yersinia* high-pathogenicity island. Int Microbiol 2:161–167

Carniel E (2002) Plasmids and pathogenicity islands of *Yersinia*. Curr Top Microbiol Immunol 264:89–108

Cavanaugh DC, Randall R (1959) The role of multiplication of *Pasteurella pestis* in mononuclear phagocytes in the pathogenesis of fleaborne plague. J Immunol 85:348–363

Clark MA, Hirst BH, Jepson MA (1998) M-cell surface beta1 integrin expression and invasin-mediated targeting of *Yersinia pseudotuberculosis* to mouse Peyer's patch M cells. Infect Immun 66:1237–1243

Cornelis GR (2002a) *Yersinia* type III secretion: send in the effectors. J Cell Biol 158:401–408

Cornelis GR (2002b) The *Yersinia* Ysc-Yop 'type III' weaponry. Nat Rev Mol Cell Biol 3:742–752

Denecker G, Declercq W, Geuijen CA, Boland A, Benabdillah R, van Gurp M, Sory MP, Vandenabeele P, Cornelis GR (2001) *Yersinia enterocolitica* YopP-induced apoptosis of macrophages involves the apoptotic signaling cascade upstream of bid. J Biol Chem 276:19706–19714

Deveraux QL, Roy N, Stennicke HR, Van Arsdale T, Zhou Q, Srinivasula SM, Alnemri ES, Salvesen GS, Reed JC (1998) IAPs block apoptotic events induced by caspase-8 and cytochrome *c* by direct inhibition of distinct caspases. EMBO J 17:2215–2223

El-Maraghi NRH, Mair NS (1979) The histopathology of enteric infection with *Yersinia pseudotuberculosis.* Am J Clin Pathol 71:631–639

Ghosh S, Karin M (2002) Missing pieces in the NF-kappaB puzzle. Cell 109 [Suppl]:S81–96

Girardin SE, Boneca IG, Carneiro LA, Antignac A, Jehanno M, Viala J, Tedin K, Taha MK, Labigne A, Zahringer U, Coyle AJ, DiStefano PS, Bertin J, Sansonetti PJ, Philpott DJ (2003a) Nod1 detects a unique muropeptide from gram-negative bacterial peptidoglycan. Science 300:1584–1587

Girardin SE, Boneca IG, Viala J, Chamaillard M, Labigne A, Thomas G, Philpott DJ, Sansonetti PJ (2003b) Nod2 is a general sensor of peptidoglycan through muramyl dipeptide (MDP) detection. J Biol Chem 278:8869–8872

Goguen JD, Walker WS, Hatch TP, Yother J (1986) Plasmid-determined cytotoxicity in *Yersinia pestis* and *Yersinia pseudotuberculosis.* Infect Immun 51:788–794

Grassl GA, Kracht M, Wiedemann A, Hoffmann E, Aepfelbacher M, von Eichel-Streiber C, Bohn E, Autenrieth IB (2003) Activation of NF-kappaB and IL-8 by *Yersinia enterocolitica* invasin protein is conferred by engagement of Rac1 and MAP kinase cascades. Cell Microbiol 5:957–971

Haase R, Kirschning CJ, Sing A, Schrottner P, Fukase K, Kusumoto S, Wagner H, Heesemann J, Ruckdeschel K (2003) A dominant role of Toll-like receptor 4 in the signaling of apoptosis in bacteria-faced macrophages. J Immunol 171:4294–4303

Han KJ, Su X, Xu LG, Bin LH, Zhang J, Shu HB (2004) Mechanisms of the TRIF-induced interferon-stimulated response element and NF-kappaB activation and apoptosis pathways. J Biol Chem 279:15652–15661

Hanski C, Kutschka U, Schmoranzer HP, Naumann M, Stallmach A, Hahn H, Menge H (1989) Immunohistochemical and electron microscopic study of interaction of *Yersinia enterocolitica* serotype 08 with intestinal mucosa during experimental enteritis. Infect Immun 57:673–678

Hauser AR, Engel JN (1999) *Pseudomonas aeruginosa* induces type-III-secretion-mediated apoptosis of macrophages and epithelial cells. Infect Immun 67:5530–5537

Hoebe K, Du X, Georgel P, Janssen E, Tabeta K, Kim SO, Goode J, Lin P, Mann N, Mudd S, Crozat K, Sovath S, Han J, Beutler B (2003) Identification of Lps2 as a key transducer of MyD88-independent TIR signalling. Nature 424:743–748

Hsu LC, Park JM, Zhang K, Luo JL, Maeda S, Kaufman RJ, Eckmann L, Guiney DG, Karin M (2004) The protein kinase PKR is required for macrophage apoptosis after activation of Toll-like receptor 4. Nature 428:341–345

Inohara N, Ogura Y, Fontalba A, Gutierrez O, Pons F, Crespo J, Fukase K, Inamura S, Kusumoto S, Hashimoto M, Foster SJ, Moran AP, Fernandez-Luna JL, Nunez G (2003) Host recognition of bacterial muramyl dipeptide mediated through NOD2. Implications for Crohn's disease. J Biol Chem 278:5509–5512

Irmler M, Thome M, Hahne M, Schneider P, Hofmann K, Steiner V, Bodmer JL, Schroter M, Burns K, Mattmann C, Rimoldi D, French LE, Tschopp J (1997) Inhibition of death receptor signals by cellular FLIP. Nature 388:190–195

Isberg RR, Van Nhieu GT (1995) The mechanism of phagocytic uptake promoted by invasin-integrin interaction. Trends Cell Biol 5:120–124

Juris SJ, Shao F, Dixon JE (2002) *Yersinia* effectors target mammalian signalling pathways. Cell Microbiol 4:201–211

Karahashi H, Amano F (1998) Apoptotic changes preceding necrosis in lipopolysaccharide-treated macrophages in the presence of cycloheximide. Exp Cell Res 241:373–383

Karin M, Lin A (2002) NF-kappaB at the crossroads of life and death. Nat Immunol 3:221–227

Lian CJ, Hwang WS, Kelly JK, Pai CH (1987a) Invasiveness of *Yersinia enterocolitica* lacking the virulence plasmid: an in vivo study. J Med Microbiol 24:219–226

Lian CJ, Hwang WS, Pai CH (1987b) Plasmid-mediated resistance to phagocytosis in *Yersinia enterocolitica*. Infect Immun 55:1176–1183

Marra A, Isberg RR (1997) Invasin-dependent and invasin-independent pathways for translocation of *Yersinia pseudotuberculosis* across the Peyer's patch intestinal epithelium. Infect Immun 65:3412-3421

Mills SD, Boland A, Sory M-P, van der Smissen P, Kerbourch C, Finlay BB, Cornelis GR (1997) *Yersinia enterocolitica* induces apoptosis in macrophages by a process requiring functional type III secretion and translocation mechanisms and involving YopP, presumably acting as an effector protein. Proc Natl Acad Sci USA 94:12638–12643

Monack DM, Mecsas J, Bouley D, Falkow S (1998) *Yersinia*-induced apoptosis in vivo aids in the establishment of a systemic infection. J Exp Med 188:2127–2137

Monack DM, Mecsas J, Ghori N, Falkow S (1997) *Yersinia* signals macrophages to undergo apoptosis and YopJ is necessary for this cell death. Proc Natl Acad Sci USA 94:10385–10390

Monack DM, Navarre WW, Falkow S (2001) *Salmonella*-induced macrophage death: the role of caspase-1 in death and inflammation. Microbes Infect 3:1201–1212

Nakajima R, Brubaker RR (1993) Association between virulence of *Yersinia pestis* and suppression of gamma interferon and tumor necrosis factor alpha. Infect Immun 61:23–31

Navarre WW, Zychlinsky A (2000) Pathogen-induced apoptosis of macrophages: a common end for different pathogenic strategies. Cell Microbiol 2:265–273

Orth K (2002) Function of the *Yersinia* effector YopJ. Curr Opin Microbiol 5:38–43

Orth K, Palmer LE, Bao ZQ, Stewart S, Rudolph AE, Bliska JB, Dixon JE (1999) Inhibition of the mitogen-activated protein kinase superfamily by a *Yersinia* effector. Science 285:1920–1923

Orth K, Xu Z, Mudgett MB, Bao ZQ, Palmer LE, Bliska JB, Mangel WF, Staskawicz B, Dixon JE (2000) Disruption of signaling by *Yersinia* effector YopJ, a ubiquitin-like protein protease. Science 290:1594–1597

Palmer LE, Hobbie S, Galan JE, Bliska JB (1998) YopJ of *Yersinia pseudotuberculosis* is required for the inhibition of macrophage TNFα production and the downregulation of the MAP kinases p38 and JNK. Mol Microbiol 27:953–965

Pepe J, Miller VL (1993) *Yersinia enterocolitica* invasin: a primary role in the initiation of infection. Proc Natl Acad Sci USA 90:6473–6477

Perry RD, Fetherston JD (1997) *Yersinia pestis*—etiologic agent of plague. Clin Microbiol Rev 10:35–66

Plano GV, Day JB, Ferracci F (2001) Type III export: new uses for an old pathway. Mol Microbiol 40:284–293

Ramamurthi KS, Schneewind O (2002) Type III protein secretion in *Yersinia* species. Annu Rev Cell Dev Biol 18:107–133

Revell PA, Miller VL (2001) *Yersinia* virulence: more than a plasmid. FEMS Microbiol Lett 205:159–64

Ruckdeschel K, Harb S, Roggenkamp A, Hornef M, Zumbihl R, Kohler S, Heesemann J, Rouot B (1998) *Yersinia enterocolitica* impairs activation of transcription factor NF-κB: involvement in the induction of programmed cell death and in the suppression of the macrophage tumor necrosis factor α production. J Exp Med 187:1069–1079

Ruckdeschel K, Machold J, Roggenkamp A, Schubert S, Pierre J, Zumbihl R, Liautard J-P, Heesemann J, Rouot B (1997a) *Yersinia enterocolitica* promotes deactivation of macrophage mitogen-activated protein kinases extracellular signal-regulated kinase-1/2, p38, and c-Jun NH_2-terminal kinase. J Biol Chem 272:15920–15927

Ruckdeschel K, Mannel O, Richter K, Jacobi CA, Trulzsch K, Rouot B, Heesemann J (2001a) *Yersinia* outer protein P of *Yersinia enterocolitica* simultaneously blocks the nuclear factor-kappa B pathway and exploits lipopolysaccharide signaling to trigger apoptosis in macrophages. J Immunol 166:1823–1831

Ruckdeschel K, Mannel O, Schrottner P (2002) Divergence of apoptosis-inducing and preventing signals in bacteria-faced macrophages through myeloid differentiation factor 88 and IL-1 receptor-associated kinase members. J Immunol 168:4601–4611

Ruckdeschel K, Richter K (2002) Lipopolysaccharide desensitization of macrophages provides protection against *Yersinia enterocolitica*-induced apoptosis. Infect Immun 70:5259–5264

Ruckdeschel K, Richter K, Mannel O, Heesemann J (2001b) Arginine-143 of *Yersinia enterocolitica* YopP crucially determines isotype-related NF-kappaB suppression and apoptosis induction in macrophages. Infect Immun 69:7652–7662

Ruckdeschel K, Roggenkamp A, Lafont V, Mangeat P, Heesemann J, Rouot B (1997b) Interaction of *Yersinia enterocolitica* with macrophages leads to macrophage cell death through apoptosis. Infect Immun 65:4813–4821

Schesser K, Spiik A-K, Dukuzumuremyi J-M, Neurath MF, Pettersson S, Wolf-Watz H (1998) The *yopJ* locus is required for *Yersinia*-mediated inhibition of NF-κB activation and cytokine expression: YopJ contains a eukaryotic SH2-like domain that is required for its repressive activity. Mol Microbiol 28:1067–1080

Simonet M, Richard S, Berche P (1990) Electron microscopic evidence for in vivo extracellular localization of *Yersinia pseudotuberculosis* harboring the pYV plasmid. Infect Immun 58:841–845

Skurnik M, Peippo A, Ervela E (2000) Characterization of the O-antigen gene clusters of *Yersinia pseudotuberculosis* and the cryptic O-antigen gene cluster of *Yersinia pestis* shows that the plague bacillus is most closely related to and has evolved from *Y. pseudotuberculosis* serotype O:1b. Mol Microbiol 37:316–330

Straley SC, Bowmer WS (1986) Virulence genes regulated at the transcriptional level by Ca^{2+} in *Yersinia pestis* include structural genes for outer membrane proteins. Infect Immun 51:445–454

Straley SC, Cibull ML (1989) Differential clearance and host-pathogen interactions of $YopE^-$ and $YopK^-YopL^-$ *Yersinia pestis* in BALB/c mice. Infect Immun 57:1200–1210

Takeda K, Akira S (2004) Microbial recognition by Toll-like receptors. J Dermatol Sci 34:73–82

Van Antwerp DJ, Martin SJ, Kafri T, Green DR, Verma IM (1996) Suppression of TNFα-induced apoptosis by NF-κB. Science 274:787–789

Vazquez-Torres A, Fantuzzi G, Edwards CK 3rd, Dinarello CA, Fang FC (2001) Defective localization of the NADPH phagocyte oxidase to *Salmonella*-containing phagosomes in tumor necrosis factor p55 receptor-deficient macrophages. Proc Natl Acad Sci USA 98:2561–2565

Viboud GI, Bliska JB (2001) A bacterial type III secretion system inhibits actin polymerization to prevent pore formation in host cell membranes. EMBO J 20:5373–5382

Viboud GI, So SS, Ryndak MB, Bliska JB (2003) Proinflammatory signalling stimulated by the type III translocation factor YopB is counteracted by multiple effectors in epithelial cells infected with *Yersinia pseudotuberculosis*. Mol Microbiol 47:1305–1315

Wang C, Deng L, Hong M, Akkaraju GR, Inoue J, Chen ZJ (2001) TAK1 is a ubiquitin-dependent kinase of MKK and IKK. Nature 412:346–351

Wang C-U, Mayo MW, Baldwin AS Jr (1996) TNF- and cancer therapy-induced apoptosis: potentiation by inhibition of NF-κB. Science 274:784–787

Zhang Y, Bliska JB (2003) Role of Toll-like receptor signaling in the apoptotic response of macrophages to *Yersinia* infection. Infect Immun 71:1513–1519

Note added in proof

Ruckdeschel et al. have recently shown that TRIF is impertant for proapoptotic signaling in macrophages infected with *Yersinia* (J Immunol 173:3320–3328)

CTMI (2004) 289:175–184

Entamoeba histolytica Activates Host Cell Caspases During Contact-Dependent Cell Killing

D. R. Boettner[1] · W. A. Petri[2] (✉)

[1] Department of Microbiology, University of Virginia, MR4 Bldg. Room 2115, Lane Road, Charlottesville, VA 22908-1340, USA
[2] Division of Infectious Diseases and International Health, University of Virginia, MR4 Bldg. Room 2115, Lane Road, Charlottesville, VA 22908-1340, USA
wap3g@virginia.edu

Abstract *Entamoeba histolytica* is a human intestinal parasite that causes amoebic colitis as well as liver abscesses. Host tissues are damaged through a three-step process involving adherence, contact-dependent cytolysis, and phagocytosis. These three processes all contribute to the pathogenicity of this parasite. Adherence is provided by the Gal/GalNAc adherence lectin. Host cells are lysed in a contact-dependent fashion. There is evidence that suggests that this contact-dependent killing involves the induction of the host cell's apoptotic machinery. Phagocytosis can then occur, consistent with metazoan apoptotic clearance.

1 Introduction to Amoebiasis

Entamoeba histolytica is an intestinal parasite common to the socio-economically downtrodden. It infects geographically diverse populations where water treatment is poor or nonexistent (Shamsuzzaman et al. 2000). This parasite has a simple life cycle containing two stages: active trophozoites and dormant, tetranucleated cysts (Flynn 1973). Cysts are highly infectious; it is estimated that as few as ten cysts are required to establish colonization. Infection begins when cysts are ingested, travel

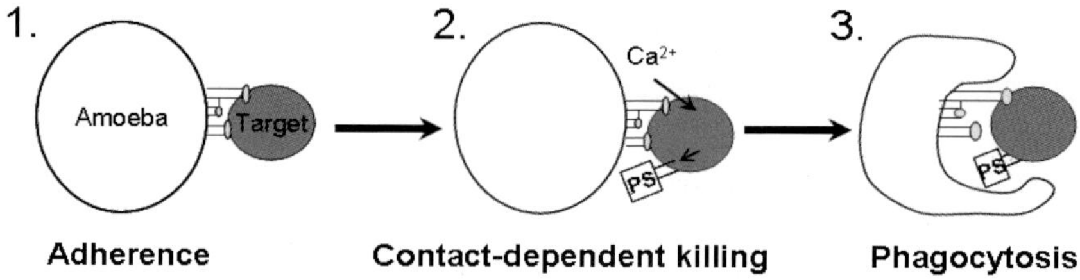

Fig. 1 *Entamoeba histolytica* pathogenesis occurs in three distinct steps. *Entamoeba histolytica* cell killing is described in three steps. *1*: The amoeba comes in contact with the cell and adheres to it in a galactose-inhibitable manner. *2*: The amoeba causes changes in the host cells; this has been seen as calcium fluxes with cells, caspase activation, and morphological changes. This step is also believed to involve the exposure of surface antigens that facilitate the uptake of the host cell. *3*: The now dying host cell is engulfed by the adherent amoeba. One surface change, phosphatidylserine exposure, has been linked to this process, but many others may have a role

to the intestine, and then excyst, each yielding four trophozoites. Although most colonized individuals are asymptomatic, clinical outcomes include colitis and liver or brain abscesses (Haque et al. 2003). The World Health Organization states that there are nearly 100,000 deaths due to infection by *Entamoeba histolytica* per annum (WHO 1997).

In 1903, Schaudin named *Entamoeba histolytica* for its ability to kill host cells. Since then, much effort has been concentrated on understanding this phenomena and how it participates in pathogenesis. Host cell killing has been described as a multistep process including adherence, contact-dependent killing, and clearance of dying host cells (Fig. 1). These steps allow the parasite to invade deeper into host tissues as well as evade detection by the immune system. There are undoubtedly many factors involved in these processes; currently there is evidence suggesting roles for genes encoding the Gal/GalNac lectin subunits (Petri et al. 2002), amoebapores (Leippe et al. 1991), and cysteine proteinases (Que and Reed 2000).

2
Overview of Contact-Dependent Cytolysis of Host Cells

Entamoeba histolytica kills multiple cell types, including neutrophils, erythrocytes (Guerrant et al. 1981), lymphocytes (Salata et al. 1987), and epithelial cells (Kobiler and Mirelman 1981). After adherence, morphological changes occur quickly. Typically between 5 and 15 min, changes appear in vitro such as the rounding up of a polarized monolayer (Li et al. 1994), the loss of tight junctions (Leroy et al. 2000), and membrane blebbing (Ravdin et al. 1988). However, this process requires intact trophozoites. Neither amoebic lysates nor conditioned media can be used to yield these effects (Ravdin et al. 1980). At 4°C amoeba adhere to host cells, but cytolysis is not evident (Hudler et al. 1983). Hence, adherence is required for cytolysis, but not sufficient.

Entamoeba histolytica adherence to host cells is 90% inhibited by the addition of galactose or *N*-acetyl galactosamine in vitro. Adherence is mediated by the Gal/GalNAc lectin of *E. histolytica* (Petri et al. 1987). This lectin is made up of three subunits: the heavy subunit (Hgl) (Mann et al. 1991; Tannich et al. 1991), the intermediate subunit (Igl) (Cheng et al. 1998), and the light subunit (Lgl) (McCoy et al. 1993). The addition of mucins, which are often galactose modified, inhibits adherence to host tissues (Chadee et al. 1987). Glycosylation-deficient CHO cell mutants, lacking terminal galactose residues on their *N*- and *O*-linked sugars, are not recognized for adherence and consequently are spared from cytolysis (Li et al. 1989). It is clear that initial adherence via the Gal/GalNac lectin is required for cytolysis.

The events following adherence are still enigmatic. After incubation with amoeba, there is a 20-fold increase in intracellular calcium in host cells (Ravdin et al. 1988). The addition of EDTA or the treatment of host cells with verapamil blocks host cell death (Ravdin et al. 1982). The importance of this event is tempered by the evidence that purified Gal/GalNAc lectin will also increase intracellular calcium in host cells without causing cytolysis (Ravdin et al. 1988). The addition of cytochalasin completely disrupts the cytotoxicity, indicating that the cytoskeleton of *E. histolytica* has a function in this process (Ravdin and Guerrant 1981). Neutralization of amoebic vesicles with 10 mM NH_4Cl, which increases the pH of vesicles to >5.7, prevents amoebic killing of CHO cells in vitro (Ravdin et al. 1986).

Other proteins have been examined in conjunction with pathogenesis. Amoebapore is a saposin-like protein, similar to NK-lysin, that has antimicrobial activity (Leippe et al. 1994). In vitro studies have also shown the ability of amoebapore to lyse host tissues. This interaction is independent of calcium, as shown by the depletion of free intracellular calcium with BAPTA and extracellular calcium with EDTA (Berninghausen and Leippe 1997a). Immunogold staining has shown that amoebapore is associated with phagosomes but is not constitutively secreted outside the cell (Leippe et al. 1995). With antisense technology, amoebapore knockdown strains were constructed. These strains have reduced virulence in the liver abscess model (Zhang et al. 2004) and reduce cytolytic activity in vitro (Bracha et al. 1999). However, it is unclear whether this is an artifact of the antibiotic selection required to harbor the antisense vector.

Unlike Amoebapore a plethora of cysteine proteases are secreted into the surrounding medium by *E. histolytica* (Schulte and Scholze 1989). These proteins have been implicated in *E. histolytica* invasion in the gut. Targets of these proteases have been found to be extracellular matrix proteins such as laminin (Li et al. 1995). Cysteine proteinases are most convincingly associated with invasion deeper into tissue and survival, but not with the cytolysis events. This may change as many queries are made into the roles of these proteins.

3 Evidence of Host Cell Apoptosis During Infection

Several attempts to classify the nature of cytolysis have been performed. Dying host cells were examined for apoptotic phenotypes by a number of labs. One lab showed that dying cells exhibited DNA laddering consistent with late apoptosis. However, this effect could not be overcome by the overexpression of Bcl-2 (Ragland et al., 1994). Another group elegantly showed that the host cells in contact with amoeba lack many hallmarks of apoptosis. Both human myeloid cells and Jurkat T-leukemia cells were killed in a fashion that led to greater cell size and loss of membrane integrity and were not TUNEL positive (Berninghausen and Leippe 1997b).

More recently, investigators have directly tested for caspase involvement in host cell killing (Fig. 2). Jurkat T-leukemia cells were preloaded with a rhodamine-G_1D_2-rhodamine caspase 3 substrate, which when cleaved separates the two probes, ending their quenching effect. Clearly, after contact with *E. histolytica*, caspase 3 was cleaved as assayed by cells becoming fluorescent. Host cell cytolysis could also be blocked by using the pan-caspase inhibitor Ac-DEVD-CHO. Sorting out the upstream regulators of caspase 3 has proven to be more difficult. Caspase 8-deficient cells were killed by *E. histolytica* without reduction, and the caspase 9 inhibitor AC-LEHD-fmk did not inhibit cell death, by ^{51}Cr release (Huston et al. 2000). This leaves many open questions as to what is principally causing this effect.

Some in vivo evidence exists indicating that there is a role for apoptosis in actual infection. Work done on the amoebic liver abscess model of

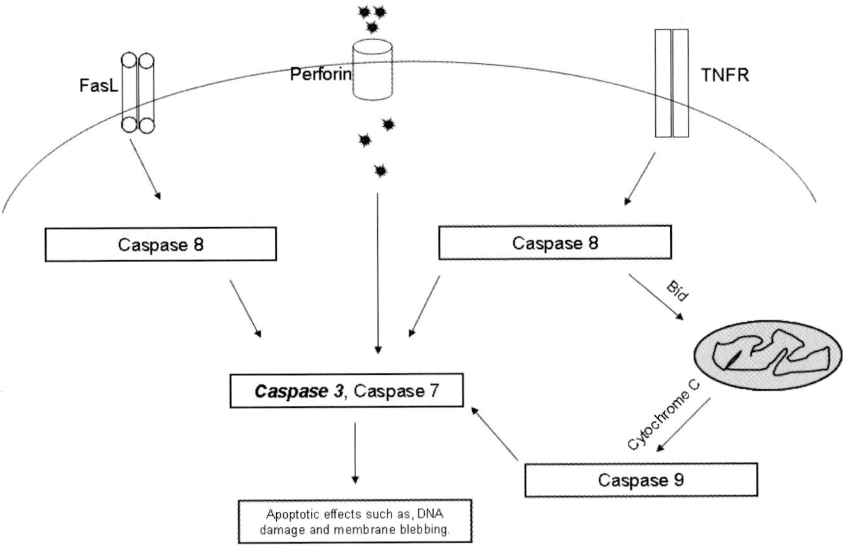

Fig. 2 Classic apoptotic pathways involve the activation of caspase 2 by caspase 8 or granzyme B. Caspase 8 can be activated directly through Fas-FasL interaction or downstream of TNF receptor activation. Caspase 3 can also be activated directly upon granzyme B inside the cell

infection showed that tissue from the site of infection was positive for DNA laddering. Also, hepatocytes near the abscess were TUNEL positive (Seydel and Stanley 1998). This was also corroborated by sections of tissue from the intestinal model of amoebiasis, which revealed that not only is there massive tissue damage surrounding the ulcerated mucosal epithelium but the amoeba appear to be ingesting intact TUNEL-positive cells (Huston et al. 2003). More evidence for a role of apoptosis in *E. histolytica* infection was attained by treating mice with Z-VAD-fmk, a pan-caspase inhibitor, which yielded protection from liver abscess formation, showing a direct involvement of apoptosis in *E. histolytica* infection (Yan and Stanley 2001). There are multiple pathways that initiate apoptosis (Fig. 2), such as through Fas-Fas ligand interaction or TNF receptor signaling. Unfortunately, efforts to discern whether this was Fas/FasL- or TNF receptor mediated provided no clear answer. Both Fas-deficient and TNF receptor-deficient mice had no altered phenotype (Seydel and Stanley 1998).

4 Phagocytosis

In metazoans, the eventual fate of apoptotic cells is ingestion by phagocytes (Ravichandran 2003). When speaking of a noted phagocyte, *Entamoeba histolytica*, it is impossible to ignore this final stage of apoptosis. Earlier work had shown that phagocytosis-deficient amoeba were avirulent in the hamster liver abscess model (Orozco et al. 1983). The next question was simply, Is killing necessary to ingest host cells? An inextricable link was made between phagocytosis and apoptosis when living Jurkat T-leukemia cells and cells made apoptotic by either UV exposure or actinomycin D treatment were incubated with amoeba simultaneously in a phagocytosis assay. The apoptotic cells were ingested, but healthy cells were not (Huston et al. 2003). In individual experiments the number of phagocytosis-positive amoeba was higher when incubated with apoptotic cells compared to healthy cells when cell killing was reduced by adding NH_4Cl.

This was a red flag that perhaps one purpose for activating host cell caspases was to utilize surface changes for ingestion of the dying cells. One of these major changes is phosphatidylserine (PS) exposure. The plasma membrane is organized into a bilayer with two asymmetric lipid leaflets (Bretscher 1972). The membrane is kept asymmetric by the activity of an ATP-dependent transporter that selectively redistributes PS to the inner leaflet (Seigneuret and Devaux 1984). During apoptosis this transporter is deactivated and PS diffuses out of the inner leaflet and throughout the plasma membrane. This process can also be expedited by the activation of a scramblase, a protein that facilitates the indiscriminate flipping of phospholipids in the membrane (Williamson et al. 1992). This change is recognized by many metazoan receptors as a signal for ingestion (Fadok and Henson 2003). Early experiments had shown that *E. histolytica* responds to PS liposomes, leading to actin polymerization in the amoeba (Bailey et al. 1987). This was examined further by the addition of PS liposomes to healthy Jurkat cells, which directly increased the ingestion rate of the cells (Huston et al. 2003). These experiments have led to a belief that PS is recognized and allows the host cells to be ingested by *E. histolytica*, in a manner similar to that of metazoans.

5 Conclusions and Future Directions of Studies

It is still not clear whether apoptosis is the sole or even the primary method of killing host cells. What is known is that DNA fragmentation has been seen in vitro as well as in vivo, caspase 3 is activated after contact, cytolysis as well as infection can be blocked by the addition of caspase inhibitors, and cells are only ingested by these trophozoites after a killing event. It is also clear that *Entamoeba histolytica* recognizes PS flipping on the outer leaflet of the host cells to ingest these cells.

There are still more questions than answers, but it is clear that the host cell has some role in its own death. Caspase 3 is undoubtedly activated; however, it remains unclear whether this is a direct activation, perhaps by amoebic proteases, or whether it is downstream of an untested pathway. The use of Z-VAD-fmk to block the liver abscess model also illustrates a role for apoptosis in the progression of disease, but it is disappointing that more specific inhibitors have not been identified. Finally, although there is evidence suggesting the role of apoptotic factors in the death of host cells, no one refutes that there are signs of necrosis. How these two are related is still not very well understood.

Future efforts in this field will address more clearly what is happening during the contact event. The fact that neither caspase 8 nor caspase 9 is activated begs the question of what is activating caspase 3. One theory involves a role for calcium (Fig. 3). The calcium flux that occurs in dying cells perhaps acts as a death signal through calpain to activate caspase 3. The role of calcium in apoptosis is very much debated, but calpain and caspase 3 have been shown to have common substrates such as calpstatin (Porn-Ares et al. 1998), PARP, and pro-caspase-3 (McGinnis et al. 1999). This could fit with what we already know, because verapamil inhibits calpain activity and makes host cells more resistant to contact-dependent killing. It may also be possible that caspase 3 is being directly activated by a secreted protein. Another scenario is similar to that of the perforin and granzyme B story. These are cosecreted by immune cells producing a portal for entry and an enzymatic execution of a cell (Russell 1983). Although there does not appear to be a homolog of granzyme B in the *E. histolytica genome*, there are a large family of cysteine proteinases that may substitute for this activity. These possibilities are under active investigation and will provide a window on the exploitation of mammalian apoptotic mechanisms by this parasite.

Another question that still needs to be answered is whether or not this activation of apoptosis is being used as an anti-inflammatory in the

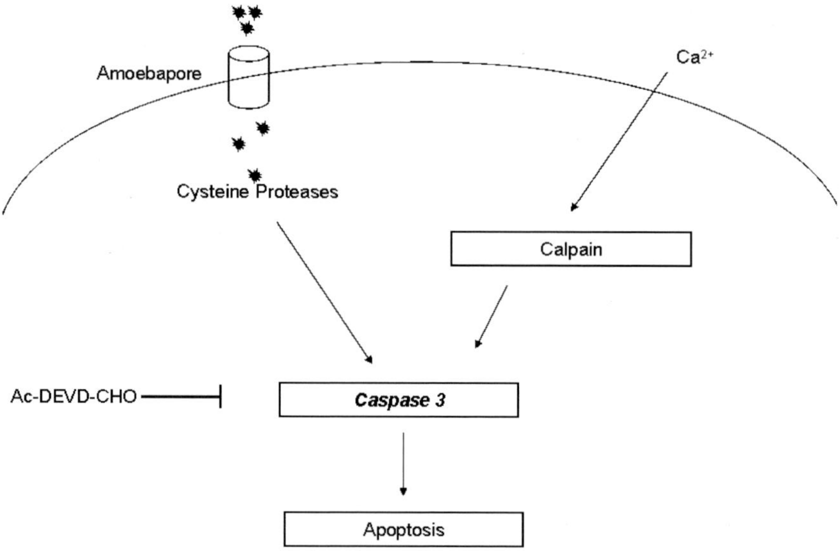

Fig. 3 Proposed models of caspase 3 activation by contact with *Entamoeba histolytica*. Interaction with amoeba causes calcium fluxes inside cells. This can lead to activation of calpain and cleavage of caspase 3. It is also possible that amoebapore and cysteine proteases may serve to directly activate caspase 3

host. Typically human disease presents with very little inflammation, which is odd considering the amount of tissue damage that occurs (Brandt and Tamayo 1970). This has prompted the belief that *E. histolytica* has coevolved with humans to take advantage of this pathway. If cells that are being killed show signs of apoptosis, they will be cleared without the attention of the adapted immune response. This gives the parasite an added advantage to establishing a chronic infection. This may even explain the ingestion of host cells that contain iron and other things that are toxic to the amoeba.

Future work in these areas will help in better understanding and dealing with these infections. We are still at the beginning of answering these questions, but with developing tools such as the Genome Project and RNAi, we should be making large strides that outpace our past.

References

Bailey GB, Day DB, Nokkaew C, Harper CC (1987) Infect Immun 55:1848–1853
Berninghausen O, Leippe M (1997a) Arch Med Res 28 Spec No:158–160
Berninghausen O, Leippe M (1997b) Infect Immun 65:3615–3621

Bracha R, Nuchamowitz Y, Leippe M, Mirelman D (1999) Mol Microbiol 34:463–472
Brandt H, Tamayo RP (1970) Hum Pathol 1:351–385
Bretscher MS (1972) Nat New Biol 236:11–12
Chadee K, Petri WA Jr, Innes DJ, Ravdin JI (1987) J Clin Invest 80:1245–1254
Cheng XJ, Tsukamoto H, Kaneda Y, Tachibana H (1998) Parasitol Res 84:632–639
Fadok VA, Henson PM (2003) Curr Biol 13:R655–657
Flynn R (1973) Parasites of Laboratory Animals. Iowa State University Press, Ames, IA
Guerrant RL, Brush J, Ravdin JI, Sullivan JA, Mandell GL (1981) J Infect Dis 143:83–93
Haque R, Huston CD, Hughes M, Houpt E, Petri WA Jr (2003) N Engl J Med 348:1565–1573
Hudler H, Stemberger H, Scheiner O, Kollaritsch H, Wiedermann G (1983) Tropenmed Parasitol 34:248–252
Huston CD, Boettner DR, Miller-Sims V, Petri WA Jr. (2003) Infect Immun 71:964–972
Huston CD, Haupt ER, Mann BJ, Hahn CS, Petri WA (2000) Cell microbiol 2:617–625
Kobiler D, Mirelman D (1981) J Infect Dis 144:539–546
Leippe M, Andra J, Nickel R, Tannich E, Muller-Eberhard HJ (1994) Mol Microbiol 14:895–904
Leippe M, Ebel S, Schoenberger OL, Horstmann RD, Muller-Eberhard HJ (1991) Proc Natl Acad Sci U S A 88:7659–7663
Leippe M, Sievertsen HJ, Tannich E, Horstmann RD (1995) Parasitology 111 (Pt 5): 569–574
Leroy A, Lauwaet T, De Bruyne G, Cornelissen M, Mareel M (2000) FASEB J 14:1139–1146
Li E, Becker A, Stanley SL Jr. (1989) Infect Immun 57:8–12
Li E, Stenson WF, Kunz-Jenkins C, Swanson PE, Duncan R, Stanley SL Jr (1994) Infect Immun 62:5112–5119
Li E, Yang WG, Zhang T, Stanley SL Jr (1995) Infect Immun 63:4150–4153
Mann BJ, Torian BE, Vedvick TS, Petri WA Jr (1991) Proc Natl Acad Sci U S A 88:3248–3252
McCoy JJ, Mann BJ, Vedvick TS, Pak Y, Heimark DB, Petri WA Jr (1993) J Biol Chem 268:24223–24231
McGinnis KM, Gnegy ME, Park YH, Mukerjee N, Wang KK (1999) Biochem Biophys Res Commun 263:94–99
Orozco E, Guarneros G, Martinez-Palomo A, Sanchez T (1983) J Exp Med 158:1511–1521
Petri WA Jr, Haque R, Mann BJ (2002) Annu Rev Microbiol 56:39–64
Petri WA Jr, Smith RD, Schlesinger PH, Murphy CF, Ravdin JI (1987) J Clin Invest 80:1238–1244
Porn-Ares MI, Samali A, Orrenius S (1998) Cell Death Differ 5:1028–1033
Que X, Reed SL (2000) Clin Microbiol Rev 13:196–206
Ragland BD, Ashley LS, Vaux DL, Petri WA Jr (1994) Exp Parasitol 79:460–467
Ravdin JI, Croft BY, Guerrant RL (1980) J Exp Med 152:377–390
Ravdin JI, Guerrant RL (1981) J Clin Invest 68:1305–1313
Ravdin JI, Moreau F, Sullivan JA, Petri WA Jr, Mandell GL (1988) Infect Immun 56:1505–1512

Ravdin JI, Schlesinger PH, Murphy CF, Gluzman IY, Krogstad DJ (1986) J Protozool 33:478–486
Ravdin JI, Sperelakis N, Guerrant RL (1982) J Infect Dis 146:335–340
Ravichandran KS (2003) Cell 113:817–820
Russell JH (1983) Immunol Rev 72:97–118
Salata RA, Cox JG, Ravdin JI (1987) Parasite Immunol 9:249–261
Schulte W, Scholze H (1989) J Protozool 36:538–543
Seigneuret M, Devaux PF (1984) Proc Natl Acad Sci U S A 81:3751–3755
Seydel KB, Stanley SL Jr (1998) Infect Immun 66:2980–2983
Shamsuzzaman SM, Haque R, Hasin SK, Petri WA Jr, Hashiguchi Y (2000) Southeast Asian J Trop Med Public Health 31:399–404
Tannich E, Ebert F, Horstmann RD (1991) Proc Natl Acad Sci U S A 88:1849–1853
WHO (1997) Epidemiol Bull 18:13–14
Williamson P, Kulick A, Zachowski A, Schlegel RA, Devaux PF (1992) Biochemistry 31:6355–6360
Yan L, Stanley SL Jr (2001) Infect Immun 69:7911–7914
Zhang X, Zhang Z, Alexander D, Bracha R, Mirelman D, Stanley SL Jr (2004) Infect Immun 72:678–683

CTMI (2005) 289:185–218

Interactions Between Malaria and Mosquitoes: The Role of Apoptosis in Parasite Establishment and Vector Response to Infection

H. Hurd (✉) · V. Carter · A. Nacer

Centre for Applied Entomology and Parasitology, School of Life Sciences, Keele University, Staffordshire, ST5 5BG, UK
h.hurd@keele.ac.uk

Abstract Malaria parasites of the genus *Plasmodium* are transmitted from host to host by mosquitoes. Sexual reproduction occurs in the blood meal and the resultant motile zygote, the ookinete, migrates through the midgut epithelium and transforms to an oocyst under the basal lamina. After sporogony, sporozoites are released into the mosquito haemocoel and invade the salivary gland before injection when next the mosquito feeds on a host. Interactions between parasite and vector occur at all stages of the establishment and development of the parasite and some of these result in the death of parasite and host cells by apoptosis. Infection-induced programmed cell death occurs in patches of follicular epithelial cells in the ovary, resulting in follicle resorption and thus a reduction in egg production. We argue that fecundity reduction will result in a change in resource partitioning that may benefit the parasite. Apoptosis also occurs in cells of the midgut epithelium that have been invaded by the parasite and are subsequently expelled into the midgut. In addition, the parasite itself dies by a process of programmed cell death (PCD) in the lumen of the midgut before invasion has occurred. Caspase-like activity has been detected in the cytoplasm of the ookinetes, despite the absence of genes homologous to caspases in the genome of this, or any, unicellular eukaryote. The putative involvement of other cysteine proteases in ancient apoptotic pathways is discussed. Potential signal pathways for induction of apoptosis in the host and parasite are reviewed and we consider the evidence that nitric oxide may play a role in this induction. Finally, we consider the hypothesis that death of some parasites in the midgut will limit infection and thus prevent vector death before the parasites have developed into mature sporozoites.

1 Introduction

In many parts of the world, malaria is re-emerging as a disease of major medical and economic importance. Transmission of the malaria parasite (*Plasmodium* spp.) from host to host is dependent on an anopheline mosquito vector and successful control methods have traditionally been directed against the mosquito. Such measures are now failing, largely due to the rapid spread of insecticide resistance.

The pressing need to find new control strategies has come at a time when we are uniquely equipped to make a fresh assessment of the biology of the malaria parasite and its mosquito vector. A resurgence in research interest in mosquito–malaria interactions has occurred over the past decade, much of it fuelled by the objective of genetically manipulating mosquitoes to be incompetent vectors. These investigations have already been aided by the publication in 2002 of both the *Anopheles* and *Plasmodium* genomes (Gardner et al. 2002; Holt et al. 2002). Here we intend to review the current understanding of the interactions that occur between the malaria parasite and its mosquito vectors and our very recent recognition that apoptosis is induced both in the parasite and in mosquito cells as a result of infection.

2
Plasmodium Life Cycle

2.1
The Asexual Cycle in the Vertebrate

The disease malaria is caused by apicomplexan parasites of the genus *Plasmodium*. Different species infect a variety of primates, rodents, birds and reptiles. Of the four species that infect humans, *Plasmodium vivax* and *Plasmodium falciparum* predominate. Because of the difficulties of performing laboratory investigation on mosquitoes infected with human malarias, much use has been made of rodent models such as *Plasmodium berghei* and *Plasmodium yoelii* and the avian malaria *Plasmodium gallinaceum*. *Plasmodium* parasites are transmitted to their vertebrate host via the bite of an infected mosquito. The infective sporozoite stage of the parasite migrates to the liver to invade hepatocytes and produce merozoites. Merozoites are released into the bloodstream when the liver cells burst and red blood cells (RBCs) are invaded. Rounds of mitotic division produce new merozoites, each of which is capable of infecting a new RBC when the erythrocyte ruptures. Some parasites differentiate to form male or female gametes instead of merozoites and arrest their cell cycle until ingested by a mosquito, marking the beginning of the sporogonic cycle.

2.2
The Sporogonic Cycle in the Mosquito

On ingestion with a blood meal, intracellular gametocytes escape from RBCs, fertilisation takes place within the midgut lumen and the resultant zygote develops into a motile ookinete. The ookinete traverses the chitinous peritrophic matrix surrounding the blood bolus to invade the midgut wall, whereupon it transforms into a sessile oocyst, resting between the midgut epithelium and basal lamina (Paul et al. 2002). Here the oocyst undergoes rounds of nuclear division, producing and subsequently releasing thousands of sporozoites. Migrating through the body cavity, the sporozoites' final destination is the mosquito salivary gland, where they are injected into a vertebrate host as the mosquito takes a subsequent blood meal (Simonetti 1996; Paul et al. 2002). Interactions between the various stages of parasite development and the vector are detailed below. Figure 1 outlines the stages of *Plasmodium* development in both the vertebrate and invertebrate hosts.

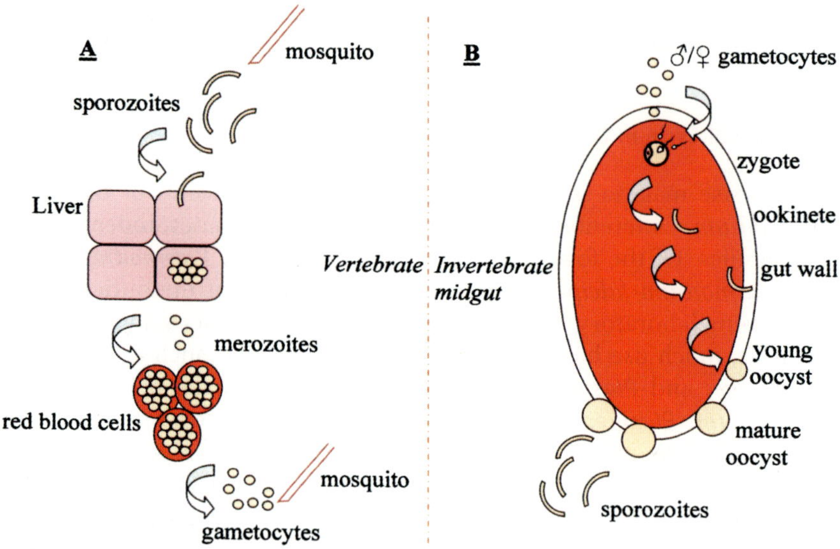

Fig. 1 Schematic life cycle of *Plasmodium* spp. **A** A mosquito takes a blood meal and injects infective sporozoites into the peripheral circulation. Within minutes, sporozoites invade hepatocytes and multiply to produce merozoite forms of the parasite. Hepatocytes rupture and release merozoites which invade red blood cells (RBCs) and undergo further multiplication. RBCs rupture and invasion of more RBCs occurs. Some parasites differentiate into gametocytes and are ingested by the mosquito during a blood meal. **B** In the mosquito digestive tract, gametocytes develop into male and female gametes and fuse to form a zygote. The zygote develops into a motile ookinete, which traverses the midgut epithelium, develops into an oocyst and replicates to produce sporozoites which ultimately accumulate in the salivary gland of the mosquito

2.2.1 Gamete Formation and Fertilisation

During gametogenesis, male gametes (microgametes) undergo three rounds of genome replication, usually within 15 min (Simonetti 1996). The content of the resulting nucleus is divided, resulting in eight haploid male gametes, each with an axoneme, a genome, and a kinetosome. Microgametes emerge from the RBC by a process known as exflagellation and move within the blood meal to fertilise a female gamete (macrogamete) (Sinden and Croll 1975). Several factors have been implicated in the induction of gametogenesis including a drop in temperature, a rise in pH from 7.4 to 7.8–8.0, cyclic AMP and a molecule termed mosquito exflagellation factor or gametocyte activating factor, which has been identified as xanthurenic acid and shown to trigger a rapid rise in cytosolic calcium in gametocytes (Billker et al. 1997, 2004; Arai et al. 2001).

2.2.2 The Zygote and Ookinete

The zygote develops within an hour of fertilisation (Simonetti 1996) and the parasite begins to take on apicomplexan characteristics; the apical complex is formed, including the polar rings, rhoptries and micronemes (Sinden 1978; Han et al. 2000). Within 10–25 h, the zygote gives rise to a motile ookinete, an invasive stage of the life cycle.

The ookinete traverses the peritrophic matrix (PM), a thick, acellular chitin-containing layer constructed de novo by the mosquito after a blood meal, separating the food bolus from the midgut epithelium (Billingsley and Lehane 1996; Langer et al. 2001; Abraham and Jacobs-Lorena 2004). The effect of the PM as a barrier to ookinete invasion has not been clearly resolved. It seems that in *P. berghei* infections, the presence or absence of the PM does not affect oocyst numbers (Billingsley and Rudin 1992). In contrast, it was observed that the PM reduced the number of oocysts on the midgut in the *P. gallinaceum-Aedes aegypti* system (Billingsley and Rudin 1992). The parasite's ability to exit the PM is crucial. Susceptibility to mosquito trypsins makes the ookinete vulnerable to digestion (Simonetti 1996; Billingsley and Sinden 1997). Paradoxically, the parasite seems to require mosquito trypsins to trigger the activation of pro-chitinase to chitinase that enables it to digest chitin in the PM and pass through it (Shahabuddin et al. 1993; Ghosh et al. 2000).

The route of midgut wall penetration has been the subject of much debate (Sinden and Billingsley 2001). Studies using the same parasite–mosquito combination have reported contradictory results with respect to an intercellular or an intracellular route. Torii et al. (1992) showed that parasites penetrated both through and between midgut cells, suggesting that initial penetration occurred through cells and that the parasite exited between cells at the basal end of the midgut. Recent work suggests a conserved invasion route, parasites penetrating midgut epithelial cells at the luminal surface, moving laterally through adjacent cells and finally reaching the basal lamina (Han et al. 2000; Zieler and Dvorak 2000; Baton and Ranford-Cartwright, in press; Vlachou et al. 2004).

Further to this debate is the contentious issue of ookinete preference for specific midgut cell types. Some authors reported that ookinetes invaded a specific group of cells, termed “Ross cells”. These cells were reported to contain high amounts of vesicular ATPases and had different staining properties and distinct morphological features (e.g. less microvilli) when examined under transmission electron microscopy in comparison to cells that were not invaded by ookinetes (Shahabuddin and Pimenta 1998). In contrast, other authors failed to detect the high levels of vesicular ATPases but instead noted that ookinetes caused extensive damage to invaded cells and possibly cell death through apoptosis. This raised the possibility that the morphological features observed in “Ross cells” were actually the result of damage to the epithelial cells (Han et al. 2000; Zieler and Dvorak 2000; Han and Barillas-Mury 2002).

2.2.2.1 Ookinete Surface Proteins

The major ookinete surface proteins interacting with the mosquito midgut fall into two categories, the P25 subfamily (including Pfs25, Pgs25, Pys25 and Pbs25) and the P28 subfamily (including Pfs28, Pgs28, Pys21 and Pbs21) (Tsuboi et al. 1998). In *P. berghei*, Pbs21 and Pbs25 may confer protection from immune factors, whilst being essential for midgut invasion and ookinete to oocyst differentiation (Sidén-Kiamos et al. 2000; Tomas et al. 2001; Arrighi 2002). Pbs21 is first expressed 3 h after the induction of gametogenesis and persists until 2–3 days after an infective blood meal (Simonetti et al. 1993; Blanco et al. 1999). Although Sidén-Kiamos et al. (2000) reported that neither single- nor double-knockout parasites for Pbs21/Pbs25 affected parasite motility or invasion, the authors recorded a 90%–95% reduction in oocyst formation in vitro. Pbs21 appears to be involved in binding to components of the basal lamina (Arrighi and Hurd 2002). Further proteins expressed in the ookinete stage which are thought to be important for motility, invasion and transformation include circumsporozoite- and thrombospondin-related adhesive protein (CTRP), von Willebrand factor type A domain-related protein (WARP) and secreted apical ookinete protein (SOAP). The first of these, CTRP, is expressed from 10 h post-fertilisation, predominantly at the apical end of the ookinete, and remains for at least 24 h. CTRP has also been localised to the micronemes, secretory organelles of the apicomplexans (Dessens et al. 1999; Yuda et al. 2001; Limviroj et al. 2002). Targeted disruption and knockout studies have shown that parasites lacking CTRP fail to develop into oocysts, consequently blocking mosquito transmission of these parasites (Dessens et al. 1999; Templeton et al. 2000; Yuda et al 1999). The soluble micronemal protein expressed by *P. berghei* ookinetes, WARP, is thought to have a crucial role in motility and binding to the midgut (Yuda et al. 2001). It contains an adhesive domain (the A domain) that is known to be implicated in cell–cell and cell–matrix interactions, where PbWARP may interact with PbCTRP during midgut invasion (Colombatti et al. 1993; Cruz et al. 1995; Lee et al. 1995; Yuda et al. 2001). SOAP is thought to play an important part in ookinete invasion of the epithelium and was shown to strongly interact with mosquito laminin. Gene disruption studies showed that parasites lacking SOAP were impaired in their ability to cross the midgut wall and form oocysts (Dessens et al. 2003).

2.2.3 The Oocyst and Sporozoite

Once the parasite arrives at the basal lamina it transforms into a spherical sessile oocyst characterised by resorption of the apical complex and secretory organelles. The oocyst grows from 2–3 μm to 40 μm in 10–14 days after blood feeding as it undergoes rounds of nuclear division for sporozoite formation (Janse and Waters 2002; Thathy et al. 2002). These polarised sporozoites possess the characteristic features of invasive stages of the apicomplexa. They contain three membranes: two inner membranes that form an inner membrane complex that seems to be constructed de novo and an outer pellicle derived from the oocyst plasma membrane (Thathy et al. 2002; Kappe et al. 2004). Underlying the inner membranes is a network of longitudinal microtubules that are involved in motility and may also maintain sporozoite rigidity and shape. Sporulation begins as the oocyst plasma membrane retracts from the capsule and invaginates, creating several sporoblasts from which sporozoites bud in successive waves, a process that requires the circumsprozoite protein (CSP) (Simonetti et al. 1993; Thathy et al. 2002). CSP is a major sporozoite surface protein that has been implicated in gliding motility (Kappe et al. 2004), binding and invasion of salivary glands and liver cells (Potocnjak et al. 1980; Yoshida et al. 1981; Sidjanski et al. 1997). It is expressed within the oocyst and persists until 30 h post-invasion of hepatocytes (Aikawa et al. 1981). The thrombospondin-related adhesive protein, TRAP, is also implicated in sporozoite motility (Kappe et al. 2004).

When sporozoite formation is complete the oocysts burst, releasing thousands of sporozoites into the haemocoel. Sporozoites appear to be passively carried by the mosquito's circulatory system and eventually bind to and invade the distal lobes of the salivary glands, a process mediated by CSP (Sidjanski et al. 1997; Myung et al. 2004). They then migrate to the salivary ducts before injection into a new host when the mosquito next blood feeds.

3 Interactions with the Vector

3.1 Triggering the Immune Response

Infection with *Plasmodium* triggers a variety of innate immune responses in the mosquito that contribute to limiting infection. Parasite killing is initiated in the midgut lumen by nitric oxide (Luckhart et al. 1998, 2003). Invasion initiates the upregulation of many innate immunity genes and a further wave of upregulation occurs when sporozoites are released from the oocysts into the haemolymph (Dimopoulos et al. 1998). Both systemic and local responses are initiated with the fat body tissue, haemocytes, midgut epithelial cells and salivary glands being involved. The resultant response is multifaceted and involves pattern recognition receptors, components involved in opsonisation, serine protease cascade components, antimicrobial peptides and, in a refractory mosquito strain, a melanisation cascade is initiated (see reviews by Barillas-Mury et al. 2000; Dimopoulos et al. 2001). Our current understanding of the extent to which these responses result in parasite killing is meagre. Those innate immune responses initiated in susceptible mosquitoes clearly are not fully effective, but they may serve to limit infection intensity. However, gene disruption technology has demonstrated that at least one innate immune response, namely defensin upregulation, is ineffective (Blandin et al. 2002).

3.2 Infection and Nutrients

During its prolonged stay in the mosquito, the malaria parasite must compete with its vector for nutrients. The parasite's requirements for certain free amino acids have been determined by different authors who compared the composition of mosquito homogenate or haemolymph after feeding on glucose solutions, uninfected mice, or infected mice with the *P. berghei-Anopheles stephensi* model (Gad et al. 1979; Mack et al. 1979). After an uninfected blood meal, free amino acid concentrations rose between 60% and 70%. In contrast, when fed on blood infected with*P. berghei*, amino acid concentrations only rose between 15% and 25% (Mack et al. 1979). Mack et al. (1979) analysed the amino acid concentrations of the haemolymph of mosquitoes 4 and 11 days after an infected or uninfected blood meal and recorded the subtle variations in this context. They reported that infected mosquitoes had greater increases in arginine, and greater decreases in valine and histidine concentrations, with the total disappearance of methionine. The limiting effects of the depletion of one or more specific amino acids are unknown but could contribute to reduced vector fitness.

4 Fitness Costs Associated with Infection

During any parasite relationship the host will incur some degree of cost associated with infection. However, conflicting selection pressures exerted on both partners in these symbioses are likely to resolve such that the parasite maximises its fitness by extracting resources from its host without jeopardising its transmission success. Parasites such as malaria, which require a blood-sucking insect to complete their life cycle, are dependent on the ability of the vector insect to seek out and feed on a new host. Furthermore, the time span required for the development of infective sporozoites is surprisingly long, relative to the likely lifespan of a mosquito in the field (Koella 1999). We would expect to find that any costs imposed by *Plasmodium* have a limited effect on mosquito longevity and blood feeding behaviour. Costs that are imposed by infection will drive the evolution of countermeasures in the host such that avoidance of infection or defence against infection will arise, unless the costs of these defences outweigh the gain achieved by being infection free. When examining the mosquito–malaria relationship it is therefore essential to consider responses made by both organisms. It has recently become apparent that programmed cell death has a role to play in the resolution of these conflicts.

4.1 Reproductive Fitness

The costs imposed by malaria infection on mosquito longevity are not clear-cut (Ferguson and Read 2002) and may be moderated by the trade-off that is predicted to exist between longevity and fecundity

(Sheldon and Verhulst 1996). However, in common with other insect infections, malaria has been conclusively shown to reduce mosquito reproductive fitness (Hurd 1990). This has been demonstrated to occur in avian, rodent and human malaria species (reviewed in Hurd 2003). Approximately a quarter of the developing eggs within an egg batch from an infected female fail to develop and a proportion of those eggs that are oviposited fail to hatch, despite the presence of sperm within the female spermatheca (Ahmed et al. 1999). The resulting fecundity reduction has been shown to occur in *P. yoelii nigeriensis* infections of both *An. stephensi* and *An. gambiae* (Hogg and Hurd 1995; Ahmed et al. 1999) and in *P. falciparum* infections of *An. gambiae* (Hogg and Hurd 1997).

Two aspects of the gonotrophic cycle are affected by *Plasmodium* infection. In the ovary, growth of the oocytes is retarded and many of the follicles that begin development are resorbed before ovulation (Carwardine and Hurd 1997). Oocyte growth occurs via the incorporation of the yolk proteins or vitellogenins (Vgs) that are synthesised in the fat body. A parasite-induced reduction in Vg mRNA in the fat body of *An. gambiae* occurs after infection with *P. y. nigeriensis* and again after a second blood meal, when oocysts are present. Vg circulating in the haemolymph is initially reduced but, during the second cycle, this titre increases because of impaired uptake by the ovary (Ahmed et al. 2001). These effects of infection result in fewer resources being devoted to reproduction. Thus, by altering the partitioning of nutrients between somatic maintenance and reproduction, sufficient nutrients may be made available for the repair of damage caused by invasion, the mounting of an innate immune defense against the parasite and the nutrient requirements of the developing oocysts (Hurd 2001, 2003). Much of this alteration of resource management appears to be initiated by the programmed cell death of patches of follicular epithelial cells within the terminal follicles of the ovary (Hopwood et al. 2001).

The Vg that is synthesised in the fat body gains access to the oocyte via the follicular epithelial cells that surround it. After a blood meal the follicular epithelial cells of the terminal follicle undergo a conformational change and spaces develop between them to facilitate passage of the yolk proteins. Each follicle contains one oocyte and seven nurse cells (Sokolova 1994), the latter supplying RNA and other macromolecules to the oocyte's cytoplasm. By 48–50 h post blood meal, oocyte maturation is complete, the follicular epithelial cells have secreted the chorion and the nurse cells have degenerated.

4.1.1
Apoptosis and Follicular Resorption

Many insects have been observed to adjust the size of their egg batch in accordance with physiological or environmental conditions. This adjustment is achieved by the resorption of some or all of the developing ovarian follicles. The mechanism underlying follicle, or capsule, resorption has been studied in *Drosophila*. Starvation, or failure to obtain the sex peptide acquired during mating, will result in premature nurse cell apoptosis, and thus capsule resorption, via changes in the titre of two hormones. Juvenile hormone protects the early vitellogenic oocytes whereas apoptosis is induced by 20-hydroxyecdysone. As a result of their studies, Bownes and her colleagues (Soller et al. 1999) concluded that control of oocyte development beyond a critical stage operates via a balance of these two hormones.

Apoptotic-like death of nurse cells is a normal process during *Drosophila* oogenesis. However, this usually occurs near the end of oogenesis and is accompanied by the "dumping" of the cytoplasmic contents of the nurse cells into the oocyte (Cavaliere et al. 1998; Soller et al. 1999; reviewed by Buszczak and Cooley 2000). The *Drosophila* caspase homologue, *dcp-1,* has been shown to be essential for the completion of normal oogenesis (Foley and Cooley 1998). Expression of the activators of apoptosis *reaper (rpr)* and *head involution defective (hid)* leads to a stage-specific degeneration of the nurse cells by apoptosis and expression can occur at any stage of oogenesis (Chao and Nagoshi 1999). However, neither these, nor *grim,* are required for nurse cell apoptosis during successful oogenesis. Transcripts for two further *Drosophila* caspases, *dredd* and *damm,* have been detected at high levels in nurse cells or egg chambers (Harvey et al. 2001). The ecdysone-inducible caspase DRONC appears to be upregulated just before nurse cell dumping and genes for DECAY and the Bcl-2 homologue, DEBCL, are expressed throughout oogenesis (Buszczak and Cooley 2000).

Mosquitoes have also been reported to regulate egg batch size according to the availability of nutrients; follicle resorption occurring when blood meal size is small (Clements and Boocock 1984). Under experimental conditions in which the haemoglobin content of the blood meal of mosquitoes fed on an uninfected or a malaria-infected mouse was similar, significant follicle resorption occurred in the infected mosquitoes alone (Carwardine and Hurd 1997). Follicle resorption was first detected 12 h after feeding, at a time when *P. y. nigeriensis* ookinetes are still developing within the midgut lumen of *An. stephensi,* or just beginning to invade the midgut epithelial cells (Carwardine and Hurd 1997).

Plasmodium-induced follicle resorption has been linked to the occurrence of apoptosis in the follicular epithelial cells (Hopwood et al. 2001). Patches of epithelial cells containing nuclei with condensed chromatin were detected with acridine orange staining and by the examination of ultrathin sections of ovaries (Fig. 2). In addition, an in situ terminal deoxynucleotidyl transferase-mediated dUTP-biotin nick end labelling (TUNEL) assay was used to detect DNA fragmentation in the nuclei of both follicular epithelium and nurse cells of ovaries from females 16 and 18 h after infection (Fig. 2).

In ovaries from uninfected mosquitoes the number of resorbing follicles or those containing apoptotic epithelial cells never exceeded 4% at any time point examined (16–36 h post-feeding), whereas it reached 14%–16% in infected females. The maximum proportion of follicles containing apoptotic cells occurred between 20 and 22 h post-infection but the appearance of resorbing follicles lagged behind and peaked at 24 h. This suggested that cell death initiated follicle resorption. Confirmation that apoptotic death of follicle cells was the cause of follicle resorption, rather than the result of it, was obtained by injecting mosquitoes with a caspase inhibitor, benzyloxycarbonyl-Val-Ala-Asp (OMe) fluoromethylketone (Z-VAD.fmk), immediately after they had fed on a gametocytaemic mouse. When examined 24 h later, significantly fewer resorbing follicles were detected in those mosquitoes that had been injected with a caspase inhibitor compared with non-injected or control-injected groups (Hopwood et al. 2001).

At present we do not know which apoptotic or anti-apoptotic genes are involved in follicle resorption in *Anopheles spp.* However, the *Drosophila* caspases involved in follicle resorption have orthologues in the

Fig. 2 Follicles from ovaries of malaria-infected *Anopheles stephensi* developing normally (*left*) or showing evidence of apoptosis (*right*). **A** whole mounts of individual follicles 24 h after blood feeding, viewed under UV light after staining with acridine orange. The left follicle (**Ai**) is exhibiting autofluorescence. Condensed, apoptotic nuclei are visible in the right follicle (**Aii**), stained orange (*a*). Scale bars=20 μm. **B** sections of follicles from ovaries 16 h after blood feeding, stained with methyl green and treated by TUNEL. DNA fragmentation (stained brown) can be observed in follicular epithelial (*f*) and nurse (*n*) cells on the right (**Bii**). Scale bars=10 μm. **C** Electron micrographs of a normal follicle 18 h after blood feeding (**Ci**) and a resorbing

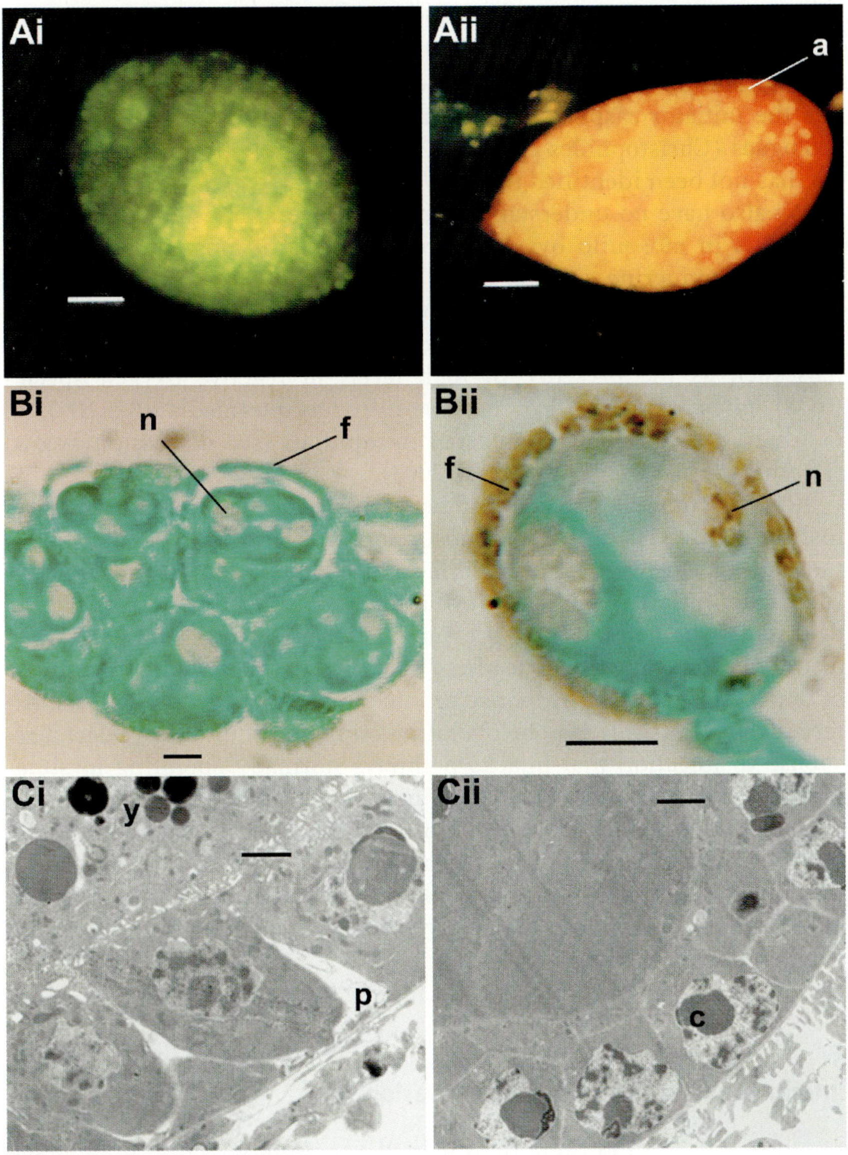

follicle 22 h after blood feeding (**Cii**). Patency (*p*) has developed in follicular epithelial cells of the normal follicle and yolk spheres (*y*) have been deposited whereas, in the resorbing follicle, no yolk is visible, epithelial cells are not patent and their nuclei contain condensed chromatin (*c*). Scale bars=2 μm. (Reproduced with permission from Hopwood et al. 2001)

An. gambiae genome. The *Drosophila* short-prodomain caspase, DCP-1, has been grouped with the mosquito caspase S8, the short-prodomain caspase DAMM was found to be most similar to *Anopheles* S9 and S10 and the initiator caspases DREDD and DRONC also have mosquito orthologues (Christophides et al. 2002). However, the pro-apoptotic gene, *rpr*, has not been identified in *Anopheles*.

We also have no understanding of the triggers that induce or inhibit apoptosis in mosquito ovaries. Several aspects of mosquito oogenesis, including endocrine control, differ from *Drosophila* and, thus, the role of hormone titres in the induction of apoptosis in the ovaries of infected mosquitoes is open to conjecture. This pathway could be initiated by signals such as the presence of high levels of reactive oxygen species (ROS), nitric oxide (NO) or reactive nitrogen intermediates (RNIs), all of which are abundant in infected mosquitoes (see discussion below) (Luckhart et al. 1998; Kumar et al. 2003).

Although there is much still to be learnt about the induction of apoptosis in the ovaries of infected mosquitoes it seems probable that this is the pathway by which the malaria-induced reduction in reproductive success is initiated. Taking an evolutionary perspective, we cannot yet determine whether selection pressures have operated on the parasite to directly induce follicle cell death, or to indirectly manipulate the process of oogenesis via a host response to infection that triggers apoptosis. Alternatively, pressures may be operating on the mosquito that select a response to infection that hijacks pathways normally operating in situations of nutrient depletion. If hypotheses concerning the links between fecundity reduction and longevity are upheld, then we can assume that both the malaria parasite and the mosquito will gain by some degree of fecundity reduction as the repartitioning of nutrient reserves may enable the mosquito to live longer, lay another batch of eggs and transmit sporozoites to a new malaria host.

5 Interactions Resulting from Gut Invasion

At approximately the same time that *Plasmodium* infection is inducing cell death in the follicular epithelium, ookinetes are penetrating and traversing the midgut epithelium. Here, too, evidence has been presented to suggest that infection induces host cell death by apoptosis. Using their in vitro model to study *P. gallinaceum* invasion of the midgut of the yellow fever mosquito, *Ae. aegypti*, Zieler and Dvorak (2000) reported morphological changes occurring in midgut epithelial cells as a result of cell invasion. Nuclei were observed to swell and to move to an apical position, cell surface blebbing occurred and a decrease of refractive index took place. Cell-impermeant dyes were used to demonstrate loss of plasma membrane integrity. By preloading the epithelial cells with PhiPhiLux-G_1D_2, a cell-permeant reagent containing a caspase-3-consensus cleavage site, they located several ookinete-containing midgut cells that also contained an activated caspase-3-specific protease, concluding that midgut epithelium invasion was followed by apoptotic cell death.

Using another malaria–mosquito association, *P. berghei-An. stephensi*, Han et al. (2000) also showed that ookinete-invaded cells may die by programmed cell death. The nuclei of these cells showed signs of fragmentation and DNA condensation. They were not able to demonstrate nucleosome-like ladders when DNA profiles were examined, possibly due to the low proportion of the total cells that were undergoing apoptosis. *An. stephensi* midgut cells exhibiting signs of DNA fragmentation were also observed to be protruding into the gut lumen. It was proposed that these damaged cells were budded off into the midgut lumen and that within 48 h the gut epithelium had repaired the damage as actin fibres became involved in a "purse-string" mechanism to extrude the damaged cells (Han et al. 2000). These observations gave rise to the "time bomb theory" which proposed that ookinetes must rapidly move out of invaded cells before the cells die. Similar observations have now been made in *P. falciparum-An. stephensi* infections where ookinetes that failed to escape in time are seen within cells that are extruded into the gut lumen (Baton and Ranford-Cartwright, in press). Apoptotic cell death of invaded epithelial cells in *An. stephensi* and *An. gambiae* has also been demonstrated, and distinguished from necrotic cells, by using the nucleic acid stain YO-PRO1, which passes through the plasma membrane of apoptotic cells, and propidium iodide, which only diffuses through the damaged membranes of necrotic cells. Clusters of apoptotic cells were observed with differing stages of apoptosis, thus giving an indication of the invasion pathway (Vlachou et al. 2004).

In addition to morphological markers, caspase activity has been linked to ookinete invasion. Ancaspase-7 is one of three apoptosis-related molecules identified in cDNA libraries enriched for sequences expressed immediately after ookinete invasion of the mosquito midgut (Abraham et al. 2004). Ancaspase-7 shares 40% identity and 60% similarity with the *Drosophila* caspase, DECAY, and 54% similarity with mammalian caspase-7. Ancaspase-7 demonstrated a threefold increase in transcription from a basal level on day 1 post-infection compared with transcription in mosquitoes bloodfed but not infected. A ~33-kDa protein was identified with an anti-ancaspase-7 antibody. This antibody also recognised a putative proteolytic product of ~19 kDa that was only present in gut cells after invasion with very high numbers of parasites, thus suggesting that the caspase was activated on cell invasion (Abraham et al. 2004). These results suggest that very heavy infections are required to induce apoptosis but the result may be a function of the lack of sensitivity of the assay. The occurrence of gut cell apoptosis in natural *P. falciparum* infections, where parasite burden is very low, has not yet been demonstrated. In contrast, the presence of mid- or late-stage oocysts was not shown to induce apoptosis-promoting gene expression but was linked with expression of putative anti-apoptotic genes (Srinivasan et al. 2004).

At this early stage in our understanding of the process of parasite-induced midgut cell death, it is difficult to ascribe a role for apoptosis in the establishment of infection, other than as a mechanism for eliminating damaged cells in a manner that prevents the leakage of haemolymph from the haemocoel to the midgut lumen. Indeed Vlachou et al. (2004) suggest that the hood formed by lamellipodia from adjacent midgut cells may be stimulated by the parasite. Much work is still to be done before we can answer such fundamental questions as, do cell death signals come from the vector or the parasite?

In answer to this question, it has been proposed that the *P. berghei* cell surface protein, Pbs21, could act as an initiator of the apoptotic cascade as it is shed from ookinetes in transit through midgut epithelial cells (Han and Barillas-Mury 2002). Alternatively, apoptosis may be triggered by a localised increase in NO concentration. Han and colleagues (2000) reported that, in the gut of infected *An. stephensi*, cells protruding into the lumen stain heavily for nitric oxide synthase (*As*NOS). This localised staining suggests that transcriptional upregulation of *AsNOS* is initiated by invasion. However, the observation that programmed cell death is confined to cells that have been invaded is surprising, given that NO diffuses rapidly and is highly toxic. The occurrence of larger clusters of apoptotic cells around the site of invasion would be more predictable if NO operates as a cell death signal. Surprisingly, ookinetes appear to escape intact from damaged cells and are not affected by the killing mechanism operating in the host cells (Han et al. 2000). This is far from the case reported for the fate of ookinetes in the midgut lumen.

5.1 Parasite Death in the Midgut Lumen

The midgut lumen is a hostile environment in which few developing parasites succeed in becoming motile ookinetes and invading the midgut. Losses occur during gametogenesis, fertilisation may fail to occur and developing ookinetes are vulnerable to the action of digestive enzymes (Billingsley and Sinden 1997; Ghosh et al. 2000). Thus few of the gametocytes that were originally imbibed survive to form oocysts on the midgut wall (Vaughan et al. 1994; Alavi et al. 2003). Until recently, attention has focused on killing mechanisms that operate in the midgut epithelial cells, particularly of mosquitoes selected to be refractory to malaria (Collins et al. 1986; Vernick et al. 1995). It has now been established that approximately 50% of *P. berghei* ookinetes that mature in the midgut lumen of a susceptible strain of *An. stephensi* are dying before they invade the epithelium (Al-Olayan et al. 2002b). Ookinetes of *P. berghei* can be grown in vitro with a well-established technique (Sinden 1997; Al-Olayan et al. 2002a). A slightly smaller proportion of parasites developing in culture also showed signs of death (Al-Olayan et al. 2002b). Both ookinetes and zygotes of this rodent malaria exhibit several morphological characteristics indicative of programmed cell death by apoptosis. Nuclei were shown to contain condensed chromatin, as detected by acridine orange staining and by observation of ultrathin sections. TUNEL was used to detect DNA fragmentation in apoptotic nuclei and translocation of phosphatidylserine onto the external leaflet of the ookinete outer membrane was observed after staining with annexin V-CY3 (see Fig. 3). Membrane blebbing and the formation of apoptotic bodies appeared to indicate later stages of the apoptotic process (Al-Olayan et al. 2002b).

5.1.1 Caspase-Like Activity in the Ookinete

A large proportion of ookinetes incubated with a carboxyfluorescine derivative of benzyloxycarbonyl valyalanylaspartic acid fluoromethyl ketone (fam-VAD-fmk) exhibit fluorescent activity in the cytoplasm that is indicative of caspase-like activity (Al-Olayan et al. 2002b). This particular compound is a broad-spectrum caspase inhibitor. The caspase-3 subfamily inhibitor, Z-DEVD.fmk, and another general caspase-inhibitor, boc-ASP.fmk, both inhibited the process of apoptosis, whereas the caspase-1 family inhibitor, Z-YVAD.cmk, did not. Interestingly, the cas-

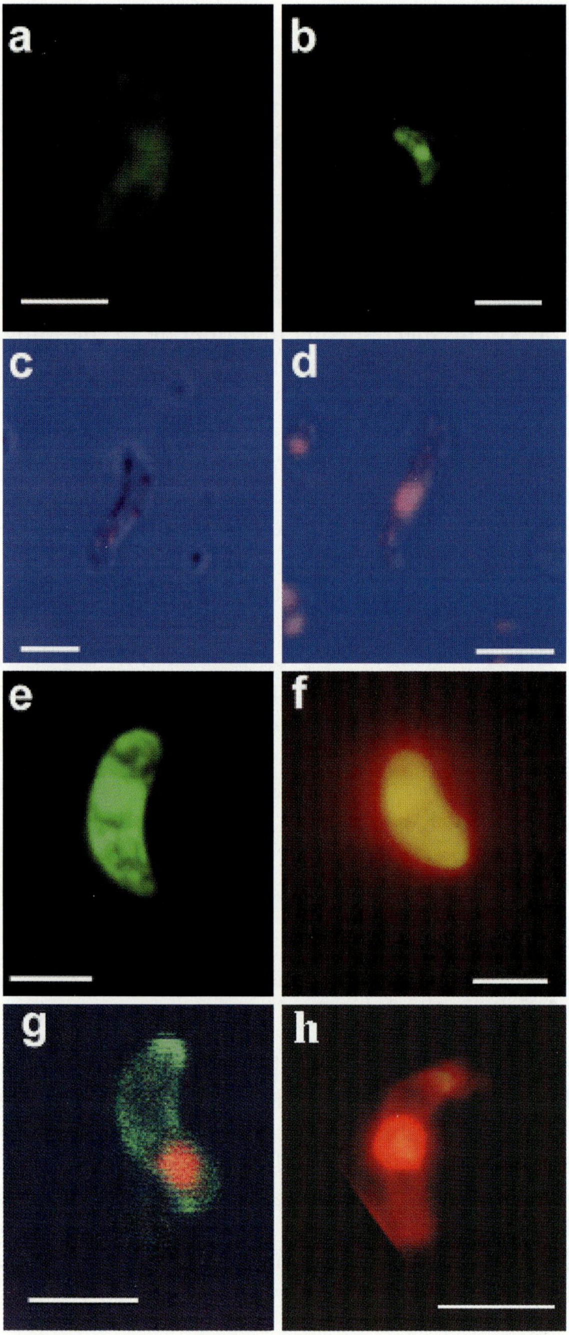
a
b
c
d
e
f
g
h

pase-1 family is considered to act more frequently as an inflammatory caspase than to play a part in apoptotic pathways in the mammalian system. The proportion of in vitro-grown *P. berghei* ookinetes undergoing apoptosis at 18, 20 and 24 h post-infection was reduced by 89%, 51% and 71%, respectively, in the presence of a further inhibitor, Z-VAD.fmk. Z-VAD.fmk did not have a significant effect on the initiation of apoptosis in zygotes at 18 or 20 h post-culture but it was effective on zygotes that were still present at 24 h. Introduction of caspase inhibitors to an infective blood meal resulted in a similar inhibition of apoptosis in zygotes and ookinetes present in the midgut lumen. Once apoptosis was inhibited in vivo, additional ookinetes survived and were able to invade the midgut and transform into oocysts. A supplement of 0.1 mM Z-VAD.fmk to a blood meal doubled the number of oocysts developing on the midgut wall (Al-Olayan et al. 2002b).

Despite the experimental evidence that suggests a functional role for caspase-like proteins in the apoptotic cell death of *Plasmodium*, no caspase homologues have been identified in the published *P. falciparum* genome (Gardner et al. 2002). The study by Al-Olayan et al. (2002b) observed caspase-like activity associated with two protein subunits of ~45 kDa and ~28 kDa in the lysate of ookinetes that developed in vitro, but the identity of these proteins is, as yet, unknown. Interpretation of findings regarding the identification of substrates of caspase inhibitors may need to be re-examined in the light of a recent report that they are not as specific as previously claimed. They will also inactivate cathepsins to varying degrees (Rozman-Pungercar et al. 2003).

Fig. 3a–h *P. berghei* ookinetes exhibiting morphological and biochemical features typical of metazoan apoptosis. **a, b** Ookinetes stained with acridine orange: **a** ookinete with intact nuclear material, **b** apoptotic ookinete with condensed chromatin (*green*). **c, d** Ookinetes treated with an apoTag TUNEL kit: **c** non-apoptotic parasite, **d** ookinete with a nucleus containing fragmented DNA (*pink*). **e, f** Ookinetes treated with apoptosis detection kit Annexin V-CY3: **e** viable ookinete also stained with 6-carboxyfluorescein diacetate (*green*), **f** ookinete with annexin-Cy3.18 bound to phosphatidylserine on the outer plasma membrane (*red*). **g, h** Ookinetes treated with a CaspaTag™ kit: **g** ookinete in early stages of apoptosis with a nucleus visualised with Hoechst stain (*red*) and caspase-positive cytoplasm (*green*), **h** propidium iodide-stained nucleus in a later-stage apoptotic ookinete with a compromised membrane (*red*). Scale bars: **a, b**=10 μm, **c–h**=5 μm. (Reproduced with permission from Al-Olayan et al. 2002a)

It is likely that an ancient apoptotic pathway operates in *Plasmodium* and other unicellular eukaryotes (Aravind et al. 2001). Inhibitors of cysteine proteases have been reported to react with other unicellular eukaryotes undergoing programmed cell death, including the kinetoplastids *Trypanosoma cruzi, Leishmania major, Leishmania donovani* (Piacenza et al. 2001), the apicomplexan *Cryptosporidium parvum* (Ojcius et al. 1999) and two parasitic protozoans that lack mitochondria, *Trichomonas vaginalis* and *Giardia intestinalis* (Chose et al. 2003). No caspases have been identified in the genomes of any of these parasites.

Other proteases of the cysteine protease Clan CD share a primary sequence including the catalytic diad, histidine and cysteine and also their secondary structure. Members of this family are found in virtually all non-metazoan eukaryote lineages (Szallies et al. 2002). In addition to calpain, metacaspases and paracaspase have been identified in the *P. falciparum* genome (Uren et al. 2000; Mottram et al. 2003). The metacaspase gene (gene ID PF13_0298 Ref to plasmoDB) has been cloned from a *P. falciparum* cDNA library and encodes a putative catalytic diad composed of His404 and Cys460 (Deponte and Becker 2004). Two metacaspase-like genes, *mc1* and *mc2*, that code for proteins of 593 and 1454 amino acids, respectively, have been identified in the *P. berghei* genome.

The yeast caspase-1 is a 52-kDa metacaspase (Yor197w) that is activated by apoptosis inducers, cleaves in a typical caspase manner and has proteolytic activity (Madeo et al. 2002). It is encoded by the gene *YCA1*. A range of apoptosis markers have been attributed to the yeast, *Saccharomyces cervisiae* (Frohlich and Madeo 2000). Caspase-like activity was reduced in organisms with a mutation in the conserved cysteine 297 of the *YCA1* gene and the normal cleavage of the C-terminal fragment of the molecule was blocked, suggesting that this cleavage enables proteolytic activity (Madeo et al. 2002). Madeo and colleagues suggest that YCA1 might be concentrated, processed and activated in an apoptosome and that it requires a stimulus such as H_2O_2 for activation.

Although five metacaspase genes have been identified in *Trypanosoma brucei (TbMCA1–TbMCA5)* their activity has not yet been associated with apoptosis. Over-expression of *TbMCA4* did, however, lead to mitochondrial dysfunction, growth inhibition and cell death (Szallies et al. 2002). Features typical of apoptosis have been described in *T. brucei rhodesiense* including DNA fragmentation into oligonucleosomal repeat-sized fragments of 193±5 bp whilst cell membranes remain intact, the migration of condensed chromatin to the periphery of the nuclear membrane occurs and surface membrane vesiculation occurs (Welburn et al. 1996). Features of nuclear morphology and DNA fragmentation typical of apoptosis have also been described in *T. cruzi* (Ameisen et al. 1995) and *L. amazonensis* (Moreira et al. 1996).

Calpain is an additional candidate protease for induction of apoptotic pathways (Wang 2000). Arnoult et al. (2002a) identified a calpain-like sequence in the kinetoplastid database. They propose these proteases to be putative executors in *Leishmania* apoptosis. In addition, cathepsin B- and L-like cysteine proteases induce death in *L. major* (Selzer et al. 1999).

Widespread accord has now been reached in support of the existence of programmed cell death pathways within lineages of the parasitic protozoans and morphological markers indicate that the process is a form of apoptosis (reviewed in Arnoult et al. 2002a). Despite this progress, we are still some way from unravelling the execution pathways that initiate this process in *Plasmodium*, or other unicellular eukaryotes, and several candidate cysteine proteases must be investigated. It is possible that no common pathways will be identified, and that a mechanism for controlled cell death has arisen several times.

What does seem likely at the present time is that apoptosis can be induced by several triggers and that these may differ amongst different protozoans. Many investigators have seen apoptotic features after applying treatments that organisms may not have encountered during their evolution, for example, anti-leishmanial drugs in *L. donovani* (Lee et al. 2002) and staurosporine in *L. major* (Arnoult et al. 2002b). Similarly, Picot and colleagues (1997) reported that chloroquine treatment of the blood stages of *P. falciparum* resulted in oligonucleosomal DNA fragmentation, as demonstrated by ladder formation. Recently, Deponte and Becker (2004) reported unpublished findings that link the treatment of *P. falciparum* with anti-malarial drugs to the occurrence of DNA fragmentation in cultures of erythrocytic stages. They also report that H_2O_2, a molecule more likely to be encountered, causes DNA fragmentation. These findings suggest that apoptosis can be induced by several stressors.

Stimuli likely to be met during a parasite's life cycle have also been implicated as apoptotic triggers. Induction of apoptosis by heat shock has been reported in *L. amazonensis* (Moreira et al. 1996). Lectins present in the tsetse midgut may initiate cell death as the lectin ConA has been shown to stimulate apoptosis in *T. b. rhodesiense* (Welburn and Murphy 1998) and serum-induced apoptosis has been reported in *T. cruzi* (Piacenza et al. 2001). The ROS H_2O_2 causes several apoptotic features to occur in *L. donovani* promastigotes, including an increase in activation of caspase-like protease (Das et al. 2001).

5.1.2
Signals that Induce Parasite Apoptosis

Investigations that definitively identify triggers that induce apoptosis in the mosquito stages of malaria have yet to be performed. Nevertheless, indirect evidence is accumulating that points the finger at conditions naturally occurring in the midgut of mosquitoes that have fed on gametocytaemic blood, as over 50% of *P. berghei* ookinetes undergo apoptosis in this location without investigator manipulation. This cell death could be inhibited by the addition of N^{ω}-nitro-L-arginine methyl ester (L-NAME) to the infectious blood meal (as measured with acridine orange staining). L-NAME inhibits the activity of the enzyme NOS that catalyses the conversion of arginine to citrulline and the subsequent production of NO; the D form of the enzyme, D-NAME, is inactive. D-NAME had no effect on the proportion of ookinetes that were undergoing apoptosis at 18 and 20 h after feeding (Al-Olayan et al. 2002b) whereas the percentage of ookinetes exhibiting caspase-like activity was reduced from approximately 50% to 20% after L-NAME treatment (Al-Olayan and Hurd, unpublished results).

Nitric oxide is known to induce apoptosis via activation of caspases (Murphy 1999). It also readily forms damaging RNIs. For example, with superoxide it will form peroxynitrate, a known inducer of apoptosis (Sandoval et al. 1997; Brockhaus and Brune 1999; Szabo 2003) and reactions with thiols and transition metal complexes form *S*-nitrosothiols (SNOs) and metal-nitrosyl adducts, respectively. Nitrate and met-haemoglobin (met-Hb), SNO-Hb or iron-nitrosyl Hb can be formed when NO reacts with partially oxygenated haemoglobin. Iron-nitrosyl Hb is a catalyst for the formation of damaging oxygen radicals and nitroxyl (HNO), another potent oxidant and inducer of apoptosis (Bai et al. 2001).

One explanation for the observation that L-NAME inhibits apoptosis is that it is reducing NO flux in the midgut lumen by inhibiting the NOS activity that has been identified in the midgut of infected mosquitoes (Luckhart et al. 1998). Expression of an inducible NOS (*AsNOS)* has been detected in *An. stephensi* after blood feeding. Infection with *P. berghei* or *P. falciparum* caused an upregulation of expression (Luckhart et al. 1998, 2003). Differences in *AsNOS* expression were first detected 6 h post-infection and were demonstrated to occur intermittently during the first 3 days of infection and again at the time of sporozoite release (Luckhart et al. 1998,. 2003). Administration of L-NAME in the blood meal increased the burden of *P. berghei* in *An. stephensi*, whereas provision of the NOS substrate L-arginine (L-Arg) reduced it, indicating that AsNOS limits parasite development (Luckhart et al. 1998, 2003).

Approximately 30% of the ookinetes that develop in cultures of gametocytaemic blood show characteristics typical of apoptotic cell death, despite the absence of a mosquito-derived source of NO (Al-Olayan et al. 2002b). A potential source of NOS activity in these cultures is the white cell component of the rodent blood used to initiate the cultures. Inactivation of gametocyte activity has previously been associated with the presence of white blood cells (Naotunne et al. 1993; Cao et al. 1998). The removal of white blood cells before initiation of the culture resulted in a statistically significant decrease in the percentage of apoptotic ookinetes but this reduction was small relative to the action of L-NAME in vivo (Al-Olayan et al. 2002b). Clearly, apoptotic signals additional to NOS may be present in the infected rodent blood.

Several workers have demonstrated that RNIs reduce malaria transmission by affecting gametocyte infectivity (Motard et al. 1993; Naotunne et al. 1993). Inactivation has also been associated with superoxide (Harada et al. 2001; Lanz-Mendoza et al. 2002). In addition, serum, a known inducer of apoptosis, is added to the RPMI medium used to culture developing ookinetes. Ookinetes will not develop in vitro without the presence of serum (personal observations) but it is not known whether it induces apoptosis.

RNIs are present in the midgut blood at inflammatory levels, derived in part from a rodent malaria-associated increase in NOS induction in the midgut epithelium (Dimopoulos et al. 1998; Luckhart et al. 1998, 2003). They have been detected between 8 and12 h after infection and continue through to at least 24 h after feeding. Luckhart et al. (2003) expect these to include nitrate, nitrite and *S*-nitrosothiols.

An additional factor, of mammalian origin, has recently been implicated in parasite death. Mammalian transforming growth factor β1 (TGF-β1), present in the blood meal in an inactive form, is activated in the midgut, probably by haem and the redox product of NO, nitroxyl anion (Luckhart et al. 2003). Luckhart et al. showed that human TGF-β1 significantly reduced the number of *P. falciparum* oocysts developing on the midgut of *An. stephensi*. Remarkably, TGF-β1 was shown to induce *AsNOS* expression in an *An. stephensi* cell line, thus introducing the possibility that it acts as an immunomodulatory cytokine in the mosquito (Luckhart et al. 2003).

Together, these results suggest that several factors from both host and vector could contribute to parasite apoptosis in the mosquito midgut.

These include *Anopheles* NOS, mammalian NOS II, and perhaps NOS-independent factors such as ROS (Kumar et al. 2003) and TGF-β1.

6
Parasite Apoptosis as a Life History Strategy

A proportion of erythrocytic-stage *Plasmodium* die in response to adverse conditions but programmed cell death has not been detected in the course of unchallenged infections (Picot et al. 1997; Deponte and Becker 2004). In contrast, it would appear that it is a part of the life history strategy of the initial stages of infection in the mosquito. However, it must be remembered that rodent malarias can result in extremely heavy burdens of infection in the mosquito in comparison with *P. falciparum*. We have now observed that *P. falciparum* ookinetes undergo apoptotic death in the midgut lumen (Pardo, Ranford-Cartwright, Hurd, personal observations). In the tsetse fly, *T. brucei* multiply in the midgut lumen and it has been suggested that PCD may act as a mechanism to control clonal growth so that these protozoans do not overwhelm the vector (Ameisen et al. 1995). Additionally, Welburn et al. (1996) observed that *T. cruzi* epimastigotes that do not convert to the trypomastigote stage in the tsetse will die. Although malaria parasites do not multiply in the midgut lumen, considerable damage to the midgut is incurred when they invade the epithelial cells. Studies of the effect of infection on vector fitness are contradictory (Ferguson and Read 2002). We have noted that heavy infections result in a considerable death rate within the first 28 h post-infection; thus it is conceivable that, in parallel with the trypanosome–tsetse fly association, malaria PCD will limit infection intensity, enabling the vector to survive to transmit the parasite.

Alternative functions have been proposed for the existence of PCD in unicellular eukaryotes including an altruistic response to severe oxidative damage that will conserve nutrient resources for the surrounding cells of its clone (Frohlich and Madeo 2000) or the elimination of cells that become producers of large amounts of ROS (Skulachev 1996). A greater understanding of the mechanism underlying *Plasmodium* apoptosis will be required before these hypotheses can be evaluated.

These viewpoints arise from consideration of the parasite and the benefits that may accrue to the survivors from the death of what are likely to be members of the same clone. We must also consider the possibility that parasite death is an unavoidable consequence of having stimulated the mosquito immune response (Dimopoulos et al. 2002). Although susceptible mosquitoes are unable to eliminate the parasite completely, they may be able to control the infection by eliminating some of the invaders, thereby exhibiting some degree of resistance.

7 In Conclusion

The literature concerned with the role of apoptosis in the establishment and maintenance of parasitic infections is expanding rapidly. In comparison with several other protozoan parasites the malaria–mosquito association is a latecomer to this stage and our understanding of the triggers that induce apoptosis in vector tissues and in the parasite itself is rudimentary. With more knowledge, an insight into the role of this cell death mechanism in the life history of the malaria parasite should follow. Likewise, we are just at the beginning of what should prove to be an exciting path towards an understanding of the induction and execution of ancient apoptotic pathways in the parasite.

References

Abraham EG, Islam S, Srinivasan P, Ghosh AK, Valenzuela JG, Ribeiro JM, Kafatos FC, Dimopoulos G, Jacobs-Lorena M (2004) Analysis of the *Plasmodium* and *Anopheles* transcriptional repertoire during ookinete development and midgut invasion. J Biol Chem 279:5573–5580

Abraham EG, Jacobs-Lorena M (2004) Mosquito midgut barriers to malaria parasite development. Insect Biochem Mol Biol 34:667–671

Ahmed AM, Maingon R, Romans P, Hurd H (2001) Effects of malaria infection on vitellogenesis in *Anopheles gambiae* during two gonotrophic cycles. Insect Mol Biol 10:347–356

Ahmed AM, Maingon RD, Taylor PJ, Hurd H (1999) The effects of infection with *Plasmodium yoelii nigeriensis* on the reproductive fitness of the mosquito *Anopheles gambiae*. Invertebr Reprod Dev 36:217–222

Aikawa M, Yoshida N, Nussenzweig RS, Nussenzweig V (1981) The protective antigen of malarial sporozoites (*Plasmodium berghei*) is a differentiation antigen. J Immunol 126:2484–2495

Alavi Y, Arai M, Mendoza J, Tufet-Bayona M, Sinha R, Fowler K, Billker O, Franke-Fayard B, Janse CJ, Waters A, Sinden RE (2003) The dynamics of interactions between *Plasmodium* and the mosquito: a study of the infectivity of *Plasmodium berghei* and *Plasmodium gallinaceum*, and their transmission by *Anopheles stephensi, Anopheles gambiae* and *Aedes aegypti*. Int J Parasitol 33:933–943

Al-Olayan EM, Beetsma AL, Butcher GA, Sinden RE, Hurd H (2002a) Complete development of mosquito phases of the malaria parasite in vitro. Science 295:677–679

Al-Olayan EM, Williams GT, Hurd H (2002b) Apoptosis in the malaria protozoan, *Plasmodium berghei*: A possible mechanism for limiting intensity of infection in the mosquito. Int J Parasitol 32:1133–1143

Ameisen JC, Idziorek T, Billautmulot O, Loyens M, Tissier JP, Potentier A, Ouaissi A (1995) Apoptosis in a unicellular eukaryote (*Trypanosoma cruzi*)—implications for the evolutionary origin and role of programmed cell-death in the control of cell-proliferation, differentiation and survival. Cell Death Differ 2:285–300

Arai M, Billker O, Morris HR, Panico M, Delcroix M, Dixon D, Ley SV, Sinden RE (2001) Both mosquito-derived xanthurenic acid and a host blood-derived factor regulate gametogenesis of *Plasmodium* in the midgut of the mosquito. Mol Biochem Parasitol 116:17–24

Aravind L, Dixit VM, Koonin EV (2001) Apoptotic molecular machinery: vastly increased complexity in vertebrates revealed by genome comparisons. Science 291:1279–1284

Arnoult D, Akarid K, Grodet A, Petit PX, Estaquier J, Ameisen JC (2002a) On the evolution of programmed cell death: apoptosis of the unicellular eukaryote *Leishmania major* involves cysteine proteinase activation and mitochondrion permeabilization. Cell Death Differ 9:65–81

Arnoult D, Parone P, Martinou JC, Antonsson B, Estaquier J, Ameisen JC (2002b) Mitochondrial release of apoptosis-inducing factor occurs downstream of cytochrome *c* release in response to several proapoptotic stimuli. J Cell Biol 159:923–929

Arrighi RB, Hurd H (2002) The role of *Plasmodium berghei* ookinete proteins in binding to basal lamina components and transformation into oocysts. Int J Parasitol 32:91–98

Bai P, Bakondi E, Szabo E, Gergely P, Szabo C, Virag L (2001) Partial protection by poly(ADP-ribose) polymerase inhibitors from nitroxyl-induced cytotoxicity in thymocytes. Free Radic Biol Med 31:1616–1623

Barillas-Mury C, Wizel B, Han YS (2000) Mosquito immune responses and malaria transmission: lessons from insect model systems and implications for vertebrate innate immunity and vaccine development. Insect Biochem Mol Biol 30:429–442

Baton LA, Ranford-Cartwright LC (2004) *Plasmodium falciparum* ookinete invasion of the midgut epithelium of *Anopheles stephensi* is consistent with the Time Bomb model. Parasitology (in press)

Billingsley PF, Lehane MJ (eds) (1996) The Insect Midgut. Chapman and Hall

Billingsley PF, Rudin W (1992) The role of the mosquito peritrophic membrane in bloodmeal digestion and infectivity of *Plasmodium* species. J Parasitol 78:430–440

Billingsley PF, Sinden RE (1997) Determinants of malaria-mosquito specificity. Parasitol Today 13:297–301

Billker O, Dechamps S, Tewari R, Wenig G, Franke-Fayard B, Brinkmann V (2004) Calcium and a calcium-dependent protein kinase regulate gamete formation and mosquito transmission in a malaria parasite. Cell 117:503–514

Billker O, Shaw MK, Margos G, Sinden RE (1997) The roles of temperature, pH and mosquito factors as triggers of male and female gametogenesis of *Plasmodium berghei* in vitro. Parasitology 115:1–7

Blanco ARA, Paez A, Gerold P, Dearsly AL, Margos G, Schwartz RT, Barker G, Rodriguez MC, Sinden RE (1999) The biosynthesis and post-translational modification of Pbs21, an ookinete-surface protein of *Plasmodium berghei*. Mol Biochem Parasitol 98:163–173

Blandin S, Moita LF, Kocher T, Wilm M, Kafatos FC, Levashina EA (2002) Reverse genetics in the mosquito *Anopheles gambiae*: targeted disruption of the Defensin gene. EMBO Rep 3:852–856

Brockhaus F, Brune B (1999) Overexpression of CuZn superoxide dismutase protects RAW 264.7 macrophages against nitric oxide cytotoxicity. Biochem J 338:295–303

Buszczak M, Cooley L (2000) Eggs to die for: cell death during *Drosophila* oogenesis. Cell Death Differ 7:1071–1074

Cao Y-M, Tsuboi T, Torii M (1998) Nitric oxide inhibits the development of *Plasmodium yoelii* gametocytes into gametes. Parasitol Int 47:157–166

Carwardine SL, Hurd H (1997) Effects of *Plasmodium yoelii nigeriensis* infection on *Anopheles stephensi* egg development and resorption. Med Vet Entomol 11:265–269

Cavaliere V, Taddei C, Gargiulo G (1998) Apoptosis of nurse cells at the late stages of oogenesis of *Drosophila melanogaster*. Dev Genes Evol 208:106–112

Chao S, Nagoshi RN (1999) Induction of apoptosis in the germline and follicle layer of *Drosophila* egg chambers. Mech Dev 88:159–172

Chose O, Sarde CO, Gerbod D, Viscogliosi E, Roseto A (2003) Programmed cell death in parasitic protozoans that lack mitochondria. Trends Parasitol 19:559–564

Christophides GK, Zdobnov E, Barillas-Mury C, Birney E, Blandin S, Blass C, Brey PT, Collins FH, Danielli A, Dimopoulos G, Hetru C, Hoa NT, Hoffmann JA, Kanzok SM, Letunic I, Levashina EA, Loukeris TG, Lycett G, Meister S, Michel K, Moita LF, Muller HM, Osta MA, Paskewitz SM, Reichhart JM, Rzhetsky A, Troxler L, Vernick KD, Vlachou D, Volz J, von Mering C, Xu J, Zheng L, Bork P, Kafatos FC (2002) Immunity-related genes and gene families in *Anopheles gambiae*. Science 298:159–165

Clements AN, Boocock MR (1984) Ovarian development in mosquitoes: stages of growth and arrest and follicular resorption. Physiol Entomol 9:1–8

Collins FH, Sakai RK, Vernick KD, Paskewitz S, Seeley DC, Miller LH, Collins WE, Campbell CC, Gwadz RW (1986) Genetic selection of a *Plasmodium*-refractory strain of the malaria vector *Anopheles gambiae*. Science 234:607–610

Colombatti A, Bonaldo P, Dolian R (1993) Type A molecules: interacting domains found in several non-fibrillar collagens and in other extracellular matrix proteins. Matrix 13:297–306

Cruz MA, Yuan H, Lee JR, Wise RJ, Handin RI (1995) Interaction of the von Willebrand factor (vWF) with collagen. J Biol Chem 270:10822–10827

Das M, Mukherjee SB, Shaha C (2001) Hydrogen peroxide induces apoptosis-like death in *Leishmania donovani* promastigotes. J Cell Sci 114:2461–2469

Deponte M, Becker K (2004b) *Plasmodium falciparum*—do killers commit suicide? Trends Parasitol 20:165–169

Dessens JT, Beetsma AL, Dimopoulos G, Wengenlik K, Crisanti A, Kafatos F, Sinden RE (1999) CTRP is essential for mosquito infection by malaria ookinetes. EMBO J 18:6221–6227

Dessens JT, Siden-Kiamos I, Mendoza J, Mahairaki V, Khater E, Vlachou D, Xu XJ, Kafatos FC, Louis C, Dimopoulos G, Sinden RE (2003) SOAP, a novel malaria ookinete protein involved in mosquito midgut invasion and oocyst development. Mol Microbiol 49:319–329

Dimopoulos G, Christophides GK, Meister S, Schultz J, White KP, Barillas-Mury C, Kafatos FC (2002) Genome expression analysis of *Anopheles gambiae*: responses to injury, bacterial challenge, and malaria infection. Proc Natl Acad Sci USA 99:8814–8819

Dimopoulos G, Muller HM, Levashina EA, Kafatos FC (2001) Innate immune defense against malaria infection in the mosquito. Curr Opin Immunol 13:79–88

Dimopoulos G, Seeley D, Wolf A, Kafatos FC (1998) Malaria infection of the mosquito *Anopheles gambiae* activates immune-responsive genes during critical transition stages of the parasite life cycle. EMBO J 17:6115–6123

Ferguson HM, Read AF (2002) Why is the effect of malaria parasites on mosquito survival still unresolved? Trends Parasitol 18:256–261

Foley K, Cooley L (1998) Apoptosis in late stage *Drosophila* nurse cells does not require genes within the H99 deficiency. Development 125:1075–1082

Frohlich KU, Madeo F (2000) Apoptosis in yeast—a monocellular organism exhibits altruistic behaviour. FEBS Lett 473:6–9

Gad AM, Maier WA, Piekarski G (1979) Pathology of *Anopheles stephensi* after infection with *Plasmodium berghei berghei*. II. Changes in amino acid contents. Z Parasitenkd 60:263–276

Gardner MJ, Hall N, Fung E, White O, Berriman M, Hyman RW, Carlton JM, Pain A, Nelson KE, Bowman S, Paulsen IT, James K, Eisen JA, Rutherford K, Salzberg SL, Craig A, Kyes S, Chan MS, Nene V, Shallom SJ, Suh B, Peterson J, Angiuoli S, Pertea M, Allen J, Selengut J, Haft D, Mather MW, Vaidya AB, Martin DM, Fairlamb AH, Fraunholz MJ, Roos DS, Ralph SA, McFadden GI, Cummings LM, Subramanian GM, Mungall C, Venter JC, Carucci DJ, Hoffman SL, Newbold C, Davis RW, Fraser CM, Barrell B (2002) Genome sequence of the human malaria parasite *Plasmodium falciparum*. Nature 419:498–511

Ghosh A, Edwards MJ, Jacobs-Lorena M (2000) The journey of the malaria parasite in the mosquito: hopes for the new century. Parasitol Today 16:196–201

Han YS, Barillas-Mury C (2002) Implications of Time Bomb model of ookinete invasion of midgut cells. Insect Biochem Mol Biol 32:1311–1316

Han YS, Thompson J, Kafatos FC, Barillas-Mury C (2000) Molecular interactions between *Anopheles stephensi* midgut cells and *Plasmodium berghei*: the time bomb theory of ookinete invasion of mosquitoes. EMBO J 19:6030–6040

Harada M, Owhashi M, Suguri S, Kumatori A, Nakamura M, Kanbara H, Matsuoka H, Ishii A (2001) Superoxide-dependent and -independent pathways are involved in the transmission blocking of malaria. Parasitol Res 87:605–608

Harvey NL, Daish T, Mills K, Dorstyn L, Quinn LM, Read SH, Richardson H, Kumar S (2001) Characterization of the *Drosophila* caspase, DAMM. J Biol Chem 276:25342–25350

Hogg JC, Hurd H (1995) *Plasmodium yoelii nigeriensis*: the effect of high and low intensity of infection upon the egg production and bloodmeal size of *Anopheles stephensi* during three gonotrophic cycles. Parasitology 111:555–562

Hogg JC, Hurd H (1997) The effects of natural *Plasmodium falciparum* infection on the fecundity and mortality of *Anopheles gambiae* s. l. In north east Tanzania. Parasitology 114:325–331

Holt RA, Subramanian GM, Halpern A, Sutton GG, Charlab R, Nusskern DR, Wincker P, Clark AG, Ribeiro JM, Wides R, Salzberg SL, Loftus B, Yandell M, Majoros WH, Rusch DB, Lai Z, Kraft CL, Abril JF, Anthouard V, Arensburger P, Atkinson PW, Baden H, de Berardinis V, Baldwin D, Benes V, Biedler J, Blass C, Bolanos R, Boscus D, Barnstead M, Cai S, Center A, Chaturverdi K, Christophides GK, Chrystal MA, Clamp M, Cravchik A, Curwen V, Dana A, Delcher A, Dew I, Evans CA, Flanigan M, Grundschober-Freimoser A, Friedli L, Gu Z, Guan P, Guigo R, Hillenmeyer ME, Hladun SL, Hogan JR, Hong YS, Hoover J, Jaillon O, Ke Z, Kodira C, Kokoza E, Koutsos A, Letunic I, Levitsky A, Liang Y, Lin JJ, Lobo NF, Lopez JR, Malek JA, McIntosh TC, Meister S, Miller J, Mobarry C, Mongin E, Murphy SD, O'Brochta DA, Pfannkoch C, Qi R, Regier A, Remington K, Shao H, Sharakhova MV, Sitter CD, Shetty J, Smith TJ, Strong R, Sun J, Thomasova D, Ton LQ, Topalis P, Tu Z, Unger MF, Walenz B, Wang A, Wang J, Wang M, Wang X, Woodford KJ, Wortman JR, Wu M, Yao A, Zdobnov EM, Zhang H, Zhao Q, Zhao S, Zhu SC, Zhimulev I, Coluzzi M, della Torre A, Roth CW, Louis C, Kalush F, Mural RJ, Myers EW, Adams MD, Smith HO, Broder S, Gardner MJ, Fraser M, Birney E, Bork P, Brey PT, Venter JC, Weissenbach J, Kafatos FC, Collins FH, Hoffman SL (2002) The genome sequence of the malaria mosquito *Anopheles gambiae*. Science 298:129–149

Hopwood JA, Ahmed AM, Polwart A, Williams GT, Hurd H (2001) Malaria-induced apoptosis in mosquito ovaries: a mechanism to control vector egg production. J Exp Biol 204:2773–2780

Hurd H (1990) Physiological and behavioural interactions between parasites and invertebrate hosts. Adv Parasitol 29:271–318

Hurd H (2001) Host fecundity reduction: a strategy for damage limitation? Trends Parasitol 17:363–368

Hurd H (2003) Manipulation of medically important insect vectors by their parasites. Annu Rev Entomol 48:141–161

Janse CJ, Waters AP (2002) The *Plasmodium berghei* research model of malaria. http://www.lumc.nl/1040/research/malaria/model.html

Kappe SH, Buscaglia CA, Bergman LW, Coppens I, Nussenzweig V (2004) Apicomplexan gliding motility and host cell invasion: overhauling the motor model. Trends Parasitol 20:13–16

Koella JC (1999) An evolutionary view of the interactions between anopheline mosquitoes and malaria parasites. Microbes Infect 1:303–308

Kumar S, Christophides GK, Cantera R, Charles B, Han YS, Meister S, Dimopoulos G, Kafatos FC, Barillas-Mury C (2003) The role of reactive oxygen species on *Plasmodium* melanotic encapsulation in *Anopheles gambiae*. Proc Natl Acad Sci USA 100:14139–14144

Langer RC, Hayward RE, Tsuboi T, Tachibana M, Torii M, Vinetz JM (2001) Micronemal transport of *Plasmodium* ookinete chitinases to the electron-dense area of the apical complex for extracellular secretion. Infect Immun 68:6461–6465

Lanz-Mendoza H, Hernandez-Martinez S, Ku-Lopez M, Rodriguez Mdel C, Herrera-Ortiz A, Rodriguez MH (2002) Superoxide anion in *Anopheles albimanus* hemolymph and midgut is toxic to *Plasmodium berghei* ookinetes. J Parasitol 88:702–706

Lee JO, Rieu P, Arnaout MA, Liddington R (1995) Crystal structure of the A domain from the alpha subunit of integrin CR3 (CD11b/CD18). Cell 80:631–638

Lee N, Bertholet S, Debrabant A, Muller J, Duncan R, Nakhasi HL (2002) Programmed cell death in the unicellular protozoan parasite *Leishmania.* Cell Death Differ 9:53–64

Limviroj W, Yano K, Yuda M, Ando K, Chinzei Y (2002) Immuno-electron microscopic observations of *Plasmodium berghei* CTRP localisation in the midgut of the vector mosquito *Anopheles stenphensi.* J Parasitol 88:664–672

Luckhart S, Crampton AL, Zamora R, Lieber MJ, Dos Santos PC, Peterson TML, Emmith N, Lim J, Wink DA, Vodovotz Y (2003) Mammalian transforming growth factor beta1 activated after ingestion by *Anopheles stephensi* modulates mosquito immunity. Infect Immun 71:3000–3009

Luckhart S, Vodovotz Y, Cui L, Rosenberg R (1998) The mosquito *Anopheles stephensi* limits malaria parasite development with inducible synthesis of nitric oxide. Proc Natl Acad Sci USA 95:5700–5705

Mack SR, Samuels S, Vanderberg JP (1979) Hemolymph of *Anopheles stephensi* from uninfected and *Plasmodium berghei*-infected mosquitoes: 2. Free amino acids. FEBS Lett 430:59–63

Madeo F, Herker E, Maldener C, Wissing S, Lachelt S, Herlan M, Fehr M, Lauber K, Sigrist SJ, Wesselborg S, Frohlich KU (2002) A caspase-related protease regulates apoptosis in yeast. Mol Cell 9:911–917

Moreira ME, Del Portillo HA, Milder RV, Balanco JM, Barcinski MA (1996) Heat shock induction of apoptosis in promastigotes of the unicellular organism *Leishmania (Leishmania) amazonensis.* J Cell Physiol 167:305–313

Motard A, Landau I, Nussler A, Grau G, Baccam D, Mazier D, Targett GA (1993) The role of reactive nitrogen intermediates in modulation of gametocyte infectivity of rodent malaria parasites. Parasite Immunol 15:21–26

Mottram JC, Helms MJ, Coombs GH, Sajid M (2003) Clan CD cysteine peptidases of parasitic protozoa. Trends Parasitol 19:182–187

Murphy MP (1999) Nitric oxide and cell death. Biochim Biophys Acta 1411:401–414

Myung JM, Marshall P, Sinnis P (2004) The *Plasmodium* circumsporozoite protein is involved in mosquito salivary gland invasion by sporozoites. Mol Biochem Parasitol 133:53–59

Naotunne TS, Karunaweera ND, Mendis KN, Carter R (1993) Cytokine-mediated inactivation of malarial gametocytes is dependent on the presence of white blood cells and involves reactive nitrogen intermediates. Immunology 78:555–562

Ojcius DM, Perfettini JL, Bonnin A, Laurent F (1999) Caspase-dependent apoptosis during infection with *Cryptosporidium parvum.* Microbes Infect 1:1163–1168

Paul REL, Brey PT, Robert V (2002) *Plasmodium* sex determination and transmission to mosquitoes. Trends Parasitol 18:32–37

Piacenza L, Peluffo G, Radi R (2001) L-Arginine-dependent suppression of apoptosis in *Trypanosoma cruzi*: contribution of the nitric oxide and polyamine pathways. Proc Natl Acad Sci USA 98:7301–7306

Picot S, Burnod J, Bracchi V, Chumpitazi BF, Ambroise-Thomas P (1997) Apoptosis related to chloroquine sensitivity of the human malaria parasite *Plasmodium falciparum*. Trans R Soc Trop Med Hyg 91:590–591

Potocnjak P, Yoshida N, Nussenzweig RS, Nussenzweig N (1980) Monovalent fragments (Fab) of monoclonal antibodies to a sporozoite surface antigen (Pb44) protect mice against malarial infection. J Exp Med 151:1504–1513

Rozman-Pungercar J, Kopitar-Jerala N, Bogyo M, Turk D, Vasiljeva O, Stefe I, Vandenabeele P, Bromme D, Puizdar V, Fonovic M, Trstenjak-Prebanda M, Dolenc I, Turk V, Turk B (2003) Inhibition of papain-like cysteine proteases and legumain by caspase-specific inhibitors: when reaction mechanism is more important than specificity. Cell Death Differ 10:881–888

Sandoval M, Ronzio RA, Muanza DN, Clark DA, Miller MJ (1997) Peroxynitrite-induced apoptosis in epithelial (T84) and macrophage (RAW 264.7) cell lines: effect of legume-derived polyphenols (phytolens). Nitric Oxide 1:476–483

Selzer PM, Pingel S, Hsieh I, Ugele B, Chan VJ, Engel JC, Bogyo M, Russell DG, Sakanari JA, McKerrow JH (1999) Cysteine protease inhibitors as chemotherapy: lessons from a parasite target. Proc Natl Acad Sci USA 96:11015–11022

Shahabuddin M, Pimenta PFP (1998) *Plasmodium gallinaceum* preferentially invades vesicular ATPase-expressing cells in *Aedes aegypti* midgut. Proc Natl Acad Sci USA 95:3385–3389

Shahabuddin M, Toyoshima T, Aikawa M, Kaslow DC (1993) Transmission-blocking activity of a chitinase inhibitor and activation of malarial parasite chitinases by mosquito protease. Proc Natl Acad Sci USA 90:4266–4270

Sheldon BC, Verhulst S (1996) Ecological immunology: costly parasite defences and trade-offs in evolutionary ecology. Tree 11:317–321

Sidén-Kiamos I, Vlachou D, Margos G, Beetsma AL, Waters AP, Sinden RE, Louis C (2000) Distinct roles for Pbs21 and Pbs25 in the in vitro ookinete to oocyst transformation of *Plasmodium berghei*. J Cell Sci 113:3419–3426

Sidjanski SP, Vanderberg JP, Sinnis P (1997) *Anopheles stephensi* salivary glands bear receptors for region I of the circumsporozoite protein of *Plasmodium falciparum*. Mol Biochem Parasitol 90:33–41

Simonetti AB (1996) The biology of malarial parasite in the mosquito—a review. Mem I Oswaldo Cruz 97:519–541

Simonetti AB, Billingsley PF, Winger LA, Sinden RE (1993) Kinetics of expression of two major *Plasmodium berghei* antigens in the mosquito vector, *Anopheles stephensi*. Jf Eukaryotic Microbiol 40:569–576

Sinden RE (1978) Cell biology. In: Killick Kendrick R, Peters W (eds) Rodent Malaria. Academic Press, London, pp 85–168

Sinden RE (1997) Infection of mosquitoes with rodent malaria. In: The molecular biology of insect disease vectors: a methods manual. Chapman & Hall (eds), London pp 67–91

Sinden RE, Billingsley PF (2001) *Plasmodium* invasion of mosquito cells: hawk or dove? Trends Parasitol 17:209–212

Sinden RE, Croll NA (1975) Cytology and kinetics of microgametogenesis and fertilization in *Plasmodium yoelii nigeriensis*. Parasitology 70:53–65

Skulachev VP (1996) Why are mitochondria involved in apoptosis? Permeability transition pores and apoptosis as selective mechanisms to eliminate superoxide-producing mitochondria and cell. FEBS Lett 397:7–10

Sokolova MI (1994) A redescription of the morphology of mosquito (Diptera: Culicidae) ovarioles during vitellogenesis. Bull Soc Vector Ecol 19:53–68

Soller M, Bownes M, Kubli E (1999) Control of oocyte maturation in sexually mature *Drosophila* females. Dev Biol 208:337–351

Srinivasan P, Abraham EG, Ghosh AK, Valenzuela J, Ribeiro JM, Dimopoulos G, Kafatos FC, Adams JH, Fujioka H, Jacobs-Lorena M (2004) Analysis of the *Plasmodium* and *Anopheles* transcriptomes during oocyst differentiation. J Biol Chem 279:5581–5587

Szabo C (2003) Multiple pathways of peroxynitrite cytotoxicity. Toxicol Lett 140–141:105–112

Szallies A, Kubata BK, Duszenko M (2002) A metacaspase of *Trypanosoma brucei* causes loss of respiration competence and clonal death in the yeast *Saccharomyces cerevisiae*. FEBS Lett 517:144–150

Templeton TJ, Kaslow DC, Fidock DA (2000) Developmental arrest of the human malaria parasite *Plasmodium falciparum* within the mosquito midgut via CTRP gene disruption. Mol Microbiol 36:1–9

Thathy V, Fujioka H, Gantt S, Nussenzweig R, Nussenzweig V, Ménard R (2002) Levels of circumsporozoite protein in the *Plasmodium* oocyst determine sporozoite morphology. EMBO J 21:1586–1596

Tomas AM, Margos G, Dimopoulos G, van Lin LHM, de Koning-Ward TF, Sinha R, Lupetti P, Beetsma AL, Rodriguez MC, Karras M, Hager A, Mendoza J, Butcher GA, Kafatos F, Janse CJ, Waters AP, Sinden RE (2001) P25 and P28 proteins of the malaria ookinete surface have multiple and partially redundant functions. EMBO J 20:3975–3983

Torii M, Nakamura K, Seiber KP, Miller LH, Aikawa M (1992) Penetration of the mosquito (*Aedes aegypti*) midgut by ookinetes of *Plasmodium gallinaceum*. J Protozool 39:449–454

Tsuboi T, Kaslow DC, Gozar MM, Tachibana M, Cao YM, Torii M (1998) Sequence polymorphism in two novel *Plasmodium vivax* ookinete surface proteins, Pvs25 and Pvs28, that are malaria transmission-blocking vaccine candidates. Mol Med 4:772–782

Uren AG, O'Rourke K, Aravind L, Pisabarro MT, Seshagiri S, Koonin EV, Dixit VM (2000) Identification of paracaspases and metacaspases: two ancient families of caspase-like proteins, one of which plays a key role in MALT lymphoma. Mol Cell 6:961–967

Vaughan JA, Hensley L, Beier JC (1994) Sporogonic development of *Plasmodium yoelii* in five anopheline species. J Parasitol 80:674–681

Vernick KD, Fujioka H, Seeley DC, Tandler B, Aikawa M, Miller LH (1995) *Plasmodium gallinaceum*: a refractory mechanism of ookinete killing in the mosquito, *Anopheles gambiae*. Exp Parasitol 80:583–595

Vlachou D, Zimmermann T, Cantera R, Janse CJ, Waters AP, Kafatos FC (2004) Real-time, in vivo analysis of malaria ookinete locomotion and mosquito midgut invasion. Cell Microbiol 6:671–685

Wang KK (2000) Calpain and caspase: can you tell the difference? Trends Neurosci 23:20–26

Welburn SC, Dale C, Ellis D, Beecroft R, Pearson TW (1996) Apoptosis in procyclic *Trypanosoma brucei rhodesiense* in vitro. Cell Death Differ 3:229–236

Welburn SC, Murphy NB (1998) Prohibitin and RACK homologues are up-regulated in trypanosomes induced to undergo apoptosis and in naturally occurring terminally differentiated forms. Cell Death Differ 5:615–622

Yoshida N, Nussenzweig RS, Potocnjak P, Nussenzweig N, Aikawa M (1981) Biosynthesis of Pb44, the protective antigen of sporozoites of *Plasmodium berghei*. J Exp Med. 154:1225–1236

Yuda M, Sakaida H, Chinzei Y (1999) Targeted disruption of the *Plasmodium berghei* CTRP gene reveals its essential role in malaria infection of the vector mosquito. J Exp Med 190:1711–1715

Yuda M, Yano K, Tsuboi T, Torii M, Chinzei Y (2001) von Willebrand Factor A domain-related protein, a novel microneme protein of the malaria ookinete highly conserved throughout the *Plasmodium* parasites. Mol Biochem Parasitol 116:65–72

Zieler H, Dvorak JA (2000) Invasion in vitro of mosquito midgut cells by the malaria parasite proceeds by a conserved mechanism and results in death of the invaded midgut cells. Proc Natl Acad Sci USA 97:11516–11521

CTMI (2005) 289:219–238

Apoptosis and Its Modulation During Infection with *Toxoplasma gondii*: Molecular Mechanisms and Role in Pathogenesis

C. G. K. Lüder (✉) · U. Gross

Institute of Medical Microbiology, Georg-August-University, Kreuzbergring 57, 37075 Göttingen, Germany
clueder@gwdg.de

Abstract Infection with the obligate intracellular protozoan *Toxoplasma gondii* leads to lifelong persistence of the parasite in its mammalian hosts including humans. Apoptosis plays crucial roles in the interaction between the host and the parasite. This includes innate and adaptive defense mechanisms to restrict intracellular parasite replication as well as regulatory functions to modulate the host's immune response. Not surprisingly, however, *T. gondii* also extensively modifies apoptosis of its own host cell or of uninfected bystander cells. After infection, apoptosis is triggered in T lymphocytes and other leukocytes, thereby leading to suppressed immune responses to the parasite. T cell apoptosis may be largely mediated by Fas engagement but also occurs independently of Fas under certain conditions. Depending on the magnitude of T cell apoptosis, it is either associated with unrestricted parasite replication and severe pathology or facilitates a stable parasite-host-interaction. However, *T. gondii* has also evolved strategies to inhibit host cell apoptosis. Apoptosis is blocked by indirect mechanisms in uninfected bystander cells, thereby modulating the inflammatory response to the parasite. In contrast, inhibition of apoptosis in infected host cells by direct interference with apoptosis-signaling cascades is thought to facilitate the intracellular development of *T. gondii*. Blockade of apoptosis by intracellular parasites may be achieved by different means including interference with the caspase cascade, increased expression of antiapoptotic molecules by infected host cells, and a decreased activity of the poly(ADP-ribose) polymerase. The intriguing dual activity of *T. gondii* to both promote and inhibit apoptosis requires a tight regulation to promote a stable parasite host-interaction and establishment of persistent toxoplasmosis.

1 Introduction

Toxoplasma gondii is an obligate intracellular protozoan parasite that infects a broad range of warm-blooded animals including approximately 30% of the human population worldwide (Tenter et al. 2000). Infection is orally acquired by ingestion of infectious sporozoites that have been released by parasitized cats via the feces or by ingestion of persisting parasites contained in raw or undercooked meat from intermediate hosts. Furthermore, transmission can occur vertically by transplacental transmission to the offspring (see below). During primary infection, disseminating parasites, the so-called tachyzoites, are able to actively invade and rapidly replicate in nearly any cell type investigated so far. In the face of an adequate immune response, these parasites differentiate into metabolically dormant cyst-forming bradyzoites that are able to persist predominantly in neural and muscular tissue. Infection of immunocompetent individuals is mostly asymptomatic or leads to mild symptoms only but gives rise to chronic infection that persists throughout the host's life. However, *T. gondii* is a major opportunistic pathogen of newborns from recently infected mothers (Petersen et al. 2001) and of immunocompromised patients, i.e. those with AIDS or under immunosuppressive therapy (Ammassari et al. 1996). In these individuals, the premature or suppressed immune system is not able to control parasite multiplication, eventually leading to life-threatening disease.

The importance of T cell-mediated immunity for antiparasitic activity has been supported by experimental infections in mice showing that $CD8^+$ and $CD4^+$ T lymphocytes are required to avoid unrestricted parasite multiplication (Suzuki and Remington 1988; Gazzinelli et al. 1992). Whereas $CD8^+$ T cells appear to represent the most important effector cells against *T. gondii*, $CD4^+$ T cells rather fulfill important regulatory functions (Denkers and Gazzinelli 1998). During the early phase of infection, natural killer (NK) cells may fulfill T cell-independent effector functions against the parasite (Sher et al. 1993). Production of IFN-γ

and additional proinflammatory cytokines, e.g. TNF-α, activate infected cells to exert antiparasitic activity and are the major mediators of resistance against *T. gondii* (Suzuki et al. 1988; Sibley et al. 1991). Furthermore, lysis of infected target cells by $CD8^+$, $CD4^+$, and NK cells may also contribute to parasite control at least under certain conditions (Denkers et al. 1997; Hu et al. 1999; Curiel et al. 1993). Despite the induction of these potentially effective immune responses against *T. gondii*, the host is nevertheless unable to clear the infection.

A variety of factors have been described that may contribute to the parasite's ability to establish and maintain a persistent infection in its immunocompetent host. Accumulating evidence indicates that this also includes alterations of apoptosis in distinct host cell populations (Lüder et al. 2001; Heussler et al. 2001). This is not surprising because apoptosis is known to play a critical role in the regulation of the immune response (Opferman and Korsmeyer 2003), as an effector mechanism of NK cells and cytotoxic T lymphocytes (CTL) to eliminate infected target cells (Lieberman 2003), and as an innate response of cells after infection by intracellular pathogens (Williams 1994). More interestingly, *T. gondii* both promotes and inhibits apoptosis. Inhibition of host cell apoptosis may allow undisturbed intracellular development, thereby facilitating parasite survival. Increased apoptosis of immune cells after infection, on the other hand, is thought to partially downregulate effective immune responses against *T. gondii* leading to immune evasion. However, host cell apoptosis and its modulation by the parasite not only benefit *T. gondii*, but are also required to restrict tissue destruction by inflammatory responses, i.e. immunopathology.

Here we review the current knowledge on the molecular mechanisms of the dual activity of *T. gondii* in apoptosis. We also focus on the role of apoptosis and its modulation by the parasite for the course of toxoplasmosis and for the pathogenesis of disease.

2
Infection with *T. gondii* as a Trigger of Apoptosis

Acute infection of both humans and mice with *T. gondii* induces a state of transient immunosuppression as determined by decreased antibody and T lymphocyte responses to homologous and heterologous antigens (Strickland and Sayles 1977; Wing et al. 1983; Luft et al. 1984; Yano et al. 1987). Among other factors, apoptosis of T lymphocytes triggered by *T. gondii* may restrict the immune response to the parasite (Khan et al.

1996; Liesenfeld et al. 1997; Wei et al. 2002). Indeed, high levels of apoptosis in splenocytes have recently been associated with unrestricted parasite multiplication leading to high parasite burdens in various tissues of the host (Mordue et al. 2001; Gavrilescu and Denkers 2001). Cell death within the spleen was not restricted to certain populations but was detected in $CD4^+$ and $CD8^+$ T lymphocytes, B lymphocytes, NK cells, and granulocytes (Gavrilescu and Denkers 2003). It must be stressed, however, that the level of splenocyte apoptosis was markedly determined by the genotype (Grigg et al. 2001) and the virulence of the parasite. This raises the possibility that parasite-triggered extensive apoptosis of leukocytes is not a general phenomenon after infection with *T. gondii* but rather represents a characteristic determinant of the course of toxoplasmosis. Importantly, lymphocyte apoptosis may also influence the local immune response after natural parasite transmission via the gut, because oral infection with *T. gondii* led to apoptosis in Peyer's patch T cells (Liesenfeld et al. 1997). Such cell death again appears to depend on the course of infection, being observed only in inbred mouse strains susceptible to severe disease (McLeod et al. 1996).

Apoptosis of $CD4^+$ T cells was preceded by a state of unresponsiveness to antigenic or mitogenic stimulation (Khan et al. 1996). Because T cell activation markers could be readily detected, this is reminiscent of a similar condition described by Lopes et al. (1995) during experimental Chagas disease, i.e. activation-induced cell death (AICD). This form of apoptosis is initiated by the interaction of Fas and FasL, augmented by IL-2, and is counteracted by Bcl-2 or Bcl-X_L (Van Parijs and Abbas 1996). Infection with *T. gondii* indeed led to the upregulation of Fas expression in Peyer's patch T cells (Liesenfeld et al. 1997) as well as splenocytes and ocular tissue (Hu et al. 1999). Furthermore, induction of apoptosis by *T. gondii* was abolished in mutant mice lacking a functional Fas-FasL system (Liesenfeld et al. 1997; Gavrilescu and Denkers 2003). Expression of Fas as well as Fas-FasL-mediated apoptosis in *T. gondii*-infected mice appear to be regulated by the secretion of proinflammatory cytokines, IL-12 and IFN-γ, and may be counterbalanced by activation of NF-κB_2 (Caamano et al. 2000). These results clearly suggest a crucial role of the interaction of Fas and its ligand in *T. gondii*-triggered apoptosis of T cells.

Recently, human dendritic cells infected with viable, but not nonviable, parasites have been shown to induce T lymphocyte apoptosis in a contact-dependent and Fas-independent fashion (Wei et al. 2002). Beside AICD, T cell death in the absence of Fas ligation (Van Parijs and Abbas 1996) may thus also contribute to T lymphocyte dysfunction during toxoplasmosis. Whether such form of apoptosis operates in vivo, however, is yet unknown. It nevertheless raises the possibility that both Fas-dependent and -independent cell death deplete T cells during toxoplasmosis. Further experiments are needed to unravel the relative contribution of these different forms of cell death for the course of infection. Furthermore, its overall impact on the transient immunosuppression during toxoplasmosis also awaits further clarification.

3
Inhibition of Host Cell Apoptosis by *T. gondii*

Besides increased apoptosis of distinct cell populations after infection, *T. gondii* has been shown clearly to also decrease host cell death (Hisaeda et al. 1997; Nash et al. 1998; Goebel et al. 1999, 2001; Channon et al. 2002; Payne et al. 2003). Parasite-mediated resistance against apoptosis was observed in both murine and human cell lines treated with diverse inducers of apoptosis, including CTL-mediated cytotoxicity, irradiation, growth factor withdrawal, TNF-α, and/or several toxic agents (Nash et al. 1998; Goebel et al. 2001; Payne et al. 2003). Furthermore, *T. gondii* also led to decreased apoptosis in primary cells cultured ex vivo after growth factor withdrawal (Hisaeda et al. 1997; Channon et al. 2002). Importantly, inhibition of apoptosis has recently also been shown to occur in vivo after intraperitoneal infection of mice with *T. gondii* (Orlowsky et al. 1999, 2002). This suggests that interference of *T. gondii* with the suicide program of host cells may modify the course of toxoplasmosis.

Three major pathways are known to trigger apoptosis in response to external or internal stimuli: (a) binding of death ligands to their specific cell surface receptors, such as Fas or TNF receptor I (death receptor pathway; Tibbetts et al. 2003), (b) release of cytochrome *c* from the mitochondria into the cytosol induced by irradiation, toxic agents, cellular stress, or growth factor withdrawal (mitochondrial pathway; Green and Reed 1998), and (c) release of perforin and granzymes by NK cells and CTL (granule-mediated cytotoxicity; Lieberman 2003). Although caspase-independent cell death may occur under certain conditions (Lieberman 2003), the apoptosis-regulating pathways mostly converge at the level of activated caspase 3 and other effector caspases, which finally lead to those cellular changes associated with apoptosis (Kaufmann and Hengartner 2001). Because *T. gondii* targets apoptosis induced by a wide range of different stimuli (see above) it might be hypothesized that this is achieved by parasite interference with a component common to all

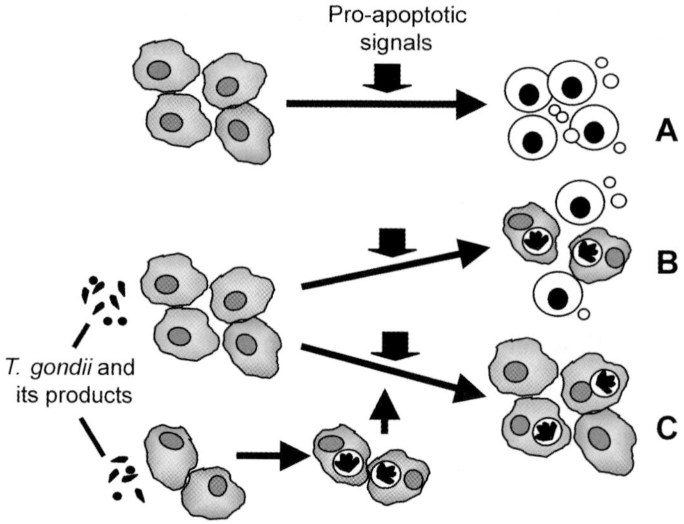

Fig. 1 Different mechanisms of *T. gondii* contribute to inhibition of host cell apoptosis during infection. **A** In the absence of the apoptosis-blocking features of the parasite, proapoptotic signals either related or unrelated to infection induce apoptosis. **B** In the presence of the parasite and its products, proapoptotic signals induce apoptosis in parasite-negative cells only, whereas direct interference of the parasite with signaling in parasite-positive host cells blocks apoptosis. **C** Apoptosis induced by proapoptotic signals may also be inhibited by indirect mechanisms, i.e., production of host-derived antiapoptotic molecules by *Toxoplasma*-infected cells. These molecules in turn block apoptosis in parasite-infected cells and uninfected bystander cells

pathways. To date, however, experimental evidence rather suggests that *T. gondii* inhibits apoptosis of host cells by different mechanisms. Whereas direct inhibition of host cell apoptosis by *T. gondii* is restricted to parasite-positive host cells, indirect mechanisms protect both infected and uninfected host cells against apoptosis (Fig. 1).

3.1 Indirect Mechanisms That Inhibit Apoptosis in Host Cells During Infection

Granulocyte colony-stimulating factor (G-CSF) and granulocyte-macrophage CSF (GM-CSF) secreted by *T. gondii*-infected human fibroblasts increased expression of the antiapoptotic Bcl-2 family member Mcl-1 and abolished apoptosis in neutrophils in vitro (Fig. 1C; Channon et al. 2002). A similar mechanism may also operate in vivo, because the inflammatory response to *T. gondii* after intraperitoneal infection of mice is accompanied by increased levels of A1, an antiapoptotic protein similar to Mcl-1 (Orlofsky et al. 1999). Importantly, parasite-induced expression of A1 led to increased numbers of peritoneal macrophages and neutrophils, possibly by inhibiting apoptosis of these cells. Although it has not been directly addressed, increased expression of A1 may result from parasite-driven secretion of inflammatory cytokines such as GM-CSF (Orlofsky et al. 1999). Expression of heat shock protein (HSP) 65 in inflammatory macrophages after infection with *T. gondii* strains of low virulence also prevents apoptosis of these cells (Hisaeda et al. 1997). Furthermore, depletion of $\gamma\delta$ T cells abolished parasite-driven HSP65 expression and induced apoptosis in macrophages from *Toxoplasma*-infected mice (Hisaeda et al. 1997). This indicates that priming of $\gamma\delta$ T lymphocytes by *Toxoplasma* or its products under certain conditions regulates apoptosis of inflammatory macrophages in an indirect fashion. Whether the increase in HSP65 expression is mediated by secretion of inflammatory cytokines in vivo awaits further clarification.

3.2
Direct Inhibtion of Host Cell Apoptosis by *T. gondii*

In addition to indirect mechanisms, *T. gondii* has evolved strategies to directly inhibit host cell apoptosis (Nash et al. 1998; Goebel et al. 1999, 2001; Payne et al. 2003). Such inhibition requires the presence of intracellular parasites that may directly interfere with signaling cascades of the host cell (Fig. 1B). It is therefore restricted to parasite-positive host cells and is not observed in parasite-negative bystander cells (Goebel et al. 1999). Direct inhibition of host cell apoptosis has been predominantly investigated in vitro after treatment of host cells with proapoptotic stimuli to obtain high-level apoptosis. However, it has recently also been observed in peritoneal exudate macrophages from *Toxoplasma*-infected mice, indicating that it operates in vivo as well (Orlofsky et al. 2002).

Efforts have been undertaken to unravel the underlying mechanisms of this parasite-host cell interaction (Fig. 2). The results show that *Toxoplasma* interferes with activation of the caspase cascade, thereby abolishing cleavage of nuclear target proteins (Goebel et al. 2001; Payne et al. 2003). Although activation of caspase 9 and caspase 3 via the mitochondrial pathway was unequivocally inhibited by *T. gondii*, clear evidence is still lacking that the death receptor pathway involving caspase 8 activation is also targeted by the parasite. Inhibition of caspase 8 activity by *T. gondii* has been demonstrated in murine fibroblasts by Payne et al.

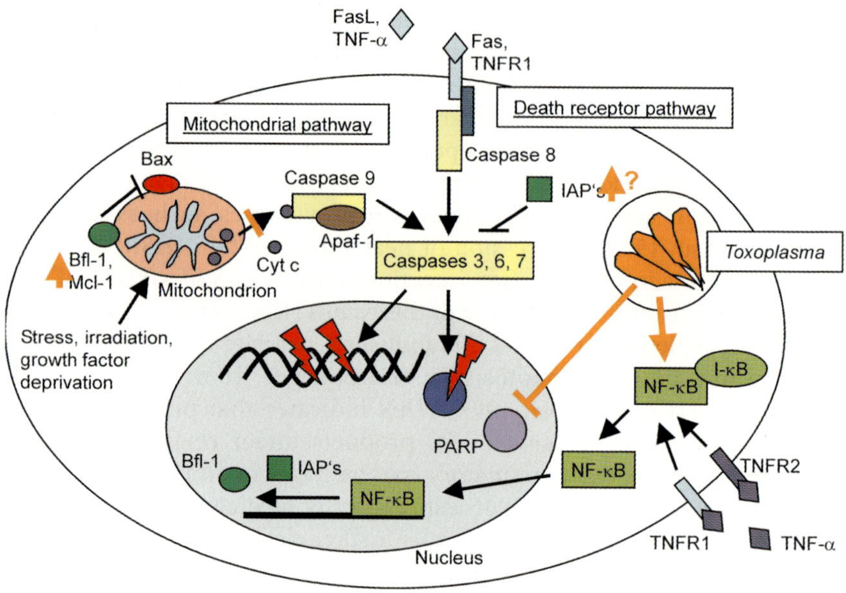

Fig. 2 Direct interference of intracellular *T. gondii* with signaling cascades regulating host cell apoptosis. Cell death may be initiated via the mitochondrial pathway or the death receptor pathway. Both pathways converge at the level of activated effector caspases, which then cleave cellular target proteins, leading to DNA fragmentation and apoptosis. A third major pathway initiated by the release of perforin and granzymes by cytotoxic NK cells and lymphocytes is not depicted. Intracellular parasites exert diverse mechanisms that are thought to contribute to direct inhibition of host cell apoptosis (indicated in *orange*): (i) Increased expression of antiapoptotic members of the Bcl-2 protein family, e.g. Bfl-1 or Mcl-1, (ii) inhibition of the cytochrome *c* release from mitochondria into the cytosol leading to decreased activation of the caspase 9 cascade, (iii) upregulation of IAPs possibly further inhibiting activity of caspases, (iv) activation of NF-κB by *T. gondii* in distinct cell types or under distinct conditions, thereby inducing the transcription of genes encoding antiapoptotic molecules, including Bfl-1 and IAPs, and (v) decreased cellular levels of PARP. The different antiapoptotic activities of *T. gondii* may be only partially related to each other

(2003) but was not found to occur in human histiocytic cells by others (Goebel et al. 2001), possibly indicating species- or cell type-specific differences. Furthermore, because the death receptor pathway partially overlaps with the mitochondrial pathway, at least in certain cell types (Kuwana et al. 1998; Scaffidi et al. 1999), inhibition of caspase 8 activity by *T. gondii* as described (Payne et al. 2003) does not necessarily exclude the involvement of the mitochondrial pathway. The effect of *T. gondii* on the death receptor signaling pathway thus remains to be established.

Several mechanisms have been described that accompany decreased caspase activation in the presence of intracellular *T. gondii.* Inhibition of caspase 9 and caspase 3 activity clearly correlated with decreased cytochrome *c* release from mitochondria into the cytosol of *T. gondii*-infected human-derived tumor cells (Fig. 2; Goebel et al. 2001). Such interference may be at least partially mediated by increased levels of antiapoptotic proteins of the Bcl-2 family, because expression of Mcl-1 and Bfl-1/A1, but not other antiapoptotic Bcl-2 proteins, was increased after parasitic infection (Goebel et al. 2001; Molestina et al. 2003). Although this remains to be established, increased levels of these proteins may reduce the activity of proapoptotic Bax, Bak, and Bok to induce the cytochrome *c* release from mitochondria of infected cells (Adams and Cory 1998; Kaufmann and Hengartner 2001). In addition to increased levels of antiapoptotic Bcl-2 family members, inhibition of apoptosis by *T. gondii* may also be related to a parasite-driven increase of inhibitors of apoptosis (IAP) proteins, namely, NAIP1, IAP1, and IAP2 (Fig. 2; Blader et al. 2001; Molestina et al. 2003). Because IAPs are known to block apoptosis by direct inhibition of distinct caspases, such a mechanism would operate downstream of cytochrome *c* release and activation of caspase 8 (Deveraux et al. 1998). Degradation of the poly(ADP-ribose) polymerase (PARP) in the presence of intracellular parasites as described by Goebel et al. (2001) is also possibly involved in the inhibition of apoptosis (Fig. 2). PARP is well known as a target of activated caspases but also promotes cell death under certain conditions (Tanaka et al. 1995; Jacobson and Jacobson 1999). Although direct evidence is still lacking, it thus appears plausible that diminished PARP levels in *Toxoplasma*-infected cells may inhibit apoptosis in a caspase-independent fashion (Alano et al. 2004). In conclusion, at least three different mechanisms have been described that possibly abolish host cell apoptosis by intracellular *T. gondii.* Further analyses are clearly required to determine whether they represent redundant mechanisms that are of functional significance for inhibition of apoptosis in *T. gondii*-infected cells. It will also be of major interest to determine whether these different mechanisms are linked to each other or are independently regulated. In this context, it is noteworthy that blockade of apoptosis in murine fibroblasts by *T. gondii* required the activation of NF-κB (Payne et al. 2003). NF-κB regulates proinflammatory responses during microbial infection and also functions as a cellular prosurvival pathway (Wang et al. 1998; Van Antwerp et al. 1998). This is mediated by activating the transcription of genes en-

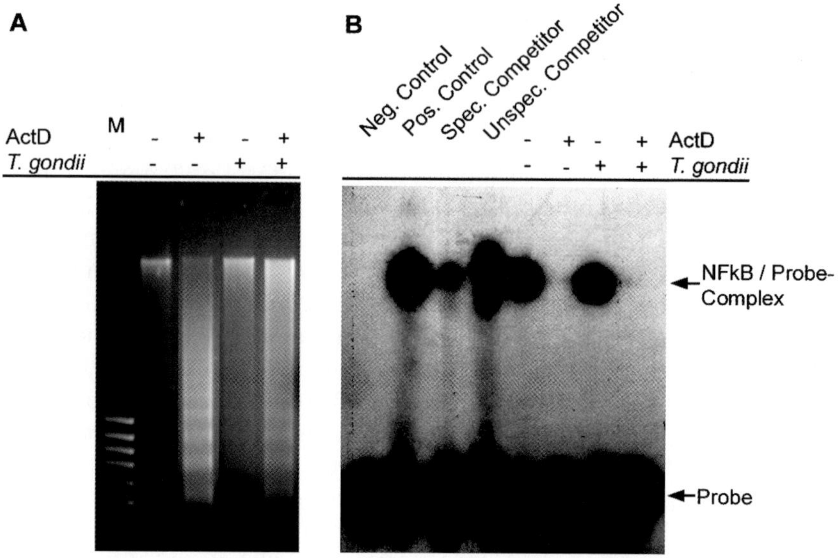

Fig. 3 Fragmentation of genomic DNA and DNA-binding activity of NF-κB in human-derived promyelocytic HL-60 cells after infection of *T. gondii*. HL-60 cells were infected with *T. gondii* or left uninfected as indicated and were then treated with actinomycin D (*ActD*) to induce apoptosis or left untreated. Eight hours after infection, genomic DNA was isolated and analyzed by agarose gel electrophoresis (**A**). In parallel, the binding activity of nuclear extracts to a radiolabeled NF-κB-specific oligonucleotide probe was assessed by electromobility shift assay (**B**). Note that ActD-induced DNA fragmentation was inhibited by *T. gondii* in the absence of DNA-binding activity of NF-κB

coding antiapoptotic molecules, including Bcl-2 and IAP proteins. Regulation of the antiapoptotic activity of *T. gondii* therefore appears to be partially linked. However, it should be noted that inhibition of apoptosis by *T. gondii* does not necessarily rely on activation of NF-κB. For example, in human-derived promyelocytic HL-60 cells, *T. gondii* inhibited actinomycin D-induced apoptosis in the absence of DNA-binding activity of NF-κB (Fig. 3). Furthermore, several groups have shown that *T. gondii* does not activate NF-κB and even inhibits activation of NF-κB induced after treatment of murine macrophages or fibroblasts with lipopolysaccharide (Butcher et al. 2001; Shapira et al. 2002). Therefore, NF-κB activation seems not to represent a general trait of intracellular *T. gondii*. Further studies are thus urgently required to unambiguously clarify the role of NF-κB in the inhibition of apoptosis by *T. gondii*.

Knowledge of the parasite molecule(s) that interfere with apoptosis-regulating signaling cascades of the host cell is still limited. Inhibtion of apoptosis required the presence of intracellular parasites, whereas intracellular replication of *T. gondii* was dispensable (Goebel et al. 1999). Notably, infection by a single viable parasite was thus sufficient to block apoptosis of the host cell. In addition, inhibition of apoptosis by *T. gondii* was reversible because it was abolished after killing the parasite (Nash et al. 1998). This indicates that the production and/or secretion of a *T. gondii* molecule are required to block apoptosis. *T. gondii* resides intracellularly in a specialized membrane-bound compartment, the so-called parasitophorous vacuole. It may therefore be hypothesized that the parasite molecule is either small enough to diffuse through pores within this membrane (Schwab et al. 1984) or is inserted into or even translocated across the membrane by yet unknown transport pathways (Cesbron-Delauw 1994; Beckers et al. 1994). Interestingly, Molestina et al. (2003) recently reported that the inhibitor of NF-κB activation, IκB, accumulates on the vacuolar membrane surrounding intracellular *T. gondii* in murine fibroblasts. This confirms the hypothesis that parasite molecules with access to the host cell cytoplasm may indeed interfere with signaling cascades of the host cell, thereby abolishing apoptosis.

3.3 Significance of Decreased Host Cell Apoptosis for Intracellular Survival

To grow and replicate, *T. gondii* relies on the sustained viability of its host cell. To date, however, the consequences of potential host cell apoptosis on the development of the parasite have not been directly addressed in detail. This question is not trivial because host cell apoptosis not only can disturb the intracellular development of the parasite but can also facilitate dissemination of certain pathogens (Gao and Kwaik 2000). Yamashita et al. (1998) reported that induction of apoptosis in *T. gondii*-infected target cells by CTL-mediated cytotoxicity does not lead to parasite death in vitro. In vivo, however, engulfment of apoptotic bodies containing viable parasites by phagocytic cells and subsequent elimination of the parasite may considerably contribute to parasite death due to apoptosis. Furthermore, whether such parasites are still able to invade new host cells also remains questionable. Because *T. gondii* actively invades its host cell and does not rely on the host cell's phagocytic machinery (Joiner and Dubremetz 1993), it appears unlikely that apoptosis significantly contributes to parasite dissemination within the host. It may thus be hypothesized that interference of *T. gondii* with apoptosis of its host cell facilitates the intracellular development of the parasite and increases parasitemia (Fig. 1B). This view is supported by Orlofsky et al. (2002), who reported a decreased number of parasites in apoptotic macrophage populations. Furthermore, it has been shown that several viruses rely on the inhibition of host cell apoptosis to ensure the developmental cycle (Barry and McFadden 1998). However, further investigations are urgently required to assess whether such direct inhibition of host cell apoptosis is also required for the development of *T. gondii.*

As discussed above (see Sect. 3.1), apoptosis may also be reduced in uninfected bystander cells as a result of indirect mechanisms triggered after infection with *T. gondii* (Fig. 1C). Such reduced levels of apoptosis in distinct cell populations of the host may lead to an enhanced inflammatory response to the parasite (Hisaeda et al. 1997; Orlofsky et al. 2002). Inflammatory leukocytes limit parasite replication by T cell-independent effector mechanisms (Sher at al. 1993) but can also induce host mortality due to overwhelming immunopathology (Gazzinelli et al. 1996). Enhanced survival of inflammatory cells thus likely fulfills a dual role during toxoplasmosis depending on the host and parasite strain as well as the dose and route of infection. Indeed, depletion of $\gamma\delta$ T cells in mice infected with a low-virulence strain of *T. gondii* abolished parasite-triggered inhibition of apoptosis in peritoneal macrophages and reduced host survival (Hisaeda et al. 1997). This suggests that indirect inhibition of apoptosis after infection with *T. gondii* contributes to efficient parasite control. On the contrary, reduced cell death possibly also increases the inflammatory response to highly virulent *T. gondii* strains, thereby leading to increased immunopathology and host mortality (Orlofsky et al. 2002). From these experimental studies in mice, it is supposed that *T. gondii*-mediated inhibition of apoptosis by indirect mechanisms fulfills a crucial role in the regulation of the immune response and the outcome of infection.

4 Roles of Apoptosis During Toxoplasmosis

As discussed above, apoptosis may potentially fulfill a variety of innate and adaptive effector as well as regulator functions in the response to *T. gondii.* However, given the fact that parasite infection exerts a variety of different effects on apoptosis of its host-derived cells, what are the actual roles of apoptosis during toxoplasmosis?

4.1
Apoptosis as an Effector Mechanism Against *T. gondii*

After infection with *T. gondii*, $CD8^+$ and $CD4^+$ T lymphocytes with cytolytic activity against parasite-infected target cells have been isolated (Hakim et al. 1991; Montoya et al. 1996). Because $CD8^+$ T cells represent the more relevant effector cell type in the effective control of *T. gondii* (Suzuki and Remington 1988; Gazzinelli et al. 1992), cytotoxicity via the induction of apoptosis has been thought to represent an important effector mechanism to control parasite replication. However, this issue is obscured by the fact that both $CD4^+$ and $CD8^+$ T cells not only exert cytotoxic effects but also produce the protective cytokine IFN-γ (Suzuki et al. 1988). With perforin knockout mice, granule-mediated cytotoxicity of T lymphocytes and NK cells has indeed been shown to be dispensable to control parasite replication during the acute stage of infection (Denkers et al. 1997). In contrast, perforin-mediated target cell lysis partially restricted tissue cyst development within the brain and decreased susceptibility of mice during chronic *Toxoplasma* encephalitis (Denkers et al. 1997). This possibly indicates that cells harboring latent bradyzoite-containing tissue cysts are more susceptible to apoptosis than tachyzoite-containing host cells. It raises the interesting hypothesis that bradyzoites and tachyzoites differ in their ability to interfere with signaling cascades of the host cell, with only the latter considerably blocking apoptosis. Alternatively, distinct conditions within the brains of infected mice may also lead to the higher susceptibility of tissue cyst-containing host cells to CTL-mediated cytotoxicity.

Beside granule-mediated cytotoxicity, CTL may also induce apoptosis via the death receptor pathway (Tibbetts et al. 2003). Its impact on the control of *T. gondii* has not been thoroughly investigated. However, CTL-mediated apoptosis via this pathway represents an important regulator of the immune response rather than an effector function against intracellular pathogens (Lieberman 2003). It may therefore play only a minor role in parasite control during toxoplasmosis.

Apoptosis as a suicide program of the cell in response to intracellular infection with *T. gondii* seems not to play a significant role as an innate effector mechanism against the parasite. This may be largely due to the broad antiapoptotic effects of *T. gondii* (see Sect. 3.2). However, because the effect of the latent bradyzoite stage of *T. gondii* on host cell apoptosis is unknown, it cannot be excluded that apoptosis restricts parasite development during chronic toxoplasmosis.

In conclusion, apoptosis plays only a minor role in the innate and adaptive defense against acute *T. gondii* infection but may contribute to parasite control during chronic toxoplasmic encephalitis.

4.2 Apoptosis in the Pathogenesis of Toxoplasmosis

In contrast to a limited role in combating the parasite, apoptosis plays a crucial role in the pathogenesis of toxoplasmosis. Induction of high levels of apoptosis in splenocytes (Mordue et al. 2001; Gavrilescu and Denkers 2001), Peyer's patch T cells (Liesenfeld et al. 1997), and peritoneal macrophages (Hisaeda et al. 1997) after infection of mice with *T. gondii* may lead to defective immune responses to the parasite. Importantly, extensive apoptosis was associated with high-level parasitemia and increased susceptibility of mice to death after infection (Mordue et al. 2001; Gavrilescu and Denkers 2001; Liesenfeld et al. 1996; Hisaeda et al. 1997). Splenocyte apoptosis was clearly less evident after infection of mice with *T. gondii* strains of lower virulence leading to reduced parasite burdens (Lee et al. 1999). The level of apoptosis in T lymphocytes and possibly other leukocytes thus appears to correlate with the induction of pathology during toxoplasmosis. Such apoptosis may dysregulate the immune response, thereby leading to unrestricted parasitemia and damage of host tissues at least under certain conditions. In contrast, low or intermediate levels of T cell death may contribute to the parasite's ability to establish persistent infections. Thus a tight regulation of T cell death may have a critical impact on a stable parasite-host interaction during toxoplasmosis. This view is supported by the finding that apoptosis is able to restrict the intraocular inflammation in response to *T. gondii* (Hu et al. 1999). Apoptosis of T cells may thus not only restrict the parasite-specific immune response but also immunopathological changes of host tissue at least in mice. It will be of major interest to determine whether apoptosis fulfills similar roles in pathogenesis of human toxoplasmosis.

5 Concluding Remarks

During recent years, considerable progress has been made in our knowledge of the interaction of *T. gondii* with host cell apoptosis. It emerges that apoptosis triggered by *T. gondii* after infection of mice represents a crucial factor in the pathogenesis of disease. For obvious reasons, much less is known on the role of apoptosis during human toxoplasmosis. Because the course of infection differs between mice and humans, it will be of major importance to unravel the impact of parasite-triggered apoptosis during human toxoplasmosis. Inhibition of apoptosis by *T. gondii*, on the other hand, may also represent a crucial factor for the parasite-host interaction and the course of disease. Although underlying mechanisms have been described, the exact cell biological and molecular bases and their regulation awaits further clarification. Particularly, the parasite molecules that interact with the apoptosis signaling cascades need to be characterized. This may then allow straightforward elucidation of the impact of decreased apoptosis on the course of disease. Unraveling the fascinating dual activity of *T. gondii* in host cell apoptosis and its regulation will undoubtedly further our understanding on the interaction of one of the most successful intracellular parasites.

Acknowledgements The authors gratefully acknowledge financial support by the Deutsche Forschungsgemeinschaft for their investigations. We also thank S. Goebel for substantial contributions.

References

Adams JM, Cory S (1998) The Bcl-2 protein family: arbiters of cell survival. Science 281:1322–1326

Alano CC, Ying W, Swanson RA (2004) Poly(ADP-ribose) polymerase-1 mediated cell death in astrocytes requires NAD+ depletion and mitochondrial permeability transition. J Biol Chem Feb 11 [Epub ahead of print]

Ammassari A, Murri R, Cingolani A, De Luca A, Antinori A (1996) AIDS-associated cerebral toxoplasmosis: an update on diagnosis and treatment. In: U Gross (ed) *Toxoplasma gondii*. Curr Topics Microbiol Immunol 219, Springer, Berlin, pp 209–222

Barry M, McFadden G (1998) Apoptosis regulators from DNA viruses. Curr Opin Immunol 10:422–430

Beckers CJ, Dubremetz JF, Mercereau-Puijalon O, Joiner KA (1994) The *Toxoplasma gondii* rhoptry protein ROP2 is inserted into the parasitophorous vacuole membrane, surrounding the intracellular parasite, and is exposed to the host cell cytoplasm. J Cell Biol 127:947–961

Blader IJ, Manger ID, Boothroyd JC (2001) Microarray analysis reveals previously unknown changes in *Toxoplasma gondii*-infected human cells. J Biol Chem 276:24223–24231

Butcher BA, Kim L, Johnson PF, Denkers EY (2001) *Toxoplasma gondii* tachyzoites inhibit proinflammatory cytokine induction in infected macrophages by preventing nuclear translocation of the transcription factor NF-κB. J Immunol 167:2193–2201

Caamano J, Tato C, Cai G, Villegas EN, Speirs K, Craig L, Alexander J, Hunter CA (2000) Identification of a role for NF-κB_2 in the regulation of apoptosis and in maintenance of T cell-mediated immunity to *Toxoplasma gondii*. J Immunol 165:5720–5728

Cesbron-Delauw M-F (1994) Dense-granule organelles of *Toxoplasma gondii*: their role in the host-parasite relationship. Parasitol Today 10:293–296

Channon JY, Miselis KA, Minns LA, Dutta C, Kasper LH (2002) *Toxoplasma gondii* induces granulocyte colony-stimulating factor and granulocyte-macrophage colony-stimulating factor secretion by human fibroblasts: implications for neutrophil apoptosis. Infect Immun 70:6048–6057

Curiel TJ, Krug EC, Purner MB, Poignard P, Berens RL (1993) Cloned human $CD4^+$ cytotoxic T lymphocytes specific for *Toxoplasma gondii* lyse tachyzoite-infected target cells. J Immunol 151:2024–2031

Denkers EY, Gazzinelli RT (1998) Regulation and function of T-cell mediated immunity during *Toxoplasma gondii* infection. Clin Microbiol Rev 11:569–588

Denkers EY, Yap G, Scharton-Kersten T, Charest H, Butcher BA, Caspar P, Heiny S, Sher A (1997) Perforin-mediated cytolysis plays a limited role in host resistance to *Toxoplasma gondii*. J Immunol 159:1903–1908

Deveraux QL, Roy N, Stennicke HR, van Arsdale T, Zhou Q, Srinivasula SM, Alnemri ES, Salvesen GS, Reed JC (1998) IAPs block apoptotic events induced by caspase-8 and cytochrome *c* by direct inhibition of distinct caspases. EMBO J 17:2215–2223

Gao L-Y, Kwaik YA (2000) The modulation of host cell apoptosis by intracellular bacterial pathogens. Trends Microbiol 8:306–313

Gavrilescu LC, Denkers EY (2001) IFN-γ overproduction and high level apoptosis are associated with high but not low virulence *Toxoplasma gondii* infection. J Immunol 167:902–909

Gavrilescu LC, Denkers EY (2003) Interleukin-12 p40- and Fas ligand-dependent apoptotic pathways involving STAT-1 phosphorylation are triggered during infection with a virulent strain of *Toxoplasma gondii*. Infect Immun 71:2577–2583

Gazzinelli R, Xu Y, Hieny S, Cheever A, Sher A (1992) Simultaneous depletion of $CD4^+$ and $CD8^+$ T lymphocytes is required to reactivate chronic infection with *Toxoplasma gondii*. J Immunol 149:175–180

Gazzinelli RT, Wysocka M, Hieny S, Scharton-Kersten T, Cheever A, Kühn R, Trinchieri G, Sher A (1996) In the absence of endogenous IL-10, mice acutely infected with *Toxoplasma gondii* succumb to a lethal immune response dependent on $CD4^+$ T cells and accompanied by overproduction of IL-12, IFN-γ, and TNF-α. J Immunol 157:798–805

Goebel S, Gross U, Lüder CGK (2001) Inhibition of host cell apoptosis by *Toxoplasma gondii* is accompanied by reduced activation of the caspase cascade and alterations of poly(ADP-ribose) polymerase expression. J Cell Sci 114:3495–3505

Goebel S, Lüder CGK, Gross U (1999) Invasion by *Toxoplasma gondii* protects human-derived HL-60 cells from actinomycin D-induced apoptosis. Med Microbiol Immunol 187:221–226

Green DR, Reed JC (1998) Mitochondria and apoptosis. Science 281:1309–1312

Grigg ME, Bonnefoy S, Hehl AB, Suzuki Y, Boothroyd JC (2001) Success and virulence in *Toxoplasma* as the result of sexual recombination between two distinct ancestries. Science 294:161–165

Hakim FT, Gazzinelli RT, Denkers E, Hieny S, Shearer GM, Sher A (1991) $CD8^+$ T cells from mice vaccinated against *Toxoplasma gondii* are cytotoxic for parasite-infected or antigen-pulsed host cells. J Immunol 147:2310–2316

Heussler VT, Küenzi P, Rottenberg S (2001) Inhibition of apoptosis by intracellular protozoan parasites. Int J Parasitol 31:1166–1176

Hisaeda H, Sakai T, Ishikawa H, Maekawa Y, Yasutomo K, Good RA, Himeno K (1997) Heat shock protein 65 induced by $\gamma\delta$ T cells prevents apoptosis of macrophages and contributes to host defense in mice infected with *Toxoplasma gondii.* J Immunol 159:2375–2381

Hu MS, Schwartzman JD, Yeaman GR, Collins J, Seguin R, Khan IA, Kasper LH (1999) Fas-FasL interaction involved in pathogenesis of ocular toxoplasmosis in mice. Infect Immun 67:928–935

Jacobson MK, Jakobson EL (1999) Discovering new ADP-ribose polymer cycles: protecting the genome and more. TIBS 24:415–417

Joiner KA, Dubremetz JF (1993) *Toxoplasma gondii*: a protozoan for the nineties. Infect Immun 61:1169–1172

Kaufmann SH, Hengartner MO (2001) Programmed cell death: alive and well in the new millennium. Trends Cell Biol 11:526–534

Khan IA, Matsuura T, Kasper LH (1996) Activation-mediated $CD4^+$ T cell unresponsiveness during acute *Toxoplasma gondii* infection in mice. Int Immunol 8:887–896

Kuwana T, Smith JS, Muzio M, Dixit V, Newmeyer DD, Kornbluth S (1998) Apoptosis induction by caspase-8 is amplified through the mitochondrial release of cytochrome *c*. J Biol Chem 273:16589–16594

Lee YH, Channon JY, Matsuura T, Schwartzman JD, Shin DW, Kasper LH (1999) Functional and quantitative analysis of splenic T cell responses following oral *Toxoplasma gondii* infection in mice. Exp Parasitol 91:212–221

Lieberman J (2003) The ABCs of granule-mediated cytotoxicity: new weapons in the arsenal. Nat Rev Immunol 3:361–370

Liesenfeld O, Kosek J, Remington JS, Suzuki Y (1996) Association of $CD4^+$ T cell-dependent, interferon-γ-mediated necrosis of the small intestine with genetic susceptibility of mice to peroral infection with *Toxoplasma gondii.* J Exp Med 184:597–607

Liesenfeld O, Kosek JC, Suzuki Y (1997) Gamma interferon induces Fas-dependent apoptosis of Peyer's patch T cells in mice following peroral infection with *Toxoplasma gondii.* Infect Immun 65:4682–4689

Lopes MF, da Veiga VF, Santos AR, Fonseca MEF, DosReis GA (1995) Activation-induced $CD4^+$ T cell death by apoptosis in experimental Chagas' disease. J Immunol 154:744–752

Lüder CGK, Gross U, Lopes MF (2001) Intracellular protozoan parasites and apoptosis: diverse strategies to modulate parasite-host interactions. Trends Parasitol 17:480–486

Luft BJ, Kansas G, Engleman EG, Remington JS (1984) Functional and quantitative alterations in T lymphocyte subpopulations in acute toxoplasmosis. J Infect Dis 150:761–767

McLeod R, Johnson J, Estes R, Mack D (1996) Immunogenetics in pathogenesis of and protection against toxoplasmosis. In:Gross U (ed) *Toxoplasma gondii.* Curr Topics Microbiol Immunol 219, Springer, Berlin, pp 95–112

Molestina RE, Payne TM, Coppens I, Sinai AP (2003) Activation of NF-κB by *Toxoplasma gondii* correlates with increased expression of antiapoptotic genes and localization of phosphorylated IκB to the parasitophorous vacuole membrane. J Cell Sci 116:4359–4371

Montoya JG, Lowe KE, Clayberger C, Moody D, Do D, Remington JS, Talib S, Subauste CS (1996) Human $CD4^+$ and $CD8^+$ T lymphocytes are both cytotoxic to *Toxoplasma gondii*-infected cells. Infect Immun 64:176–181

Mordue DG, Monroy F, La Regina M, Dinarello CA, Sibley LD (2001) Acute toxoplasmosis leads to lethal overproduction of Th1 cytokines. J Immunol 167:4574–4584

Nash PB, Purner MB, Leon RP, Clarke P, Duke RC, Curiel TJ (1998) *Toxoplasma gondii*-infected cells are resistant to multiple inducers of apoptosis. J Immunol 160:1824–1830

Opferman JT, Korsmeyer SJ (2003) Apoptosis in the development and maintenance of the immune system. Nat Immunol 4:410–415

Orlofsky A, Somogyi RD, Weiss LM, Prystowsky MB (1999) The murine antiapoptotic protein A1 is induced in inflammatory macrophages and constitutively expressed in neutrophils. J Immunol 163:412–419

Orlofsky A, Weiss LM, Kawachi N, Prystowsky MB (2002) Deficiency in the antiapoptotic protein A1-a results in a diminished acute inflammatory response. J Immunol 168:1840–1846

Payne TM, Molestina RE, Sinai AP (2003) Inhibition of caspase activation and a requirement for NF-κB function in the *Toxoplasma gondii*-mediated blockade of host apoptosis. J Cell Sci 116:4345–4358

Petersen E, Pollak A, Reiter-Owona I (2001) Recent trends in research on congenital toxoplasmosis. Int J Parasitol 31:115–144

Scaffidi C, Schmitz I, Zha J, Korsmeyer SJ, Krammer PH, Peter ME (1999) Differential modulation of apoptosis sensitivity in CD95 type I and type II cells. J Biol Chem 274:22532–22538

Schwab JC, Beckers CJM, Joiner KA (1994) The parasitophorous vacuole membrane surrounding intracellular *Toxoplasma gondii* functions as a molecular sieve. Proc Natl Acad Sci USA 91:509–513

Shapira S, Speirs K, Gerstein A, Caamano J, Hunter CA (2002) Suppression of NF-κB activation by infection with *Toxoplasma gondii.* J Infect Dis 185 [Suppl]:S66–S72

Sher A, Oswald IO, Hieny, S, Gazzinelli RT (1993) *Toxoplasma gondii* induces a T-independent IFN-γ response in NK cells which requires both adherent accessory cells and TNF-α. J Immunol 150:3982–3989

Sibley LD, Adams LB, Fukutomi Y, Krahenbuhl JL (1991) Tumor necrosis factor-α triggers antitoxoplasmal activity of IFN-γ primed macrophages. J Immunol 147:2340–2345

Strickland GT and Sayles PC (1977) Depressed antibody responses to a thymus-dependent antigen in toxoplasmosis. Infect Immun 15:184–190

Suzuki Y, Orellana MA, Schreiber RD, Remington JS (1988) Interferon-γ: the major mediator of resistance against *Toxoplasma gondii.* Science 240:516–518

Suzuki Y, Remington JS (1988) Dual regulation of resistance against *Toxoplasma gondii* infection by Lyt-2^+ and Lyt-1^+, L3T4^+ T cells in mice. J Immunol 140:3943–3946

Tanaka Y, Yoshihara K, Tohno Y, Kojima K, Kameoka M, Kamiya T (1995) Inhibition and down-regulation of poly(ADP-ribose) polymerase results in a marked resistance of HL-60 cells to various apoptosis-inducers. Cell Mol Biol 41:771–781

Tenter AM, Heckeroth AR, Weiss LM (2000) *Toxoplasma gondii*: from animals to humans. Int J Parasitol 30:1217–1258

Tibbetts MD, Zheng L, Lenardo MJ (2003) The death effector domain protein family: regulators of cellular homeostasis. Nature Immunol 4:404–409

Van Antwerp DJ, Martin SJ, Verma IM, Green DR (1998) Inhibition of TNF-induced apoptosis by NF-κB. Trends Cell Biol 8:107–111

Van Parijs L, Abbas AK (1996) Role of Fas-mediated cell death in the regulation of immune responses. Curr Opin Immunol 8:355–361

Wang CY, Mayo MW, Korneluk RG, Goeddel DV, Baldwin Jr AS (1998) NF-κB antiapoptosis: induction of TRAF1 and TRAF2 and c-IAP1 and c-IAP2 to suppress caspase-8 activation. Science 281:1680–1683

Wei S, Marches F, Borvak J, Zou W, Channon J, White M, Radke J, Cesbron-Delauw M-F, Curiel TJ (2002) *Toxoplasma gondii*-infected human myeloid dendritic cells induce T-lymphocyte dysfunction and contact-dependent apoptosis. Infect Immun 70:1750–1760

Williams GT (1994) Programmed cell death: a fundamental protective response to pathogens. Trends Microbiol 2:463–464

Wing EJ, Boehmer SM, Christner LK (1983) *Toxoplasma gondii*: decreased resistance to intracellular bacteria in mice. Exp Parasitol 56:1–8

Yamashita K, Yui K, Ueda M, Yano A (1998) Cytotoxic T-lymphocyte-mediated lysis of *Toxoplasma gondii*-infected target cells does not lead to death of intracellular parasites. Infect Immun 66:4651–4655

Yano A, Norose K, Yamashita K, Aosai F, Sugane K, Segawa K, Hayashi S (1987) Immune response to *Toxoplasma gondii*-analysis of suppressor T cells in a patient with symptomatic acute toxoplasmosis. J Parasitol 73:954–961

CTMI (2005) 289:239–258

Modulation of the Immune Response in the Nervous System by Rabies Virus

M. Lafon

Unité de Neuroimmunologie Virale, Département de Neuroscience, Institut Pasteur, 25 rue du Dr Roux, 75724 Paris Cedex 15, France
mlafon@pasteur.fr

Abstract Rabies virus (RABV) is a pathogen well-adapted to the nervous system, where it infects neurons. RABV is transmitted by the bite of an infected animal. It enters the nervous system via a motor neuron through the neuromuscular junction, or via a sensory nerve through nerve spindles. It then travels from one neuron to the next, along the spinal cord to the brain and the salivary glands. The virions are then excreted in the saliva of the animal and can be transmitted to another host by bite. Thus preservation of neuronal network integrity is crucial for the virus to be transmitted. Successful invasion of the nervous system by RABV seems to be the result of a subversive strategy based on the survival of infected neurons. This strategy includes protection against virus-mediated apoptosis and destruction of T cells that invade the CNS in response to infection.

Abbreviations

RABV	Rabies virus
PV	Pasteur virus
CVS	Challenge virus standard
ERA	Evelyn Rokitniki Abelseth
CNSCNS	Central nervous system
NS	Nervous system
NC	Nucleocapsid
gld	Generalized lymphoproliferative disorder
NK	Natural killer cell
CTL	Cytotoxic lymphocyte

1 Introduction

Rabies virus (RABV) is an enveloped bullet-shaped virus belonging to the *Rhabdoviridae* family, genus *Lyssavirus*. The viral particle consists of a membrane composed of host lipids and two viral proteins, G and M, surrounding a helical nucleocapsid (NC). NC is composed of a viral negative-strand RNA molecule protected by N protein, P protein, and the RNA-dependent RNA polymerase, L protein. This virus is a well-adapted pathogen of the mammalian nervous system (NS), where it infects mostly neurons forming cytoplasmic inclusions, the Negri bodies, which are pathognomonic of rabies. This virus is most often transmitted by bite. Aerosol infection has also been reported. Virus particles from infected saliva or progeny virus particles produced by muscle infection enter the NS by the neuromuscular junctions or via a sensory nerve through nerve spindles. RABV transport occurs exclusively by retrograde axonal transport and spreads to anatomically connected sites (Gillet et al. 1986; Kelly and Strick 2000). The property to be transported retrogradely into the CNS is under the control of the surface envelope G (Etessami et al. 2000; Mazarakis et al. 2001). After a primary wave of replication within motor neuron cell bodies, RABV travels from one neuron to another along the spinal cord to the brain before spreading to the salivary glands. Virions are then excreted in the saliva and transmitted to another host by bite.

The central nervous system (CNS), eyes, and testes have intrinsic mechanisms for controlling immune response compared to other organs. This can be understood as protective mechanisms against irreversible neurological damage induced by inflammatory or cytotoxic response (Griffith et al. 1995). Nevertheless, many virus infections of the

NS, RNA virus infections in particular, can be cleared from the CNS by immune mechanisms (Griffin 2003). This may indicate that on virus infection the immune privilege of the CNS is possibly not as strong as has been proposed. However, RABV may constitute an exception because once the virus has entered the CNS, RABV progression is interrupted neither by destruction of the infected neurons nor by the immune response and the issue of rabies is fatal every time. Escape of RABV from host defense mechanisms suggests that RABV has developed a subversive strategy to avoid functional neuron impairment and host immune mechanisms, which could compromise the infectious cycle of RABV in the CNS (Lafon 2004).

Viruses have developed an abundance of strategies to subvert normal cell systems and thereby promote virus replication. Preservation of cell integrity can be obtained by manipulating the pathways that activate apoptosis to increase survival of infected cells (Alcami and Koszinowski 2000). Viruses also utilize diverse mechanisms to escape host defenses. In particular, viruses can facilitate their own dissemination by developing stealth strategies to evade CTL and NK attack. Induction of apoptosis in the T cells or NK cells that are supposed to eliminate the infected cells is one of these strategies. Killing of T cells can be accomplished by the Fas/FasL apoptotic pathway. In this chapter, we describe how a RABV that is highly adapted to the CNS has adopted invasive strategies that preserve neuron integrity and evade the immune response in the CNS.

2
Mouse Models of Infection with RABV Strains

The question of how RABV has adopted a subversive strategy to evade the immune response can be answered, at least in part, with the illustration of experiments that use the mouse as the animal model for infection with RABV. Routes of infection consist of peripheral intramuscular or intranasal injection to mimic natural exposure by bite or aerosol. Several strains of RABV with different levels of pathogenicity in mice have been selected. Some, such as CVS (Challenge virus standard), ERA (Evelyn Rokitniki Abelseth) PV (Pasteur virus), CVS-N2C, and CVS-B2C, are the result of different passages in animals or cell culture. Primary sequences of ERA, PV and CVS, CVS-B2C, and N2C are different by only a few amino acids (Anilionis et al. 1981; Tordo et al. 1986; Poch et al. 1988; Morimoto et al. 1998). Other strains are escape mutants resulting in a single mutation in the G sequence, such as CVS-F3 (also named RV-194-2) or AVO-1 with an Arg→Gln mutation at position 333 affecting the antigenic site III (Dietzschold et al. 1983; Lafon et al. 1983; Seif et al. 1985) and also ts mutant (Iwasaki et al. 1977). After injection in the periphery by the intramuscular, subcutaneous, intraplantar, or intranasal route, the pathogenic virus strain CVS invades the spinal cord and almost all brain regions and causes a fatal acute encephalitis. In contrast, injected by the same routes the RABV PV, ERA, or CVS-F3 result in a nonfatal abortive disease characterized by a transient and restricted infection of the CNS followed by irreversible paralysis of the inoculated limbs (Weiland et al. 1992; Xiang et al. 1995; Hooper et al. 1998; Irwin et al. 1999; Galelli et al. 2000). The restricted CNS invasion by abortive RABV can be a result of the intrinsic property of the abortive RABV strain for progression in the CNS, due to triggering premature cell death of the infected neurons. Interruption of the neuronal network by apoptosis can be evoked for the restricted neuroinvasiveness of abortive RABV ERA and CVSB2C because these two strains can trigger apoptosis in vitro (Thoulouze et al. 1997, 2003a; Morimoto et al. 1999; Prehaud et al. 2003). In contrast, this is not the case with the abortive RABV(PV), which does not cause death of the neuronal cells it infects (Baloul and Lafon 2003). Thus restriction of neuroinvasiveness is likely the result of other mechanisms, including the control of abortive RABV infection by the immune response.

3
The Restricted Invasion of the Brain by Abortive RABV Is Controlled by T Cells

Histological studies on brain sections from RABV-infected mice revealed a perivascular infiltration of $CD4^+$ and $CD8^+$ T cells and the presence of $CD8^+$ T cells in the parenchyma (Sugamata et al. 1992; Weiland et al. 1992; Galelli et al. 2000). CNS-infiltrating cells can be isolated by Percoll gradient from the CNS of RABV-infected mice and characterized by flow cytometry analysis (Galelli et al. 2000). Very few mononuclear cells are detected in the CNS of uninfected mice. In contrast, the CNS of infected mice was infiltrated by $CD4^+$ and $CD8^+$ T cells, with $CD8^+$ T cells outnumbering $CD4^+$ T cells from day 10 onwards (Galelli et al. 2000). It is likely these infiltrating T cells migrate from the popliteal lymph node located near the site of virus injection or from the spleen, where infection triggers RABV-specific proliferation and cytolytic activity against RABV G-expressing target cells (Galelli et al. 2000). Nevertheless, migration of immune cells through the hematoencephalic barrier is likely a selective process because the proportion of cells expressing CD69, a marker of T, B, and NK cells, detected in the CNS was higher than that of those present in the popliteal lymph node population (data not shown).

The demonstration that T cells play a critical role in controlling abortive rabies has resulted from a series of experiments using different models of immunodeficient mice. Cyclophosphamide-treated mice were infected intramuscularly with abortive RABV(ERA) or RABV(tsCVS) (Iwasaki et al. 1977; Smith et al. 1982), mice lacking T cells, due to an alteration in the Foxn1 gene affecting both development of the thymus and the follicular hair growth (nude) were infected with the abortive RABV(PV) (Galelli et al. 2000), and mice lacking both T and B cells because of defects in the gene encoding recombinases needed for recombination of functional T cell receptor and immunoglobulin genes (RAG-1

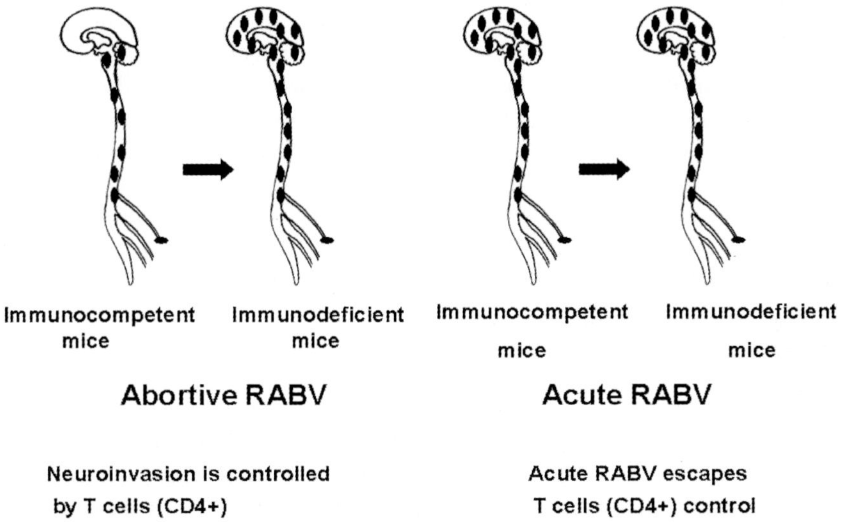

Fig. 1 Acute RABV escapes the immunosurveillance of $CD4^+$ T cells. BALB/c mice (immunocompetent mice) and nude mice (immunodeficient mice) received an intramuscular injection of 10^7 infectious particles of RABV(CVS) or (PV) into both hind legs. Invasion of BALB/c spinal cord and brain were limited after abortive RABV infection, whereas CNS was widely invaded by acute RABV. The CNS of the immunodeficient mice was extensively invaded by the abortive RABV strains, causing a complete invasion of the CNS and a severe lethal encephalitis similar to those triggered by the acute RABV(CVS) infection. In contrast, neither the neuroinvasiveness nor the lethality of acute RABV(CVS) was increased in nude mice, suggesting that acute RABV escapes the protection conferred by T cells

and RAG-2) were injected with the abortive RABV escaping mutants of CVS, CVS-F3, or RV-194-2 (Xiang et al. 1995; Hooper et al. 1998). In all instances, the CNS of the immunodeficient mice was extensively invaded by the abortive RABV strains, causing a severe lethal encephalitis similar to that triggered by the acute RABV(CVS) infection (Fig. 1). The transformation in immunodeficient mice of an abortive RABV infection into an acute RABV infection, indicates that T cells are a crucial factor in controlling RABV neuroinvasiveness. Among the T cell population, protection is conferred exclusively by $CD4^+$ T cells because fatal encephalopathogeny develops only after depletion of $CD4^+$ and not of $CD8^+$ T cells (Weiland et al. 1992; Galelli et al. 2000).

4
Apoptosis in the Abortive RABV(PV)-Infected CNS Is a T Cell-Mediated Event

T cells can participate in the clearance of infected cells by several mechanisms including cytotoxicity. The hypothesis that protective T cells trigger destruction of infected neurons was tested by comparing the number of apoptotic cells in the CNS of immunocompetent mice and T cell-deficient mice (nude mice) after infection with an abortive RABV strain. Analysis of sections of abortive RABV(PV)-infected cerebellum stained by the TUNEL technique or with Ab specific for activated caspase indicate that many cells were apoptotic and most of them were infected neurons (Galelli et al. 2000; Baloul and Lafon 2003). In contrast, despite a widespread infection of the CNS, nude mice infected with abortive RABV(PV) exhibited very few TUNEL-positive cells (Galelli et al. 2000; Baloul and Lafon 2003). This clearly indicates that neuron apoptosis in the RABV(PV)-infected CNS is a T cell-mediated event that may control virus spread in the CNS. The fact that RABV spread is controlled by T cells is consistent with the observation that neurotropic strains of RABV do not cause apoptosis of the cells they infect (Thoulouze et al. 1997, 2003a,b; Baloul and Lafon 2003; Lay et al. 2003; Prehaud et al. 2003) and that apoptosis triggered by a RABV strain inversely correlates with rabies pathogenicity (Morimoto et al. 1999, 2000; Thoulouze et al. 2003b).

The mechanisms involved in T-cell mediated neuronal apoptosis, which interrupts the propagation of RABV infection, are so far unknown. Virus-infected cells can be destroyed by classic cytotoxicity via the recognition of virus-loaded MHC class I involving $CD8^+$ T cells. Because rabies spread is controlled by CD4 T cells it is unlikely that direct cytotoxicity participates in T cell-mediated RABV clearance. Moreover, despite the large array of evidence that neurons can express class I MHC molecules after an injury, during phases of development, synaptic plasticity (Corriveau et al. 1998; Huh et al. 2000), or after infection (Pereira et al. 1994; Pereira and Simmons 1999; Kimura and Griffin 2000; Redwine et al. 2001) including RABV infection (Irwin et al. 1999), the elimination of infected neurons by direct cytotoxic mechanisms is highly questionable. Thus, with the exception of the data indicating that hippocampal neurons induced to express MHC class I mRNA become sensitive to lysis by cytotoxic T cells provided they are pulsed with MHC binding peptide (Medana et al. 2000), the demonstration that infected neurons are targets for T cell cytolysis via recognition of peptide processed with class I MHC in vivo is still lacking. Absence of cytotoxicity could be the result of the inability of neurons to load peptides into the groove of MHC class I molecules for antigen presentation (Joly and Oldstone 1992). The lack of evidence for class I MHC-mediated cytotoxicity of infected neurons may suggest that MHC class I expression during virus infection has functions that extend beyond a role in immune response.

In the CNS of mice infected with an abortive RABV strain, T cell-mediated apoptosis involves not only infected neurons but noninfected cells as well (Galelli et al. 2000). This suggests that neuronal apoptosis occurs in the absence of antigenic presentation. Such T cell-mediated activity has been described in human fetal neurons involving FasL, LFA1, and CD40 without class I MHC expression (Giuliani et al. 2003). It has also been proposed that T cells may exert clearance of infection from the CNS by secretion of antiviral cytokines such as INF-γ (Kundig et al. 1993; Binder and Griffin 2001; Chesler and Reiss 2002). T cell-dependent IFN-γ-mediated clearance of virus infection from the CNS could result from NK cell and macrophage activation or through the ability of INF-γ to induce type I nitric oxide synthase activity that can inhibit viral replication in neurons (Komatsu et al. 1996). The hypothesis that IFN-γ plays a major role in the T cell-mediated killing of RABV-infected neurons is strongly supported by the production of IFN-γ mRNAs not only in the abortive RABV(PV)-infected nervous parenchyma (Baloul and Lafon 2003) but also by migratory T cells (Galelli et al. 2000) and by the detection of NO in the course of RABV infection (Koprowski et al. 1993; Akaike et al. 1995; Van Dam et al. 1995), which was found to suppress RABV transcription (Ubol et al. 2001). A protective role for IFN-γ in RABV infection was clearly established with the observation that spread is more extensive in the brain of IFN-γ receptor-deficient mice than in normal mice (Hooper et al. 1998). In addition, antibodies play a critical role in clearance of abortive RABV from the CNS. Depletion of B cells with anti-isotype antibodies that compromise the capacity to mount an antibody response while leaving the T cell response intact demonstrated that B cells play an essential role in the clearance of the attenuated RABV(HEP) (Miller et al. 1978). Mice lacking B cells (J_{hD} knockout mice) and mice lacking T and B cells (RAG-2 mice) injected intranasally with the abortive RABV(CVS-F3) do not develop RABV-neutralizing antibodies and succumb to infection (Hooper et al. 1998). It is unknown so far by which mechanism, neutralization or antibody-dependent cell cytotoxicity (ADCC), antibodies trigger the control of CNS RABV infection.

After infection with an abortive RABV strain, normal mice and $CD8^+$ T cell-depleted mice developed high levels of IgG2a-specific RABV serum antibodies. In contrast, nude mice and $CD4^+$ T cell-depleted mice did not develop RABV-specific IgG antibodies (Weiland et al. 1992; Galelli et al. 2000). This indicates that $CD4^+$ T cells are an essential component of the protective B cell response. Thus the beneficial contribution of $CD4^+$ T cells in limiting the neuroinvasiveness of abortive RABV could be the result of the combination of providing help for B cells and of producing INF-γ.

5
T Cells Are Triggered to Death in the Acute RABV(CVS)-Infected CNS

In striking contrast to what was observed in an abortive RABV infection, intramuscular injection of the acute RABV(CVS) triggers only a limited expression of IFN-γ mRNAs in the CNS (Baloul and Lafon 2003) and only low levels of RABV-specific antibodies in the CNS (Fig. 2). Moreover, neither the neuroinvasiveness nor the lethality of acute RABV(CVS) was increased in nude mice (Camelo et al. 2000). These features suggest that T cell-mediated protection is impaired during acute rabies. However, after footpad injection, acute RABV(CVS) infection triggers RABV-specific proliferation and cytotoxic activity in popliteal lymph nodes similar to those observed after injection of abortive RABV(RV 194-2) (Irwin et al. 1999). In addition, increase in spleen cell number during infection and proliferation of splenocytes after addition of ConA were similar in the first 7 days of the acute RABV(CVS) and abortive RABV(PV) infection (Camelo et al. 2001b). Migratory mononuclear cells including cells expressing CD69 and T cells (CD3-, CD4-, and

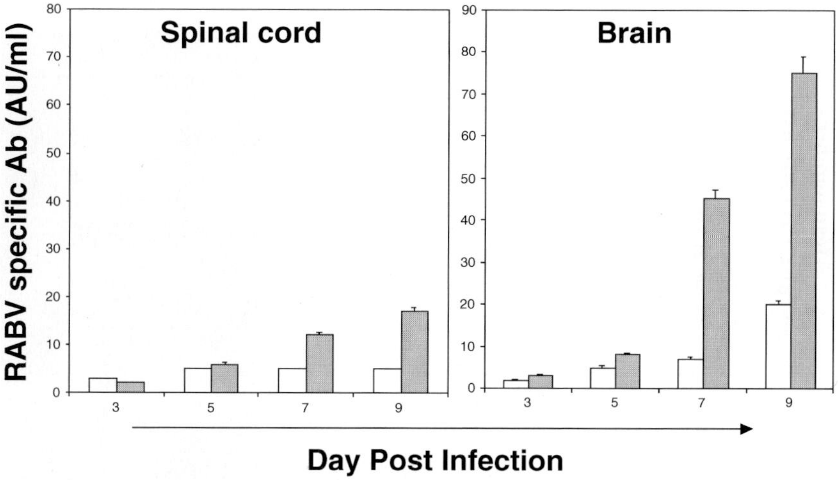

Fig. 2 Acute RABV(CVS) triggers limited amounts of RABV-specific Ab in the CNS compared to abortive RABV(PV) infection. After transcardiac perfusion, brains and spinal cord were dissected and homogenized 3, 5, 7, and 9 days p.i. with acute RABV(CVS) or abortive RABV(PV) infection. RABV-specific Ab were detected by ELISA as described previously (Lafon et al. 1990) in spinal cord or brain homogenates of RABV(CVS)-infected mice (*white histograms*) or RABV(PV)-infected mice (*gray histograms*). Results are presented as mean±SD titers of three mice. AU/ml=arbitary units per ml of homogenized NS. tissues

CD8-positive cells) were detected in the parenchyma of acute RABV(CVS)-infected mice (Camelo et al. 2000). The observation of similar levels of mononuclear cell infiltration in the two types of infection was consistent with the observation that abortive RABV(PV) and acute RABV(CVS) infections triggered similar levels of production of TNF-α (Baloul and Lafon 2003), a cytokine known to facilitate the passage of activated T and mononuclear cells across the hematoencephalic barrier (Seabrook and Hay 2001). The function of TNF-α as a mononuclear cell chemoattractant in RABV-infected CNS is illustrated by the lower influx of inflammatory cells into acute RABV(CVS)-infected mice lacking the TNF-α receptor (p55TNFR$^{-/-}$) than into the CNS of normal mice (Camelo et al. 2000).Thus the inefficiency of T cells for protection against the CVS infection is unlikely to be the result of a lack of immune activation in the periphery or of chemokine secretion by the infected CNS.

Nevertheless, despite similar migration of mononuclear cells into the CNS of acute and abortive RABV-infected mice, early in infection a ma-

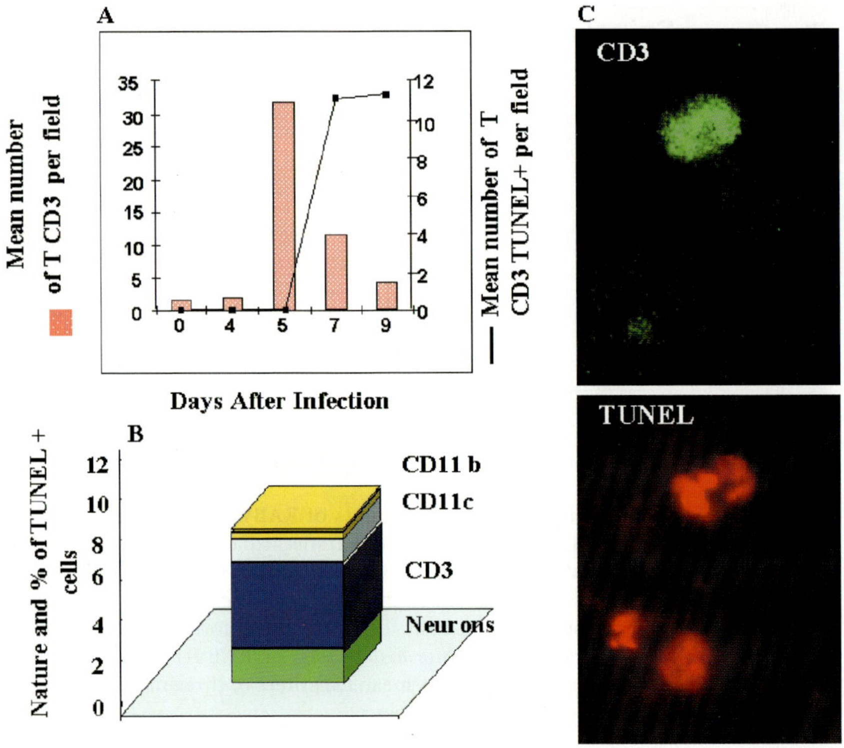

Fig. 3A–C Migratory CD3 T cells form the major group of TUNEL-positive cells in the acute RABV(CVS)-infected CNS. **A** Kinetics of CD3 T cells (*histogram*) and number of apoptotic cells (TUNEL-positive cells, *line*) per field of spinal cord sections in CVS-infected mice. Positive cells were detected by UV microscopic analysis of at least 20 different fields. **B** Mice were killed 7 days after injection with CVS into both hind legs. Spinal cord sections were double-stained with the TUNEL reagents and CD3 or CD11b and CD11c monoclonal antibody or antibody specific for RABV nucleocapsid. The number of TUNEL-positive cells and the percentages of CD11b/CD11c, nucleocapsid, or CD3 among the TUNEL-positive cells were determined in 20 microscope fields (magnification ×40) for each condition. **C** CD3-positive cells encounter apoptosis as shown by immunohistochemistry. One CD3 positive cell (*green, top panel*) is TUNEL positive (*red, bottom panel*) (×40 objective)

jor difference in the RABV(CVS) and RABV(PV) infection occurs on day 7. At that time, the number of infiltrating T cells sharply declined in the CNS of acute RABV(CVS)-infected mice (Fig. 3A), whereas it continued to increase in the CNS of abortive RABV(PV)-infected mice (Galelli et al. 2000). When the nature and the number of cells undergoing apoptosis in the CVS-infected CNS were analyzed further, it appeared that the RABV(CVS)-infected CNS contained seven times fewer apoptotic cells than the RABV(PV)-infected CNS and that in the RABV CVS infection most TUNEL-positive cells were $CD3^+$ T cells (Fig. 3B and C). Fifteen times as many $CD3^+$ T cells were apoptotic in brain sections of acute RABV(CVS)-infected mice than abortive RABV (PV)-infected mice (Baloul and Lafon 2003). Kinetic analysis of T cell mononuclear cell infiltration indicates that a decrease in the number of migratory cells coincides with the peak of apoptosis (day 7), suggesting strongly that T cell apoptosis may account for the disappearance of migratory cells in acute RABV infection.

6 $CD3^+$ T Cell Apoptosis Is a Fas/FasL-Mediated Mechanism

Fas (also called CD95 or Apo-1) is a molecule of the tumor necrosis factor-α (TNF-α) receptor family. It contains a cytoplasmic death domain. The ligand of Fas (FasL or CD95L) is a 40-kDa type II cell surface glycoprotein that belongs to the TNF family (Suda and Nagata 1994). In contrast to Fas, expression of FasL is restricted to a few cell types including cells of immune-privileged sites such as CNS. Binding of FasL to Fas results in a rapid caspase-dependent apoptosis of Fas-bearing cells (Green and Ferguson 2001) (Fig. 4A).

Because T cells are known to express Fas, the possibility that apoptosis was induced in migratory T cells by the Fas/FasL pathway in acute RABV(CVS) infection was investigated. It was expected that FasL is upregulated in early stages of the infection in RABV(CVS)-infected but not in RABV(PV)-infected CNS. Indeed, acute RABV infection induces an early production of FasL mRNA (day 5 p.i) in the spinal cord, whereas PV infection does not (Fig. 4B). FasL mRNA expression is followed by FasL protein expression, mainly on RABV(CVS)-infected neurons but also on a few uninfected cells (Baloul and Lafon 2003). Among the noninfected cells some were not of neuron origin (NeuN negative). The possibility that some of these cells are migratory T cells, which were found to produce FasL as well as Fas, cannot be excluded. However, most of the nonneuron cells of the rabies virus-infected NS that produce FasL do not have a lymphocyte morphology. They may be astrocytes or glial cells, known to express FasL (Bonetti et al. 1997; Bechmann et al. 1999; Choi et al. 1999; Kohji and Matsumoto 2000; Shin et al. 2002; Dockrell 2003). The production of FasL by uninfected cells may be a normal pro-

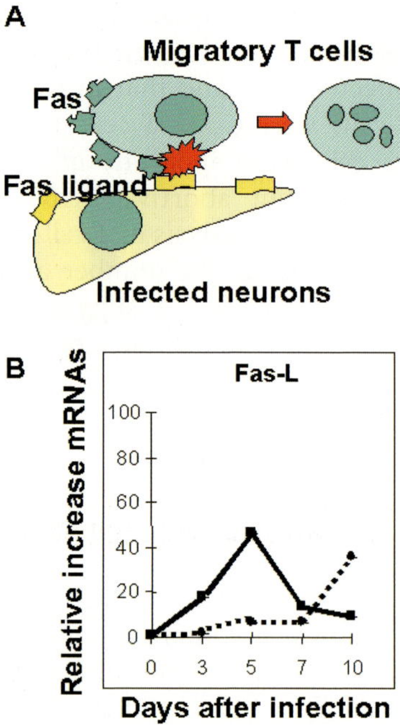

Fig. 4A, B FasL is expressed in the CNS of acute RABV(CVS)-infected mice. **A** Mechanism leading to the destruction of activated migratory T cells expressing Fas by RABV-infected neurons expressing FasL. **B** Kinetics of FasL in the CNS of RABV-infected mice. Relative FasL mRNAs expression was evaluated at day 0, 3, 5, 7, and 10 p.i. in spinal cord (left panel) of either CVS-infected (*solid line*, *black square*) or PV-infected mice (*dotted line*, *black circle*). Each point represents the arithmetic mean of results from 3–4 mice

cess in the context of immune privilege (Bechmann et al. 1999; Shin et al. 2002).

During acute RABV infection, FasL protein production was upregulated and concomitantly migratory $CD3^+$ T cells underwent apoptosis. The involvement of Fas/FasL as a pathway of T cell death was investigated in mice lacking a functional FasL (gld mice). Groups of gld mice and their normal counterparts (B6 mice) were infected with RABV(CVS). Despite similar levels of brain infection (data not shown), three times as many CD3/TUNEL double-positive cells were present in B6 mice than in gld mice infected with CVS (Fig. 5A). The lower levels of T cell death in mice lacking functional FasL strongly supports the notion that most of the T

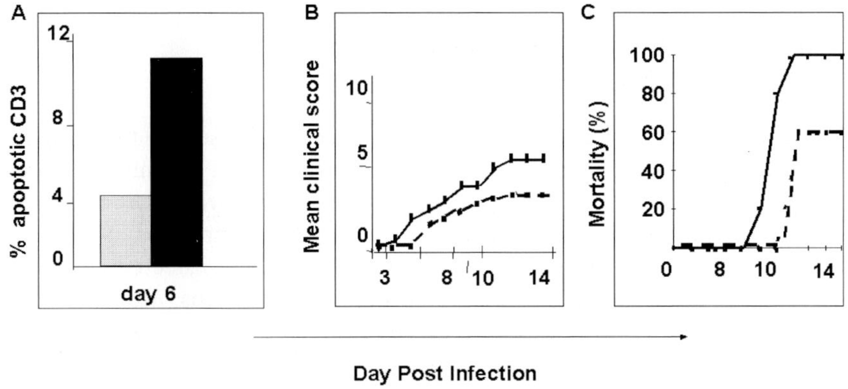

Fig. 5A–C Absence of FasL reduces severity of acute RABV disease in mice. B6 and B6 gld mice were injected i.m. with 10^7 PFU of acute RABV(CVS) in each hind leg. Percentage of TUNEL-positive CD3 cells was counted in brain sections of gld mice (*gray bar*) and B6 mice (*black bar*) on day 6 p.i. (**A**) and RABV(CVS) morbidity (**B**) and mortality (**C**) were compared in the two groups of mice. *Solid line* and *dashed lines* represent mortality and morbidity of B6 and gld mice, respectively. Experiments were performed twice, with *n*=5 and *n*=8, respectively

cells that undergo apoptosis in the CVS-infected CNS follow the Fas/FasL apoptotic pathway.

The destruction of T cells by FasL expression by neurons has been described in vitro (Medana et al. 2001) and also in nonviral inflammation of the CNS (Flugel et al. 2000). The destruction of recruited inflammatory cells by apoptosis has been observed in the CNS of mice infected with Sindbis virus (Havert et al. 2000) and Theiler's virus (Schlitt et al. 2003) and in retina infected with HCMV (Raftery et al. 2001) or RABV (Camelo et al. 2001a). The observation in RABV infection that fewer infiltrating CD3 T cells undergo apoptosis in gld than in B6 mice provides the first evidence that the apoptosis of infiltrating CD3 T cells during viral infection of the CNS occurs by the Fas/FasL pathway (Baloul et al. 2004).

Infection by the parasite *Trypanozoma cruzii* triggers cell death of B cells via Fas/FasL (Zuniga et al. 2002).The possibility that in acute RABV CNS infection, neuronal FasL upregulation could also kill infiltrating B cells has not been addressed. Because B cells play a central role in protection against RABV infection, this question deserves further attention.

7
Severity of Acute RABV(CVS) Disease Is Decreased in Mice Lacking FasL (gld Mice)

The involvement of FasL in CVS pathogenesis was assessed by comparing the progression of acute RABV(CVS) infection in B6 and gld mice from day 3 until day 12. The first signs of hindlimb weakness appeared as early as day 5 p.i. and paralysis of one hind leg appeared as early as day 6 p.i. in B6 mice. In contrast, in acute RABV(CVS)-infected gld mice, there was a 2-day delay in the development of symptoms. Moreover, the paralysis that developed was less severe in infected gld mice than in their normal counterparts. Overall, mean clinical score was significantly decreased in gld mice compared to normal mice (Fig. 5B). On day 14 p.i. all infected B6 mice were dead, whereas only 60% of the infected gld mice had died (Fig. 5C). These data indicate that acute RABV(CVS) infection is less severe in FasL-deficient mice than in normal mice and that the deletion of FasL favors the survival of mice infected with acute RABV. These data strongly support the hypothesis that FasL upregulation is a key factor in RABV pathogenesis. RABV-infected neurons can evoke T cell death as a mechanism of neuroprotection and a means of successful neuroinvasiveness.

8
How Does Acute RABV Infection Increase Neuronal FasL Expression?

The upregulation of FasL by a virus has been observed in HIV/SIV-infected macrophages and lymphocytes (Badley et al. 1996; Xu et al. 1997; Dockrell et al. 1998), in EBV-infected neutrophils (Tanner and Alfieri 1999), as well as in HCMV-infected human retinal pigment epithelial cells and dendritic cells (Chiou et al. 2001; Raftery et al. 2001). FasL transcription can be controlled in several ways (Li-Weber and Krammer 2003). Some evidence has been provided that Tat and Nef proteins of HIV-1 directly upregulate the expression of FasL (Xu et al. 1997; Li-Weber et al. 2000; Ghorpade et al. 2003). Hepatitis B virus X protein was also found to increase the expression of FasL (Lee et al. 2002). How RABV upregulates FasL has not been investigated yet. Direct regulation can take place, but it cannot be ruled out that CVS induces FasL production by an indirect mechanism, with the virus triggering FasL in cells other than those that it infects because noninfected cells express FasL.

9
Conclusion

Together, these observations indicate that RABV uses a subversive strategy based on the preservation of the neuronal network and the destruction of T cells that invade the CNS in response to infection. This is achieved in part because RABV cannot induce apoptosis of the neuron it infects, and in part because it can trigger cell death in the migratory infiltrating "protective" T cells. The Fas/FasL apoptotic pathway is thought to be involved in the disappearance of T cells because fewer CD3 T cells undergo apoptosis in gld mice (which have a *FasL* gene containing a point mutation resulting in a nonfunctional form of FasL). The upregulation of FasL is hypothesized to play a role in RABV pathogenesis because acute RABV-infected gld mice display lower rates of mortality, with death and symptoms occurring later, than acute RABV-infected normal mice. This suggests that the immunosubversive strategy of RABV involves the upregulation of FasL production to reduce the protective immune response. On the basis of our results, we propose a model to account for the success of acute RABV infection. After the injection of CVS into the hindlimbs, progressive infection of the spinal cord and the brain is accompanied by the production of the inflammatory cytokines and chemokines that attract lymphocytes, resulting in their migration across the hematoencephalitic barrier. However, acute RABV upregulates FasL. The protective T cells specifically triggered in the periphery are inefficient at controlling acute rabies virus encephalitis because they are destroyed by apoptosis shortly after their entry into the CNS. This facilitates the propagation of the neuroinvasive RABV through the CNS to the brain, resulting in the completion of the RABV infectious cycle in the host. A similar mechanism could be evoked to explain why mice lacking functional FasL showed delayed and reduced mortality after infection with the Murray Valley encephalitis flavivirus (Licon Luna et al. 2002). Because upregulation of FasL could be a critical factor that contributes to virus pathogenesis and because FasL transcription can be modulated (Cippitelli et al. 2002), it is tempting to propose that transcriptional regulation of FasL gene expression may eventually be manipulated to fulfill a therapeutic purpose.

Acknowledgements This work was supported by institutional grants from the Institut Pasteur, Paris, France. The contribution of Leïla Baloul, Serge Camelo, and Anne Galelli in the exploration of RABV immunosubversive strategy is greatly acknowledged.

References

Akaike T, Weihe E, Schaefer M, Fu ZF, Zheng YM, Vogel W, Schmidt H, Koprowski H, Dietzschold B (1995) Effect of neurotropic virus infection on neuronal and inducible nitric oxide synthase activity in rat brain. J Neurovirol 1:118–125

Alcami A, Koszinowski UH (2000) Viral mechanisms of immune evasion. Trends Microbiol 8:410–418

Anilionis A, Wunner WH, Curtis PJ (1981) Structure of the glycoprotein gene in rabies virus. Nature 294:275–278

Badley AD, McElhinny JA, Leibson PJ, Lynch DH, Alderson MR, Paya CV (1996) Upregulation of Fas ligand expression by human immunodeficiency virus in human macrophages mediates apoptosis of uninfected T lymphocytes. J Virol 70:199–206

Baloul L, Lafon M (2003) Apoptosis and rabies virus neuroinvasion. Biochimie 85:777–788

Baloul L, Camelo S, Lafon M (2004) Up-regulation of FasL in the CNS: a mechanism of immune evasion by rabies virus. Journal of Neurovirology (in press)

Bechmann I, Mor G, Nilsen J, Eliza M, Nitsch R, Naftolin F (1999) FasL (CD95L, Apo1L) is expressed in the normal rat and human brain: evidence for the existence of an immunological brain barrier. Glia 27:62–74

Binder GK, Griffin DE (2001) Interferon-gamma-mediated site-specific clearance of alphavirus from CNS neurons. Science 293:303–306

Bonetti B, Pohl J, Gao YL, Raine CS (1997) Cell death during autoimmune demyelination: effector but not target cells are eliminated by apoptosis. J Immunol 159:5733–5741

Camelo S, Castellanos J, Lafage M, Lafon M (2001a) Rabies virus ocular disease: T-cell-dependent protection is under the control of signaling by the p55 tumor necrosis factor alpha receptor, p55TNFR. J Virol 75:3427–3434

Camelo S, Lafage M, Galelli A, Lafon M (2001b) Selective role for the p55 Kd TNF-alpha receptor in immune unresponsiveness induced by an acute viral encephalitis. J Neuroimmunol 113:95–108

Camelo S, Lafage M, Lafon M (2000) Absence of the p55 Kd TNF-alpha receptor promotes survival in rabies virus acute encephalitis. J Neurovirol 6:507–518

Chesler DA, Reiss CS (2002) The role of IFN-gamma in immune responses to viral infections of the central nervous system. Cytokine Growth Factor Rev 13:441–454

Chiou SH, Liu JH, Hsu WM, Chen SS, Chang SY, Juan LJ, Lin JC, Yang YT, Wong WW, Liu CY, Lin YS, Liu WT, Wu CW (2001) Up-regulation of Fas ligand expression by human cytomegalovirus immediate-early gene product 2: a novel mechanism in cytomegalovirus-induced apoptosis in human retina. J Immunol 167:4098–4103

Choi C, Park JY, Lee J, Lim JH, Shin EC, Ahn YS, Kim CH, Kim SJ, Kim JD, Choi IS, Choi IH (1999) Fas ligand and Fas are expressed constitutively in human astrocytes and the expression increases with IL-1, IL-6, TNF-alpha, or IFN-gamma. J Immunol 162:1889–1895

Cippitelli M, Fionda C, Di Bona D, Di Rosa F, Lupo A, Piccoli M, Frati L, Santoni A (2002) Negative regulation of CD95 ligand gene expression by vitamin D3 in T lymphocytes. J Immunol 168:1154–1166
Corriveau RA, Huh GS, Shatz CJ (1998) Regulation of class I MHC gene expression in the developing and mature CNS by neural activity. Neuron 21:505–520
Dietzschold B, Wunner WH, Wiktor TJ, Lopes AD, Lafon M, Smith CL, Koprowski H (1983) Characterization of an antigenic determinant of the glycoprotein that correlates with pathogenicity of rabies virus. Proc Natl Acad Sci USA 80:70–74
Dockrell DH (2003) The multiple roles of Fas ligand in the pathogenesis of infectious diseases. Clin Microbiol Infect 9:766–779
Dockrell DH, Badley AD, Villacian JS, Heppelmann CJ, Algeciras A, Ziesmer S, Yagita H, Lynch DH, Roche PC, Leibson PJ, Paya CV (1998) The expression of Fas ligand by macrophages and its upregulation by human immunodeficiency virus infection. J Clin Invest 101:2394–2405
Etessami R, Conzelmann KK, Fadai-Ghotbi B, Natelson B, Tsiang H, Ceccaldi PE (2000) Spread and pathogenic characteristics of a G-deficient rabies virus recombinant: an in vitro and in vivo study. J Gen Virol 81:2147–2153
Flugel A, Schwaiger FW, Neumann H, Medana I, Willem M, Wekerle H, Kreutzberg GW, Graeber MB (2000) Neuronal FasL induces cell death of encephalitogenic T lymphocytes. Brain Pathol 10:353–364
Galelli A, Baloul L, Lafon M (2000) Abortive rabies virus central nervous infection is controlled by T lymphocyte local recruitment and induction of apoptosis. J Neurovirol 6:359–372
Ghorpade A, Holter S, Borgmann K, Persidsky R, Wu L (2003) HIV-1 and IL-1 beta regulate Fas ligand expression in human astrocytes through the NF-kappa B pathway. J Neuroimmunol 141:141–149
Gillet JP, Derer P, Tsiang H (1986) Axonal transport of rabies virus in the central nervous system of the rat. J Neuropathol Exp Neurol 45:619–634
Giuliani F, Goodyer CG, Antel JP, Yong VW (2003) Vulnerability of human neurons to T cell-mediated cytotoxicity. J Immunol 171:368–379
Green DR, Ferguson TA (2001) The role of Fas ligand in immune privilege. Nat Rev Mol Cell Biol 2:917–924
Griffin DE (2003) Immune responses to RNA-virus infections of the CNS. Nat Rev Immunol 3:493–502
Griffith TS, Brunner T, Fletcher SM, Green DR, Ferguson TA (1995) Fas ligand-induced apoptosis as a mechanism of immune privilege. Science 270:1189–1192
Havert MB, Schofield B, Griffin DE, Irani DN (2000) Activation of divergent neuronal cell death pathways in different target cell populations during neuroadapted sindbis virus infection of mice. J Virol 74:5352–5356
Hooper DC, Morimoto K, Bette M, Weihe E, Koprowski H, Dietzschold B (1998) Collaboration of antibody and inflammation in clearance of rabies virus from the central nervous system. J Virol 72:3711–3719
Huh GS, Boulanger LM, Du H, Riquelme PA, Brotz TM, Shatz CJ (2000) Functional requirement for class I MHC in CNS development and plasticity. Science 290:2155–2159

Irwin DJ, Wunner WH, Ertl HC, Jackson AC (1999) Basis of rabies virus neurovirulence in mice: expression of major histocompatibility complex class I and class II mRNAs. J Neurovirol 5:485–494

Iwasaki Y, Gerhard W, Clark HF (1977) Role of host immune response in the development of either encephalitic or paralytic disease after experimental rabies infection in mice. Infect Immun 18:220–225

Joly E, Oldstone MB (1992) Neuronal cells are deficient in loading peptides onto MHC class I molecules. Neuron 8:1185–1190

Kelly RM, Strick PL (2000) Rabies as a transneuronal tracer of circuits in the central nervous system. J Neurosci Methods 103:63–71

Kimura T, Griffin DE (2000) The role of $CD8^+$ T cells and major histocompatibility complex class I expression in the central nervous system of mice infected with neurovirulent Sindbis virus. J Virol 74:6117–6125

Kohji T, Matsumoto Y (2000) Coexpression of Fas/FasL and Bax on brain and infiltrating T cells in the central nervous system is closely associated with apoptotic cell death during autoimmune encephalomyelitis. J Neuroimmunol 106:165–171

Komatsu T, Bi Z, Reiss CS (1996) Interferon-gamma induced type I nitric oxide synthase activity inhibits viral replication in neurons. J Neuroimmunol 68:101–108

Koprowski H, Zheng YM, Heber-Katz E, Fraser N, Rorke L, Fu ZF, Hanlon C, Dietzschold B (1993) In vivo expression of inducible nitric oxide synthase in experimentally induced neurologic diseases. Proc Natl Acad Sci USA 90:3024–3027

Kundig TM, Hengartner H, Zinkernagel RM (1993) T cell-dependent IFN-gamma exerts an antiviral effect in the central nervous system but not in peripheral solid organs. J Immunol 150:2316–2321

Lafon M (2004) Subversive neuroinvasive strategy of rabies virus. Archives of Virology S18:149–159

Lafon M, Edelman L, Bouvet JP, Lafage M, Montchatre E (1990) Human monoclonal antibodies specific for the rabies virus glycoprotein and N protein. J Gen Virol 71:1689–1696

Lafon M, Wiktor TJ, Macfarlan RI (1983) Antigenic sites on the CVS rabies virus glycoprotein: analysis with monoclonal antibodies. J Gen Virol 64:843–851

Lay S, Prehaud C, Dietzschold B, Lafon M (2003) Glycoprotein of nonpathogenic rabies viruses is a major inducer of apoptosis in human jurkat T cells. Ann NY Acad Sci 1010:577–581

Lee MO, Choi YH, Shin EC, Kang HJ, Kim YM, Jeong SY, Seong JK, Yu DY, Cho H, Park JH, Kim SJ (2002) Hepatitis B virus X protein induced expression of interleukin 18 (IL-18): a potential mechanism for liver injury caused by hepatitis B virus (HBV) infection. J Hepatol 37:380–386

Licon Luna RM, Lee E, Mullbacher A, Blanden RV, Langman R, Lobigs M (2002) Lack of both Fas ligand and perforin protects from flavivirus-mediated encephalitis in mice. J Virol 76:3202–3211

Li-Weber M, Krammer PH (2003) Function and regulation of the CD95 (APO-1/Fas) ligand in the immune system. Semin Immunol 15:145–157

Li-Weber M, Laur O, Dern K, Krammer PH (2000) T cell activation-induced and HIV tat-enhanced CD95(APO-1/Fas) ligand transcription involves NF-kappaB. Eur J Immunol 30:661–670

Mazarakis ND, Azzouz M, Rohll JB, Ellard FM, Wilkes FJ, Olsen AL, Carter EE, Barber RD, Baban DF, Kingsman SM, Kingsman AJ, O'Malley K, Mitrophanous KA (2001) Rabies virus glycoprotein pseudotyping of lentiviral vectors enables retrograde axonal transport and access to the nervous system after peripheral delivery. Hum Mol Genet 10:2109–2121

Medana I, Li Z, Flugel A, Tschopp J, Wekerle H, Neumann H (2001) Fas ligand (CD95L) protects neurons against perforin-mediated T lymphocyte cytotoxicity. J Immunol 167:674–681

Medana IM, Gallimore A, Oxenius A, Martinic MM, Wekerle H, Neumann H (2000) MHC class I-restricted killing of neurons by virus-specific $CD8^+$ T lymphocytes is effected through the Fas/FasL, but not the perforin pathway. Eur J Immunol 30:3623–3633

Miller A, Morse HC, 3rd, Winkelstein J, Nathanson N (1978) The role of antibody in recovery from experimental rabies. I. Effect of depletion of B and T cells. J Immunol 121:321–326

Morimoto K, Foley HD, McGettigan JP, Schnell MJ, Dietzschold B (2000) Reinvestigation of the role of the rabies virus glycoprotein in viral pathogenesis using a reverse genetics approach. J Neurovirol 6:373–381

Morimoto K, Hooper DC, Carbaugh H, Fu ZF, Koprowski H, Dietzschold B (1998) Rabies virus quasispecies: implications for pathogenesis. Proc Natl Acad Sci USA 95:3152–3156

Morimoto K, Hooper DC, Spitsin S, Koprowski H, Dietzschold B (1999) Pathogenicity of different rabies virus variants inversely correlates with apoptosis and rabies virus glycoprotein expression in infected primary neuron cultures. J Virol 73:510–518

Pereira RA, Simmons A (1999) Cell surface expression of H2 antigens on primary sensory neurons in response to acute but not latent herpes simplex virus infection in vivo. J Virol 73:6484–6489

Pereira RA, Tscharke DC, Simmons A (1994) Upregulation of class I major histocompatibility complex gene expression in primary sensory neurons, satellite cells, and Schwann cells of mice in response to acute but not latent herpes simplex virus infection in vivo. J Exp Med 180:841–850

Poch O, Tordo N, Keith G (1988) Sequence of the 3386 3′ nucleotides of the genome of the AVO1 strain rabies virus: structural similarities in the protein regions involved in transcription. Biochimie 70:1019–1029

Prehaud C, Lay S, Dietzschold B, Lafon M (2003) Glycoprotein of nonpathogenic rabies viruses is a key determinant of human cell apoptosis. J Virol 77:10537–10547

Raftery MJ, Schwab M, Eibert SM, Samstag Y, Walczak H, Schonrich G (2001) Targeting the function of mature dendritic cells by human cytomegalovirus: a multilayered viral defense strategy. Immunity 15:997–1009

Redwine JM, Buchmeier MJ, Evans CF (2001) In vivo expression of major histocompatibility complex molecules on oligodendrocytes and neurons during viral infection. Am J Pathol 159:1219–1224

Schlitt BP, Felrice M, Jelachich ML, Lipton HL (2003) Apoptotic cells, including macrophages, are prominent in Theiler's virus-induced inflammatory, demyelinating lesions. J Virol 77:4383–4388

Seabrook TJ, Hay JB (2001) Intracerebroventricular infusions of TNF-alpha preferentially recruit blood lymphocytes and induce a perivascular leukocyte infiltrate. J Neuroimmunol 113:81–88

Seif I, Coulon P, Rollin PE, Flamand A (1985) Rabies virulence: effect on pathogenicity and sequence characterization of rabies virus mutations affecting antigenic site III of the glycoprotein. J Virol 53:926–934

Shin DH, Lee E, Kim HJ, Kim S, Cho SS, Chang KY, Lee WJ (2002) Fas ligand mRNA expression in the mouse central nervous system. J Neuroimmunol 123:50–57

Smith JS, McCelland CL, Reid FL, Baer GM (1982) Dual role of the immune response in street rabiesvirus infection of mice. Infect Immun 35:213–221

Suda T, Nagata S (1994) Purification and characterization of the Fas-ligand that induces apoptosis. J Exp Med 179:873–879

Sugamata M, Miyazawa M, Mori S, Spangrude GJ, Ewalt LC, Lodmell DL (1992) Paralysis of street rabies virus-infected mice is dependent on T lymphocytes. J Virol 66:1252–1260

Tanner JE, Alfieri C (1999) Epstein-Barr virus induces Fas (CD95) in T cells and Fas ligand in B cells leading to T-cell apoptosis. Blood 94:3439–3447

Thoulouze MI, Lafage M, Montano-Hirose JA, Lafon M (1997) Rabies virus infects mouse and human lymphocytes and induces apoptosis. J Virol 71:7372–7380

Thoulouze MI, Lafage M, Yuste VJ, Baloul L, Edelman L, Kroemer G, Israel N, Susin SA, Lafon M (2003a) High level of Bcl-2 counteracts apoptosis mediated by a live rabies virus vaccine strain and induces long-term infection. Virology 314:549–561

Thoulouze MI, Lafage M, Yuste VJ, Kroemer G, Susin SA, Israel N, Lafon M (2003b) Apoptosis inversely correlates with rabies virus neurotropism. Ann NY Acad Sci 1010:598–603

Tordo N, Poch O, Ermine A, Keith G, Rougeon F (1986) Walking along the rabies genome: is the large G-L intergenic region a remnant gene? Proc Natl Acad Sci USA 83:3914–3918

Ubol S, Hiriote W, Anuntagool N, Utaisincharoen P (2001) A radical form of nitric oxide suppresses RNA synthesis of rabies virus. Virus Res 81:125–132

Van Dam AM, Bauer J, Man AHWK, Marquette C, Tilders FJ, Berkenbosch F (1995) Appearance of inducible nitric oxide synthase in the rat central nervous system after rabies virus infection and during experimental allergic encephalomyelitis but not after peripheral administration of endotoxin. J Neurosci Res 40:251–260

Weiland F, Cox JH, Meyer S, Dahme E, Reddehase MJ (1992) Rabies virus neuritic paralysis: immunopathogenesis of nonfatal paralytic rabies. J Virol 66:5096–5099

Xiang ZQ, Knowles BB, McCarrick JW, Ertl HC (1995) Immune effector mechanisms required for protection to rabies virus. Virology 214:398–404

Xu XN, Screaton GR, Gotch FM, Dong T, Tan R, Almond N, Walker B, Stebbings R, Kent K, Nagata S, Stott JE, McMichael AJ (1997) Evasion of cytotoxic T lymphocyte (CTL) responses by nef-dependent induction of Fas ligand (CD95L) expression on simian immunodeficiency virus-infected cells. J Exp Med 186:7–16

Zuniga E, Motran CC, Montes CL, Yagita H, Gruppi A (2002) *Trypanosoma cruzi* infection selectively renders parasite-specific IgG^+ B lymphocytes susceptible to Fas/Fas ligand-mediated fratricide. J Immunol 168:3965–3973

CTMI (2005) 289:259–292

Apoptotic Cells at the Crossroads of Tolerance and Immunity

M. Škoberne (✉) · A.-S. Beignon · M. Larsson · N. Bhardwaj

Cancer Institute, NYU School of Medicine, 550 First Avenue, MSB507, New York, NY 10016, USA
mojca.skoberne@med.nyu.edu

Abstract Clearance of apoptotic cells by phagocytes can result in either anti-inflammatory and immunosuppressive effects or prostimulatory consequences through presentation of cell-associated antigens to T cells. The differences in outcome are due to the conditions under which apoptosis is induced, the type of phagocytic cell, the nature of the receptors involved in apoptotic cell capture, and the milieu in which phagocytosis of apoptotic cells takes place. Preferential ligation of specific receptors on professional antigen-presenting cells (dendritic cells) has been proposed to induce potentially tolerogenic signals. On the other hand, dendritic cells can efficiently process and present antigens from pathogen-infected apoptotic cells to T cells. In this review, we discuss how apoptotic cells manipulate immunity through interactions with dendritic cells.

Abbreviations

APC	Antigen-presenting cell
CR	Complement receptor
CMV	Cytomegalovirus
DC	Dendritic cell
DTH	Delayed type hypersensitivity
GM-CSF	Granulocyte-macrophage colony-stimulating factor
HSP	Heat shock protein
HSV	Herpes simplex virus
IC	Immune complex
Ig	Immunoglobulin
IFN	Interferon
IL	Interleukin
LPS	Lipopolysaccharide
LPC	Lysophosphatidylcholine
MHC	Major histocompatibility complex
MBL	Mannose-binding lectin
mDC	Myeloid dendritic cells
LOX-1	Oxidized low-density receptor 1
PGE_2	Prostaglandin E_2
PAF	Platelet activating factor
PS	Phosphatidylserine
PSR	Phosphatidylserine receptor
pDC	Plasmacytoid dendritic cells
SR-A	Scavenger receptor A
SLE	Systemic lupus erythematosus
TLR	Toll-like receptor
TGF	Transforming growth factor
TNF	Tumor necrosis factor

1 Introduction

Apoptosis is a physiological form of cell death occurring in normal tissue turnover, during embryogenesis, and after infection or inflammation of tissues. Uptake of apoptotic cells by surrounding phagocytes offers their safe disposal and prevents activation of bystander cells and tissue damage after the release of dying cell contents. Cells undergoing apoptosis are characterized by morphological as well as biochemical changes such as altered distribution of membrane lipids and exposure of modified carbohydrates on the plasma membrane. These changes enable recognition of apoptotic cells by specific receptors on phagocytes and their rapid uptake. Elimination of apoptotic cells is most expeditiously mediated by resident macrophages, but immature dendritic cells (DCs) also phagocytose apoptotic cells, albeit less efficiently. Additionally, nonprofessional phagocytes such as epithelial cells [206], fibroblasts [14], glial cells in the brain [29], mesangial cells in the kidney [83], or hepatocytes in the liver [43] contribute to the elimination of apoptotic cells or their derivatives, e.g., apoptotic blebs. The failure of apoptotic cells to frequently elicit clinically significant autoimmune responses is considered to be an active process whereby phagocytes are rendered immunosuppressive. In contrast, apoptotic cells, when delivered to DCs together with inflammatory signals, are an excellent source of antigen for stimulating of effector T cells. In this review we discuss the characteristics of apoptotic cell uptake by DCs and the circumstances leading to tolerance vs. immunity.

2 Apoptotic Cells and Their Receptors

The search for receptors and molecules mediating uptake of apoptotic cells began with the identification of the $\alpha_v\beta_3$ *vibronectin receptor* on macrophages [173]. Savill and colleagues later demonstrated that $\alpha_v\beta_3$ vibronectin receptor cooperates with the scavenger receptor *CD36*, and uses a bridging molecule, thrombospondin, to bind apoptotic cells [171]. Competitive binding of apoptotic cells and oxLDL by macrophages [23, 205] led to the discovery that apoptotic, but not live cells express oxidized moieties that are structurally analogous to those that enable recognition of oxLDL by scavenger receptors [30] such as scavenger receptor A (*SRA*), macrosialin or *CD68*, and oxidized low-density receptor-1 (*LOX-1*) [145, 153, 168].

The second important group of receptors consists of members recognizing exposed phosphatidylserine (PS). PS, which is usually confined to the cytoplasmic side of the plasma membrane by a phospholipid translocase, is translocated to the outer part of the membrane because of malfunction of the enzyme in cells undergoing apoptosis [217]. Receptors recognizing PS include *PSR*, which binds it directly or possibly via an intermediate molecule, annexin I [11], *β_2-GPI receptor*, the receptor tyrosine kinase *MER*, and $\alpha_v\beta_3$ vibronectin receptor [166, 180], which use the bridging proteins β_2-GPI, Gas6, and MFGE8, respectively [13, 6, 67, 89]. Recently, serum protein S was identified as yet another ligand of PS that expedites uptake of apoptotic cells by macrophages. Protein S is partially homologous to Gas6, but its receptor on phagocytic cells remains to be identified [6].

CD14 [39] and CD91 [143] are two additional molecules of the receptor array recognizing apoptotic cells. *CD14*, also known to bind lipopolysaccharide (LPS), has not clearly been connected to a discrete molecule on apoptotic cells; however, ICAM-3, which is expressed on apoptotic cells, has been identified as one molecule recognized by CD14 [132]. Additional candidates include PS [49] and altered carbohydrates, because CD14 possesses lectinlike activity [64]. Recognition of apoptotic cells by the α_2 macroglobulin receptor (molecule CD91) is more complex. Apparently, *CD91* associates with *calreticulin* and recognizes apoptotic cells through bridging molecules: either the C1q component of the complement system or defense collagens such as mannose-binding lectin (MBL) [143], surfactant protein A and D, and bovine conglutinin [175, 215].

Association with CD91 is not the only occasion where the complement system is involved in the uptake of apoptotic cells. Two complement receptors, *CR3* and *CR4*, that belong to the β_2 integrin family and recognize complement fragment iC3b have been identified on macrophages and/or DCs. iC3b enhances the clearance of apoptotic cells by macrophages and DCs, by coating apoptotic cells [216]. The primary role of complement receptors still remains to be shown, as it is not yet clear whether they contribute significantly to the internalization of apoptotic cells or whether they assist other receptors in increased uptake, perhaps by docking apoptotic cells to phagocytes ([216] and Škoberne and Bhardwaj, unpublished data).

Finally, *ABC*, an ATP-binding cassette transporter, has been proposed to play a role in capturing apoptotic cells [123]. However, little is known about the mechanism by which this transporter mediates phagocytosis.

Apoptotic cell-phagocyte interactions are facilitated by the deactivation of CD31 (PECAM-1). CD31 is expressed on both viable and phagocytic cells and provides a mutual "repulsion" signal that normally prevents ingestion of viable cells. However, during the process of apoptosis, CD31 is modified so that it no longer mediates detachment, thereby permitting apoptotic cells to dock to scavenger receptors [26].

Recently it was shown that phagocytic cells reach the sites where extensive apoptotic cell death occurs by chemotaxis to lysophosphatidylcholine (LPC) [110], a molecule that is first exposed on the surface of apoptotic cells in a PLA_2 -dependent manner [99] and then released by apoptotic cells to function as a chemoattractant. LPC is recognized by

G2A, a recently identified receptor on phagocytes [96]. It is tempting to speculate that LPC-G2A receptor interaction may also function to take up apoptotic cells, as G2A was shown to be internalized after binding LPC [96]. Without doubt, LPC is only one of the factors that facilitate phagocyte-apoptotic cell interactions, and several more examples will become apparent in the future.

Human DCs express several of the receptors discussed above, but so far only a few have been examined for their role in the phagocytosis of apoptotic cells. Rubartelli et al. [165] showed that $\alpha_v\beta_3$ mediates engulfment of apoptotic cells, and Albert et al. [3] could achieve up to 50% inhibition of apoptotic cell uptake by DCs previously exposed to specific antibodies against the receptors $\alpha_v\beta_5$ and CD36 . DC scavenger receptor Lox-1 binds heat shock proteins (HSPs) and cross-presents antigens bound to HSPs to T cells [36]; thus this receptor may take up not only apoptotic cells but also the cell-associated debris released from necrotic cells. This dual ability of Lox-1 may partly account for the capacity of DCs to present antigens from both apoptotic and necrotic cells [106]. In macrophages, CD91 binds HSPs and mediates the cross-presentation of associated antigens, in addition to mediating the binding of apoptotic cells [16]. Indirect recognition of apoptotic cells via Fc receptors might also play a role in vivo, as IgG-opsonized tumor apoptotic cells have been shown to be a good source of antigens in a vaccination study when charged to DCs [2]. Recently, complement receptors have also been associated with recognition of apoptotic cells by DCs, although characterization of their role needs further investigation [134, 190, 216].

Mice that lack certain receptors or factors that mediate uptake of apoptotic cells such as Mer [33], PSR [117], G2A [113], C1q [203], or IgM [25, 47] exhibit profound defects in phagocytosis of apoptotic cells and develop syndromes resembling lupus (Mer, C1q, and IgM) or are rapidly fatal (PSR). Such findings are not surprising, as failure to adequately remove apoptotic cells would have serious consequences, such as secondary necrosis and consequent inflammation. On the other hand, a loss of other receptors that also contribute to removal of C1q-opsonized apoptotic cells (e.g., CD93) does not result in development of autoimmune disease [141], nor does the loss of $\alpha_v\beta_5$, $\alpha_v\beta_3$, or CD36 receptors. In fact, these mice remain capable of cross-presenting antigens encoded by apoptotic cells to T cells [21, 179]. A multiple receptor-based system ensures a fail-safe elimination of apoptotic cells, but it seems that certain receptors are essential for the prevention of autoimmunity.

3
Inhibitory Effects of Apoptotic Cells

Several mechanisms have evolved to prevent autoimmunity to self-antigens that are contained within apoptotic cells. Most self-reactive T cell responses are eliminated early in life by central tolerance. Remaining autoimmune T cells are controlled by peripheral tolerance, through deletion, anergy or by induction of regulatory CD4+ or CD8+ T cells that actively suppress self-reactive responses. In vivo, the tissue-resident antigen-presenting cells (APCs) take up the surrounding apoptotic cells during physiological cell turnover and migrate to lymph nodes where self-antigens are processed and presented to T cells [88, 130]. In the absence of inflammation, apoptotic cells may interfere with DC maturation as well as macrophage activation [172, 210]. Therefore, DCs may reach T cell areas of lymph nodes in an immature state and induce tolerance rather than immune responses [40, 82, 161, 177] (Fig. 1).

The immunosuppressive effects of apoptotic cells were first identified on macrophages, where production of several proinflammatory factors such as TNF-α, IL-1β, IL-8, GM-CSF, and thromboxane B_2 was inhibited on their ingestion [50]. In addition, ligation of PSR by apoptotic cells stimulated TGF-β1 secretion and inhibited LPS-induced TNF-α release in macrophages [51]. Binding of CR by specific antibodies or natural ligands also inhibited IL-12 and IFN-γ production by monocytes [125] or even induced IL-10 production [226]. Similarly, ligation of CD36 induced IL-10 production and inhibited TNF-α and IL-12 production in response to LPS [219].

It is interesting that pathogens often mimic apoptotic cells and take advantage of the same receptors to enter phagocytes and to promote immunosuppressive effects. For example, infectious stages of *Leishmania major* bind CR3 and suppress IL-12 production by macrophages [63, 124, 128]. Also, *Bordetella pertussis* binds CR3, induces IL-10, and inhibits IL-12 production by a macrophage cell line on exposure to LPS and IFN-γ [129].

DCs were long believed to stay immunologically inert after ingestion of apoptotic cells, but data now suggest that they also play an active role in immunosuppression [200]. In this respect, the possible involvement of complement receptors, CD36 and PSR has received the greatest attention.

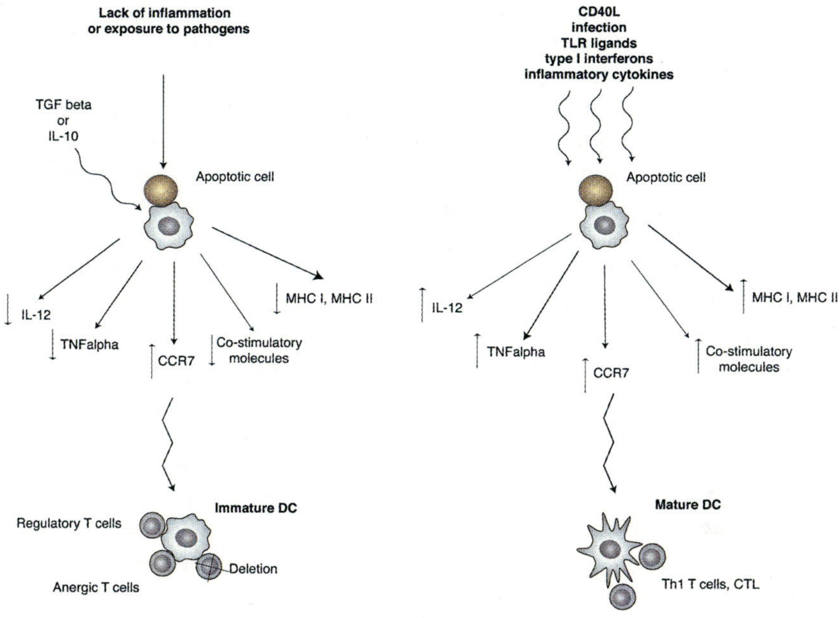

Fig. 1 Maturation state of DC influences T cell responses. In the periphery, DCs continuously capture physiologically dying apoptotic cells and process their antigens. In the absence of maturation signals, these DCs retain low expression of MHC and costimulatory molecules while they upregulate their expression of CCR7. They can then migrate to the draining lymph nodes, present antigens derived from apoptotic cells to T cells, and induce anergy, deletion, or tolerance of antigen-specific T cells. In addition, these DCs can be further skewed to a tolerizing activity by the cytokines, e.g., IL-10 or TGF-β, that are produced by neighboring macrophages that have ingested apoptotic cells, by tumor cells, or by microbes. In contrast, when DCs undergo maturation, the presentation of antigens derived from apoptotic cells results in the priming of effector T cells. Factors that contribute to full maturation of DCs include impaired clearance of apoptotic cells that is followed by secondary necrosis and release of inflammatory cell contents, presence of microbial components, inflammatory cytokines produced by infected cells or by components of preexisting immunity, such as opsonization by antibodies or activation by CD40 molecule on activated T cells

3.1
Complement Receptors CR3 and CR4

Myeloid DCs express high levels of two complement receptors, CR3 (CD11b/CD18) and CR4 (CD11c/CD18), which both recognize iC3b, a breakdown product of complement and an opsonin for apoptotic cells. Apoptotic cells activate complement via the classical pathway [101] or the alternative pathway [126, 207]. On activation, C3 breaks into C3a and C3b. The latter is deposited on the surface of apoptotic cells and is probably rapidly converted into iC3b by serum factor H or I [126, 131, 149]. Molecules on apoptotic cells responsible for activation of complement have not been identified, but exposure of PS and deposition of C3b fragments have been positively correlated [133]. In addition, IgM antibodies that activate the classic pathway of complement, bind LPC, a recently defined "find and eat me" marker of apoptotic cells (The "find and eat me" concept is nicely reviewed in [111]). In support of the role of IgM antibodies in noninflammatory removal of apoptotic cells, $IgM^{-/-}$ mice were found to develop SLE-like disorders [25, 47].

Recently, ligation of the complement receptors CR3 and CR4 has been implicated in the induction of DCs with regulatory properties [190, 216]. Verbovetski et al. used serum-deprived apoptotic Jurkat cells opsonized with complement fragment iC3b to show that engagement of CR3 and CR4 prevents DC maturation on stimulation with either LPS or CD40L. Interestingly, despite the inhibition of upregulation of various maturation markers, complement-treated DCs still heightened the expression of CCR7 (a receptor for the chemokines CCL19 and 21 expressed in the lymph node), so conceivably these DCs could migrate and induce tolerogenic T cells in the secondary lymphoid organs [216].

Further confirmation of the importance of CR in DC modulation comes from animal experimental models. Morelli et al. [134] showed that iC3b fragments exert immunomodulary functions in vivo. They demonstrated that in mice splenic marginal zone DCs use complement receptors CR3 and CR4 to phagocytose circulating apoptotic cells. Interaction of apoptotic cells with these complement receptors reduces the production of inflammatory cytokines such as IL-1α, IL-1β, and TNF-α but does not interfere with production of TGF-β1.

Sohn and colleagues used a rat model to show complement involvement in systemic tolerance induced in an immunoprivileged site. OVA-loaded APC pretreated with either polymeric iC3b or iC3b-opsonized erythrocytes induced suppression of DTH to OVA when injected into rats. These observations were supported by in vitro studies demonstrating that after exposure to iC3b-opsonized erythrocytes, OVA-loaded APCs produced TGF-β2 and IL-10. However, the authors used peritoneal exudate cells that represent a nonhomogeneous population as APCs [190].

Altogether, these studies provide support for the hypothesis that engagement of complement receptors CR3 and CR4 interferes with DCs maturation. Furthermore, they imply that such DCs could be involved in tolerance induction.

3.2 The Scavenger Receptor CD36

Plasmodium falciparum is a parasite causing malaria in humans. During its intraerythrocytic stages it expresses the protein PfEMP-1 in erythrocyte membranes that mediates binding to host cells [138]. All variants of PfEMP-1 analyzed so far bind the scavenger receptor CD36 [15]. *P. falciparum*-infected erythrocytes bind CD36 directly and the CD51/α_v integrin chain indirectly (via thrombospondin) and consequently inhibit LPS-induced expression of maturation markers on human DCs [209, 211]. Engagement of these receptors with antibodies inhibited DC maturation [210], and therefore CD36 and $\alpha_v\beta_3$ or $\alpha_v\beta_5$ are among candidates for receptors with immunosuppressive activity. Interestingly, it was recently shown that $CD8^+$ T cell responses are impaired during malaria infection [142], although the mechanisms leading to this impairment have not been identified.

3.3 The Phosphatidylserine Receptor

Although not yet shown in DCs, in macrophages the recognition of PS by PSR is crucial for a switch to an anti-inflammatory response [75] and also to resolve an already established inflammation through production of TGF-β, PGE_2, and PAF and inhibition of IL-12 production [85]. In addition to its immunosuppressive role, PSR is better known for mediating phagocytosis of apoptotic cells via a "tether-tickle mechanism." This theory proposes usage of different sets of receptors for docking the apoptotic cell to the phagocyte ("tether" signal) or providing a "tickle" signal resulting in phagocytosis of apoptotic cells. The "tickle" signal is provided by PSR as simultaneous binding of PSR converts adhesion into ingestion even by receptors that are normally not involved in phagocytosis [79]. Ligation of PSR is necessary for activation of Rac1 and Cdc42, stimulation of Arp2/3 complex-dependent actin polymerization, and phagosome formation [79]. In contrast to activated macrophages, immature DCs express Cdc42 and are constitutively macropinocytic and phagocytic [61]. As mature DCs downregulate expression of PSR, it is unlikely that PSR plays a "tickling" role [191] and receptors such as $\alpha_v\beta_5$ integrin may be self-sufficient for phagosome formation [5].

4 Cross-Presentation

Apoptotic cells were first appreciated not for their role in tolerance induction, but as a source of antigen in priming of T cell immune responses. $CD8^+$ T cells generally recognize major histocompatibility complex (MHC) class I molecules presenting peptides of 8–10 amino acids. These cells are crucial for immune responses to tumors and to pathogens such as viruses and intracellular bacteria and are primed when naive T cells encounter professional APC bearing an antigenic peptide. Two mechanisms are responsible for loading of peptides onto MHC class I molecules: the first is the endogenous or classic pathway in which the antigens are derived from proteins produced or existing in the cytosol of an APC. The second mechanism, also referred to as "cross-presentation," involves an exogenous pathway in which antigens are derived from exogenous sources (such as apoptotic cells) that have been taken up by the APC. The term "cross-priming" discriminates the priming of $CD8^+$ T cell responses to antigen derived from such sources. Cross-presentation when associated with maturation of DCs results in priming of effector $CD4^+$ and $CD8^+$ T cell responses (cross-priming) but when associated with immature or partially mature DCs can lead to "cross-tolerance" (reviewed in [20, 136, 137]).

4.1 Cell Types Responsible for Cross-Presentation

In contrast to macrophages that can cross-present but cannot cross-prime, DCs are efficient at both cross-presentation and cross-priming [27, 102, 158]. Cell types responsible for cross-presentation in vivo have been extensively studied in mouse models. DCs were always the prime candidate and their role was formally confirmed in the studies by den Haan et al. [37], where the authors showed that the $CD8^+$ subset of DCs presents cell-associated antigens to cytotoxic T cells. Further confirmation comes from Jung et al. [95], who have shown that by short-term depletion of CD11c+ DCs in mice, they could abrogate cross-priming of $CD8^+$ T cells . However, classification of DCs in mice is complex and not directly comparable to humans. In humans fewer DC subsets are known so far [the predominant are myeloid DCs (mDCs), plasmacytoid DCs (pDCs) and the Langerhans cells], whereas in mice at least six subsets of DCs have been characterized on the basis of their phenotype [72, 77, 155]. These murine DC subsets display differential and complex capacities to cross-present antigens derived from cellular sources vs. immune complexes (ICs). For example, $CD8^+$ DCs but not $CD8^-$ DCs cross-present antigens from ovalbumin (OVA)-loaded spleen cells to $CD8^+$ T cells [38]. Furthermore, $CD8^+$ DCs also present soluble OVA in vivo to $CD8^+$ T cells but $CD8^-$ DCs preferentially present soluble OVA to $CD4^+$ T cells [38]. However, the efficiency of cross-presentation of cell-associated antigens is several hundredfold greater than cross-presentation of soluble antigen [116]. Finally, both $CD8^+$ and $CD8^-$ DCs cross-present immune complexes (ICs) to $CD8^+$ T cells. However, only the $CD8^-$ DCs cross-present ICs to $CD4^+$ T cells. The loss of all three FcRs leads to loss of presentation to $CD8^+$ T cells by $CD8^-$ DCs but there is no effect on $CD8^+$ DCs—perhaps because these DCs may acquire ICs via complement fixation [38]. Another intriguing and open question is whether antigen acquisition in the peripheral tissues and transport to the draining lymph node can be segregated from the presentation of antigen to T cells. In support, recent studies by Belz et al. suggest that in addition to DCs that have migrated from the periphery, the lymph node-resident, non migrating DCs are responsible for antigen presentation after transfer of antigen from the migrating DCs [19].

Less is known about the role of pDCs in cross-presentation. Although pDCs are generally more appreciated for the production of type I interferons, studies by Jung et al [95] cannot exclude their role in cross-priming (in mice CD11c is expressed by the pDC and the mDC). Interestingly, pDCs lack (mouse) or express low levels (human) of CR3—one of the primary receptors identified so far that negatively influences DC maturation on apoptotic cell internalization [65, 77, 134, 190, 216]. However, pDCs might be less specialized for endocytosis compared to mDCs as cord blood and blood pDC are not very efficient in uptake of proteins [56, 193] and studies of tonsil or blood human pDC show that they are not very efficient in endocytosis of latex or dextran beads [65, 199]. They may be better in the transfer of membranes from live, necrotic, or apoptotic targets, although still less efficient compared to mDCs [199]. Moreover, the profile of expression of cathepsins in human mDC and pDCs might suggest that they differ in their capacity to process and present antigens [54]. Indeed, murine pDCs are less potent than mDCs in assembly of viral peptide-MHC class II complexes after in vivo or in vitro exposure to virus [104]. In addition, after intravenous injection into female mice, both CpG-matured male pDCs and mDCs were able to induce direct CTL priming against the male-specific transplantation antigen. However, contrary to mDCs, CpG-matured male pDC prepulsed with soluble OVA injected into female animals failed to cross-prime

OVA-specific $CD8^+$ T cells or prime OVA-specific $CD4^+$ T cells [167]. Further studies are required to determine whether pDCs have a direct role in cross-priming or whether they promote cross-presentation and cross-priming ability of "bystander" mDC through production of type I IFN.

4.2 Mechanism of Cross-Presentation

Presentation of exogenous antigens onto MHC class I by APCs occurs by at least two distinct mechanisms [8, 103, 116, 140, 159]. Antigens derived from various sources, e.g., soluble proteins, ICs, and protein-coated latex beads, can be conveyed from the endocytic compartment into the cytosol of APCs. In the cytosol, antigens are degraded into oligopeptides and are transported via the transporter associated with antigen processing (TAP) into the endoplasmatic reticulum (ER) for loading onto newly synthesized MHC class I molecules [57, 68, 103, 116, 140, 152, 159, 160]. In addition, cell-associated viral antigens also appear to require TAP for cross-presentation [57, 186]. Alternatively, antigens can be processed in endosomal compartments where peptides are generated with participation of cathepsin S, loaded on recycling MHC class I molecules, and transported to the cell surface for presentation [68, 152, 182]. This pathway is TAP-independent and is used by soluble proteins, multibranched lysine with attached peptides, and, under some conditions, proteins chaperoned by HSPs [8, 103, 148].

Recently, it was proposed that the process facilitating cross-presentation is carried out in a special compartment, an early ER-derived phagosome [1] that has the characteristics of both the ER and the phagosomes. After endocytosis exogenous antigens are transported from this compartment into the cytosol, possibly via the Sec61 complex, where degradation by proteasomes occurs [1, 80]. The peptides produced by proteasomes are then transported back via TAP complex into the lumen of this distinctive compartment, where antigens are finally loaded on MHC class I molecules [66, 80]. The presence of MHC class I molecules in this compartment can be explained by a target signal that they bear in their cytoplasmic domain, which directs them to endosomal compartments [120]. Conceivably, peptides generated in the cytosol through this mechanism could also access MHC class I peptides in the ER. This novel mechanism has been studied in the cross-presentation of antigen-coated latex beads by mouse DCs, and involvement of a similar process in cross-presentation of dying cells by human DCs is plausible [57].

4.3
Sources of Antigen for Cross-Presentation

In vitro, several forms of antigens can access the exogenous pathway for MHC class I presentation (reviewed in [106]), including remainders of dying cells such as apoptotic cells, apoptotic bodies or necrotic cells, HSPs, DNA- or RNA-encoded antigens, organisms, e.g., bacteria, viruses, viruslike particles, exosomes, immune complexes, and soluble proteins [135, 224] and even "bits" of live cells which "are nibbled off" by phagocytes [158, 69, 70].

Whereas in vitro data clearly show that apoptotic cells are an important source of antigens [3, 105, 107] fewer data are available regarding their role in vivo. Huang et al. [82] have shown that mouse DC subsets can acquire apoptotic cells in the intestine and transport the ingested material to mesenteric lymph nodes. Bevan's group [37] showed that the $CD8^+$ subset of murine DCs can ingest OVA-loaded cells and prime OVA-specific T cells. Although not shown, these cells were presumably dying. Iyoda et al. [90] have confirmed the above studies and state that the majority of the cells ingested by DCs were apoptotic before injection. Further confirmation for cross-presentation of apoptotic cells by DCs comes from Hugues et al. [84]. In their studies, induction of apoptosis of pancreatic β-cells led to the access of islet antigens to $CD11c^+$ and $CD11b^+$ DCs and induced the development of regulatory $CD4^+$ T cells [84]. Studies by Scheinecker et al. [177] do not address the viability of cross-presented cells but clearly show that cell-associated antigens are taken up and processed by $CD11c^+$ DCs that in turn migrate to draining lymph nodes where antigens are recognized by specific T cells. The identity of the cellular fraction that has the predominant role in cross-priming of cell-associated antigens (whether these are HSP-associated proteins, native proteins, peptides, or some other component) has only recently begun to be explored. Some answers have come from the group of Shen and Rock [183], who show that the fully processed peptides are not a major source for cross-presentation but that rather native proteins released from the dying cells are taken up by the surrounding APCs. Partially supporting these observations Norbury et al. [139] demonstrated that proteasomal substrates are the prime form of antigens transferred from donor to recipient cells. However, the contribution of HSP-peptide complexes should not be excluded on the basis of these results.

5
Cross-Priming Versus Cross-Tolerance

Under noninflammatory conditions apoptotic cells serve as a source of antigen for DCs; however, DCs fail to upregulate MHC and costimulatory molecules [118, 177] but upregulate the chemokine receptor CCR7 associated with migration to lymph nodes [216]. After uptake of apoptotic cells in the steady state, DCs reach lymph nodes in an immature state and are presumably responsible for induction of "cross-tolerance." The DCs most probably express some degree of costimulation to sufficiently activate T cells but not to generate long-term effector and memory cells [71, 118, 197]. When steady-state conditions are altered, e.g., during infection with pathogens, DCs can undergo activation, present antigens from dying cells, and prime effector T cells instead. We discuss below cross-priming to microbial antigens during infection and factors that might be involved in the switch from tolerance to immunity.

5.1
Cross-Priming to Microbial Antigens

It is unlikely that all microbes are able to infect professional APCs. In addition, microbial immune escape mechanisms involve abrogation of presentation on MHC class I molecules but nevertheless efficient long-term immunity is established, as in the case of cytomegalovirus (CMV) or herpes simplex virus (HSV) [225]. Therefore, alternative mechanisms such as cross-presentation are essential for priming of CD8+ T cells at least in the above-mentioned circumstances. In vitro, antigens of numerous viruses have been shown to be presented via cross-presentation of dying infected cells (for review see [55]), starting with pioneering studies on influenza virus by Albert and colleagues [4]. Later, the repertoire was extended to vaccinia virus [107], human CMV [9, 10, 202], Epstein-Barr virus [76, 201], HIV-1 [108, 7], canarypox virus [87], HSV [154], and measles virus [178, 181]. Apoptosis of infected cells is also a prerequisite for cross-presentation of bacteria such as *Salmonella* [227] or mycobacteria [176], whereas some other bacteria such as *Klebsiella* [93] or *Listeria* may utilize different sources of antigens [187, 208, 91]. Interestingly enough, cross-presentation of bacterial antigens not only occurs in the context of MHC class I molecules but also in the context of CD1b molecules as shown for the mycobacterial antigens [176]. Bearing in mind that, in addition to the better-studied viruses, numerous bacteria and parasites also induce apoptosis of infected cells [122, 220], the list of microorganisms to which immune responses develop via cross-presentation of apoptotic cells is expected to increase substantially.

Although studies by Sigal et al. [185] confirmed that cross-priming to viral antigens can indeed take place in vivo during infection, the relevance of cross-priming in relation to direct priming in vivo remains an open question. Yewdell and colleagues [31] have recently shown in a convincing study that cross-priming indeed plays an important role in vivo for induction of antiviral immunity. When DCs are directly infected and infection results in partial apoptosis of the APC, presumably both direct and cross-presentation will take place. However, when DCs are not infected or when the infection kills APC rapidly, cross-presentation would be the prime mechanism for activating T cells. This opinion is to some extent opposed by studies of Zinkernagel and colleagues [58] in which the authors claim that in the priming of T cell responses to tumor and viral antigens, cross-presentation is only of minor importance.

5.2 Factors That Influence the Priming Capability of Dendritic Cells

What factors distinguish the cross-presentation of microbial antigens that lead to effector cell responses and cross-presentation of self-antigens that induce tolerogenic responses? We consider the state of DCs during encounter of the antigen to set the stage. To avoid autoimmune responses, by default DCs do not mature on encounter of apoptotic cells and therefore immature DCs could induce tolerance [127, 170, 198]. In contrast, when signs of danger (e.g., infection) are present, DCs undergo maturation and switch to an active immunostimulatory role [127].

In vitro, several factors defining "danger" were shown to induce maturation and enhance cross-presentation by DCs that have ingested apoptotic cells [105, 106, 107]. However, in vivo signals responsible for the switch from tolerance to immunity of apoptotic cells are still poorly characterized. We would expect that in vivo the final outcome would depend on multiple factors, including stimulation through Toll-like receptors (TLRs), presence of proinflammatory cytokines, engagement of CD40, and presence and inhibitory activity of antigen-specific regulatory T cells, to name only a few. Below we discuss some of the factors that might be involved in turning on the cross-priming ability of DCs.

5.2.1 Microbial Factors

Microbial molecules influence the cross-priming capability of DCs on several levels. First, molecules derived from pathogens such as dsRNA, ssRNA, unmethylated CpG containing DNA (CpG DNA), LPS, or peptidoglycans, which bind an array of TLRs transmit signals that induce DCs maturation (reviewed in [97]). Second, certain TLR ligands when added to culture together with soluble OVA induce cross-presentation by bone marrow-derived $CD11c^+$ $CD4^-$ $CD8\alpha^-$ DCs [35]. The importance of TLR9 in activation of DCs after endocytosis of antigen was implied by experiments in TLR9-deficient mice in which injection of OVA covalently linked to CpG DNA induced cross-presentation but not cross-priming [74]. How stimulatory signals mediated through TLR9, and possibly other TLRs, combine with signals induced by cell-bound antigen, namely apoptotic cells, remains to be determined. Studies in DCs and macrophages indicate that this mechanism is complex, as the production of cytokines by these cells when cultured in the presence of apoptotic cells and TLR ligands, differs from that when they are exposed to either TLR ligands or apoptotic cells alone [121, 200, 216]. TLR ligation might also influence the efficiency of phagocytosis of apoptotic cells. Although not yet shown for DCs, murine macrophages upregulate the expression of apoptotic cell receptors, e.g., SR-A, Lox-1, and CD36, after triggering of certain TLRs [44]. Additionally, two groups have shown that ligation of TLR increases the uptake of antigens [24, 221]. However studies by Blander and Medzhitov [24] demonstrate that while ligation of TLR by bacteria increases their uptake and influences the maturation of phagosomes, no influence was seen on the uptake and digestion of simultaneously phagocytosed apoptotic cells. Notably, while uptake of infected apoptotic cells was not studied, the bacteria and uninfected apoptotic cells were confined to distinct compartments.

Finally, pathogen-derived TLR ligands such as LPS and CpG DNA were shown to induce DC maturation and overturn the inhibitory capacity of regulatory T cells in vitro, in a costimulatory molecule-independent and IL-6-dependent fashion [151]. However, this reversal of the suppressive effects of regulatory cells in vivo was either not observed [147] or was shown to require persistent TLR signaling [223]. In their paper [223], Yang and colleagues compared infection with hemagglutinin (HA)-encoding recombinant virus with injection of TNF-α-matured and HA-loaded DC in the ability to activate HA-specific $CD8^+$ T cells. Lentiviruses or vaccinia viruses succeeded to break CD8 tolerance and to protect against tumoral challenge. DC-based vaccines failed unless regulatory $CD4^+$ T cells were removed, TLR ligand was constantly added, or mice were concomitantly infected with an irrelevant virus (not encoding HA). However, ex vivo TLR-ligand-matured DCs could not break tolerance and after withdrawal of LPS, the production of proinflammatory cytokines by DCs diminished rapidly [223].

5.2.2 Endogenous Adjuvants

DCs also undergo maturation in the absence of foreign substances. Massive apoptotic cell death can result in failure of clearance of apoptotic cells and secondary necrosis [164]. Necrotic, in contrast to apoptotic cells mature DCs and lead to initiation of immune responses [22, 60, 170]. Several reports have described the endogenous activation of DCs via engagement of TLRs. Endogenous factors that are released from or are associated with necrotic cells induce DC activation (reviewed in [188]). They include immunostimulatory self DNA that binds TLR9 [48], self ssRNA that may stimulate TLR7 and TLR8 [41, 73], secondary structures of mRNA that activate TLR3 [98], HMGB-1 that signals through TLR2, TLR4 [150], or RAGE [174], and HSPs that stimulate TLR4 [144, 212–214]. HSPs are one of the prime candidates to contribute to conversion of a tolerogenic to a priming signal. They induce innate immune responses [214], they are released after necrosis, and the amount of expression correlates with maturation of DCs [192]. In addition, increased expression of HSPs by apoptotic cells also increases the immunogenicity of the latter [53]. Interestingly, feverlike thermal conditions enhance HSP expression on cell membranes and promote DC maturation and the priming of specific T cells [52, 53]. Recently, uric acid was identified as another factor associated with cell death and activation of DCs [184]. Finally, the immune system is alerted to massive cell death not only by factors released from dying cells, but also by factors emanating from disruption of tissue architecture, e.g., fibrinogen [189], oligosaccharides of hyaluronan [204], EDA-containing fibronectin [146], and heparan sulfate proteoglycan that stimulate TLR4.

5.2.3 Cytokines in Dendritic Cells' Milieu

The milieu in which DCs encounter apoptotic cells may regulate their stimulatory capacity. For example, DCs exposed in vitro to IL-10, TGF-

β, vascular endothelial growth factor (VEGF), or IL-10 homolog have diminished IL-12 production, expression of costimulatory molecules, and capacity to stimulate T cells [18, 34, 100, 194–196]. These factors can be produced by macrophages on ingestion of apoptotic cells, by tumor cells, or by microbes [46, 59, 62, 109, 195]. On the other hand, recognition of microbes by macrophages and neutrophils results in the production of inflammatory cytokines such as TNF-α, IL-6, or IL-1β, which mature DCs and contribute to their activation and priming capabilities [115]. For example, *Shigella flexneri* and *Salmonella* induce macrophage apoptosis, activation of caspase 1, and release of IL-1β and IL-18 [92, 169].

The role of type I IFNs, especially IFN-α in cross-priming has also gained substantial attention. In mice, type I IFNs, secreted either on TLR9 triggering [32] or viral infection [112], when coadministered with antigens were shown to be sufficient for cross-priming. IFN-α is produced in great amounts after interaction of pDCs with virus and not only influences other cells (e.g., NK cells), but can also induce maturation and perhaps cross-priming by bystander mDCs [56, 112]. Involvement of proinflammatory cytokines in overturning the tolerogenic potential of apoptotic cells has been demonstrated in studies by Zimmermann et al. By using an artificial model of TNF-α conjugated to apoptotic melanoma cells they could achieve full maturation of DCs on apoptotic cell ingestion and priming of partially protective CTL responses [228].

5.2.4
Opsonization of Apoptotic Cells with Antibodies

Sole cross-linking of FcγRI or FcγRII leads to NF-κB translocation and production of TNF-α by monocytes [45]. In addition, opsonization with antibodies and targeting FcγR was shown to mediate internalization and to overcome initial inhibitory effects of apoptotic cells [2]. Ligation of FcγR also interferes with expression of IL-10-inducible genes [94], showing the importance of this receptor as a "switch" to immunostimulatory DCs. Defective clearance of apoptotic cells in susceptible individuals was linked to initiation of autoimmunity (reviewed in [222]). Additionally, autoreactive antibodies were shown to target self-antigens located in apoptotic blebs or the chromatin released from apoptotic cells [28, 157, 162]. Apoptotic cells combined with lupus IgG induce generation of immune complexes that contain DNA. These complexes can simultaneously bind TLR9 (with the chromatin part) and the FcR (with the antibody) on pDC and B cells. Such binding induces IFN-α production by pDC

[17] and could also further activate B cells ([114] and reviewed in [218]), resulting in amplified autoimmune responses.

5.2.5 Ligation of CD40

Ligation of CD40 on DCs is one of the better-explored mechanisms that can override inhibitory effects of apoptotic cells. CD40 interacts with CD40 ligand (molecule CD154) that is expressed on immune cells such as activated B and T cells, but also on mononuclear phagocytes and activated platelets and during an inflammatory response on epithelial cells and smooth muscle cells among others (reviewed in [156]). Stimulation of CD40 was shown to override immune unresponsiveness induced by immature DCs [42, 86, 163]. In addition, tolerance induced by dying OVA-loaded cells in mice was abolished when agonistic anti-CD40 antibody was coinjected. Studies in mice show that CD40-deficient DCs are only capable of inducing a transient immune response that is followed by unresponsiveness to subsequent challenge with the same antigen [78]. A pivotal role of CD40 is supported by several therapeutic approaches in which blocking of CD40-CD154 interaction results in silencing of autoimmune disease in mouse models and induction of tolerance [12, 81, 119].

6 Concluding Remarks

Apoptotic cell death as a consequence of normal tissue turnover inhibits DC maturation and leads to tolerance. However, the concurrent presence of microbes, stress, or massive cell death presumably prevents the inhibitory effects of apoptotic cells and leads to DC maturation and induction of immune responses. Several receptors have already been associated with inhibition or stimulation of DC maturation. Molecular mechanisms leading to different outcomes after apoptotic cell ingestion are being evaluated. In the future, it will be necessary to study the signaling pathways triggered and to evaluate the hierarchy and cooperation of signals transduced via different receptors. This information will lead to new strategies to develop "tolerogenic" or "immunostimulatory" DCs.

Acknowledgments This work was in part supported by the Burroughs Wellcome Fund, NIH grants R01-Al44628, R21-Al55274, R01-Al061684. N.B. is an Elisabeth Glaser Scientist and Doris Duke Distinguished Clinical Scientist. MS was supported in part by AdFutura, Slovenia.

References

1. Ackerman AL, Kyritsis C, Tampe R, Cresswell P (2003) Early phagosomes in dendritic cells form a cellular compartment sufficient for cross presentation of exogenous antigens. Proc Natl Acad Sci USA 100:12889–12894
2. Akiyama K, Ebihara S, Yada A, Matsumura K, Aiba S, Nukiwa T, Takai T (2003) Targeting apoptotic tumor cells to Fc gamma R provides efficient and versatile vaccination against tumors by dendritic cells. J Immunol 170:1641–1648
3. Albert ML, Pearce SF, Francisco LM, Sauter B, Roy P, Silverstein RL, Bhardwaj N (1998) Immature dendritic cells phagocytose apoptotic cells via alphavbeta5 and CD36, and cross-present antigens to cytotoxic T lymphocytes. J Exp Med 188:1359–1368
4. Albert ML, Sauter B, Bhardwaj N (1998) Dendritic cells acquire antigen from apoptotic cells and induce class I-restricted CTLs. Nature 392:86–89
5. Albert ML, Kim JI, Birge RB (2000) Alphavbeta5 integrin recruits the CrkII-Dock180-rac1 complex for phagocytosis of apoptotic cells. Nat Cell Biol 2:899–905
6. Anderson HA, Maylock CA, Williams JA, Paweletz CP, Shu H, Shacter E (2003) Serum-derived protein S binds to phosphatidylserine and stimulates the phagocytosis of apoptotic cells. Nat Immunol 4:87–91
7. Andrieu M, Desoutter JF, Loing E, Gaston J, Hanau D, Guillet JG, Hosmalin A (2003) Two human immunodeficiency virus vaccinal lipopeptides follow different cross-presentation pathways in human dendritic cells. J Virol 77:1564–1570
8. Arnold D, Faath S, Rammensee H-G, Schild H (1995) Cross-priming of minor histocompatibility antigen-specific cytotoxic T cells upon immunization with the heat shock protein gp96. J Exp Med 182:885–889
9. Arrode G, Boccaccio C, Lule J, Allart S, Moinard N, Abastado JP, Alam A, Davrinche C (2000) Incoming human cytomegalovirus pp65 (UL83) contained in apoptotic infected fibroblasts is cross-presented to $CD8^+$ T cells by dendritic cells. J Virol 74:10018–10024
10. Arrode G, Boccaccio C, Abastado JP, Davrinche C (2002) Cross-presentation of human cytomegalovirus pp65 (UL83) to CD8+ T cells is regulated by virus-induced, soluble-mediator-dependent maturation of dendritic cells. J Virol 76:142–150
11. Arur S, Uche UE, Rezaul K, Fong M, Scranton V, Cowan AE, Mohler W, Han DK (2003) Annexin I is an endogenous ligand that mediates apoptotic cell engulfment. Dev Cell 4:587–598
12. Balasa B, Krahl T, Patstone G, Lee J, Tisch R, McDevitt HO, Sarvetnick N (1997) CD40 ligand-CD40 interactions are necessary for the initiation of insulitis and diabetes in nonobese diabetic mice. J Immunol 159:4620–7462
13. Balasubramanian K, Chandra J, Schroit AJ (1997) Immune clearance of phosphatidylserine-expressing cells by phagocytes. The role of beta2-glycoprotein I in macrophage recognition. J Biol Chem 272:31113–31117
14. Bartl MM, Luckenbach T, Bergner O, Ullrich O, Koch-Brandt C (2001) Multiple receptors mediate apoJ-dependent clearance of cellular debris into nonprofessional phagocytes. Exp Cell Res 271:130–141
15. Baruch DI, Gormely JA, Ma C, Howard RJ, Pasloske BL (1996) Plasmodium falciparum erythrocyte membrane protein 1 is a parasitized erythrocyte receptor

for adherence to CD36, thrombospondin, and intercellular adhesion molecule 1. Proc Natl Acad Sci USA 93:3497–3502
16. Basu S, Binder RJ, Ramalingam T, Srivastava PK (2001) CD91 is a common receptor for heat shock proteins gp96, hsp90, hsp70, and calreticulin. Immunity 14:303–313
17. Bave U, Magnusson M, Eloranta ML, Perers A, Alm GV, Ronnblom L (2003) Fc gamma RIIa is expressed on natural IFN-alpha-producing cells (plasmacytoid dendritic cells) and is required for the IFN-alpha production induced by apoptotic cells combined with lupus IgG. J Immunol 171:3296–3302
18. Bellinghausen I, Brand U, Steinbrink K, Enk AH, Knop J, Saloga J (2001) Inhibition of human allergic T-cell responses by IL-10-treated dendritic cells: differences from hydrocortisone-treated dendritic cells. J Allergy Clin Immunol 108:242–249
19. Belz GT, Smith CM, Kleinert L, Reading P, Brooks A, Shortman K, Carbone FR, Heath WR (2004) Distinct migrating and nonmigrating dendritic cell populations are involved in MHC class I-restricted antigen presentation after lung infection with virus. Proc Natl Acad Sci USA 101:8670–8675
20. Belz GT, Heath WR, Carbone FR (2002) The role of dendritic cell subsets in selection between tolerance and immunity. Immunol Cell Biol 80:463–468
21. Belz GT, Vremec D, Febbraio M, Corcoran L, Shortman K, Carbone FR, Heath WR (2002) CD36 is differentially expressed by CD8+ splenic dendritic cells but is not required for cross-presentation in vivo. J Immunol 168:6066–6070
22. Binder RJ, Han DK, Srivastava PK (2000) CD91: a receptor for heat shock protein gp96. Nat Immunol 1:151–155
23. Bird DA, Gillotte KL, Horkko S, Friedman P, Dennis EA, Witztum JL, Steinberg D (1999) Receptors for oxidized low-density lipoprotein on elicited mouse peritoneal macrophages can recognize both the modified lipid moieties and the modified protein moieties: implications with respect to macrophage recognition of apoptotic cells. Proc Natl Acad Sci USA 96:6347–6352
24. Blander JM, Medzhitov R (2004) Regulation of phagosome maturation by signals from toll-like receptors. Science 304:1014–1018
25. Boes M (2000) Role of natural and immune IgM antibodies in immune responses. Mol Immunol 37:1141–1149
26. Brown S, Heinisch I, Ross E, Shaw K, Buckley CD, Savill J (2002) Apoptosis disables CD31-mediated cell detachment from phagocytes promoting binding and engulfment. Nature 418:200–203
27. Carbone FR, Bevan MJ (1990) Class I-restricted processing and presentation of exogenous cell-associated antigen in vivo. J Exp Med 171:377–387
28. Casciola-Rosen L, Anhalt G, Rosen A (1994) Autoantigens targeted in systemic lupus erythematosus are clustered in two populations of surface structures on apoptotic keratinocytes. J Exp Med 179:1317–1330
29. Chan A, Magnus T, Gold R (2001) Phagocytosis of apoptotic inflammatory cells by microglia and modulation by different cytokines: mechanism for removal of apoptotic cells in the inflamed nervous system. Glia 33:87–95
30. Chang MK, Bergmark C, Laurila A, Horkko S, Han KH, Friedman P, Dennis EA, Witztum JL (1999) Monoclonal antibodies against oxidized low-density lipoprotein bind to apoptotic cells and inhibit their phagocytosis by elicited macro-

phages: evidence that oxidation-specific epitopes mediate macrophage recognition. Proc Natl Acad Sci USA 96:6353–6358
31. Chen W, Masterman KA, Basta S, Haeryfar SM, Dimopoulos N, Knowles B, Bennink JR, Yewdell JW (2004) Cross-priming of $CD8^+$ T cells by viral and tumor antigens is a robust phenomenon. Eur J Immunol 34:194–199
32. Cho HJ, Hayashi T, Datta SK, Takabayashi K, Van Uden JH, Horner A, Corr M, Raz E (2002) IFN-$\alpha\beta$ promote priming of antigen-specific $CD8^+$ and $CD4^+$ T lymphocytes by immunostimulatory DNA-based vaccines. J Immunol 168:4907–4913
33. Cohen PL, Caricchio R, Abraham V, Camenisch TD, Jennette JC, Roubey RA, Earp HS, Matsushima G, Reap EA (2002) Delayed apoptotic cell clearance and lupus-like autoimmunity in mice lacking the c-mer membrane tyrosine kinase. J Exp Med 196:135–140
34. Corinti S, Albanesi C, la Sala A, Pastore S, Girolomoni G (2001) Regulatory activity of autocrine IL-10 on dendritic cell functions. J Immunol 166:4312–4318
35. Datta SK, Redecke V, Prilliman KR, Takabayashi K, Corr M, Tallant T, DiDonato J, Dziarski R, Akira S, Schoenberger SP, Raz E (2003) A subset of Toll-like receptor ligands induces cross-presentation by bone marrow-derived dendritic cells. J Immunol 170:4102–4110
36. Delneste Y, Magistrelli G, Gauchat J, Haeuw J, Aubry J, Nakamura K, Kawakami-Honda N, Goetsch L, Sawamura T, Bonnefoy J, Jeannin P (2002) Involvement of LOX-1 in dendritic cell-mediated antigen cross-presentation. Immunity 17:353–362
37. Den Haan J, Lehar S, Bevan M (2000) $CD8^+$ but not $CD8^-$ dendritic cells cross-prime cytotoxic T cells in vivo. J Exp Med 192:1685–1696
38. Den Haan JM, Bevan MJ (2002) Constitutive versus activation-dependent cross-presentation of immune complexes by $CD8^+$ and $CD8^-$ dendritic cells in vivo. J Exp Med 196:817–827
39. Devitt A, Moffatt OD, Raykundalia C, Capra JD, Simmons DL, Gregory CD (1998) Human CD14 mediates recognition and phagocytosis of apoptotic cells. Nature 392:505–509
40. Dhodapkar MV, Steinman RM, Krasovsky J, Munz C, Bhardwaj N (2001) Antigen-specific inhibition of effector T cell function in humans after injection of immature dendritic cells. J Exp Med 193:233–238
41. Diebold SS, Kaisho T, Hemmi H, Akira S, Reise e Sousa C (2004) Innate antiviral responses by means of TLR7-mediated recognition of single-stranded RNA. Science 303:1529–1531
42. Diehl L, den Boer AT, Schoenberger SP, van der Voort EI, Schumacher TN, Melief CJ, Offringa R, Toes RE (1999) CD40 activation in vivo overcomes peptide-induced peripheral cytotoxic T-lymphocyte tolerance and augments anti-tumor vaccine efficacy. Nat Med 5:774–779
43. Dini L, Autuori F, Lentini A, Oliverio S, Piacentini M (1992) The clearance of apoptotic cells in the liver is mediated by the asialoglycoprotein receptor. FEBS Lett 296:174–178
44. Doyle SE, O'Connell RM, Miranda GA, Vaidya SA, Chow EK, Liu PT, Suzuki S, Suzuki N, Modlin RL, Yeh WC, Lane TF, Cheng G (2004) Toll-like receptors induce a phagocytic gene program through p38. J Exp Med 199:81–90

45. Drechsler Y, Chavan S, Catalano D, Mandrekar P, Szabo G (2002) FcgammaR cross-linking mediates NF-kappaB activation, reduced antigen presentation capacity, and decreased IL-12 production in monocytes without modulation of myeloid dendritic cell development. J Leukoc Biol 72:657–667
46. Dummer W, Becker JC, Schwaaf A, Leverkus M, Moll T, Brocker EB (1995) Elevated serum levels of interleukin-10 in patients with metastatic malignant melanoma. Melanoma Res 5:67–68
47. Ehrenstein MR, Cook HT, Neuberger MS (2000) Deficiency in serum immunoglobulin (Ig)M predisposes to development of IgG autoantibodies. J Exp Med 191:1253–1258
48. Elias F, Flo J, Lopez RA, Zorzopulos J, Montaner A, Rodriguez JM (2003) Strong cytosine-guanosine-independent immunostimulation in humans and other primates by synthetic oligodeoxynucleotides with PyNTTTTGT motifs. J Immunol 171:3697–3704
49. Fadok VA, Voelker DR, Campbell PA, Cohen JJ, Bratton DL, Henson PM (1992) Exposure of phosphatidylserine on the surface of apoptotic lymphocytes triggers specific recognition and removal by macrophages. J Immunol 148:2207–2216
50. Fadok VA, Bratton DL, Frasch SC, Warner ML, Henson PM (1998) The role of phosphatidylserine in recognition of apoptotic cells by phagocytes. Cell Death Differ 5:551–562
51. Fadok VA, Bratton DL, Rose DM, Pearson A, Ezekewitz RA, Henson PM (2000) A receptor for phosphatidylserine-specific clearance of apoptotic cells. Nature 405:85–90
52. Feng H, Zeng Y, Graner MW, Katsanis E (2002) Stressed apoptotic tumor cells stimulate dendritic cells and induce specific cytotoxic T cells. Blood 100:4108–4115
53. Feng H, Zeng Y, Graner MW, Likhacheva A, Katsanis E (2003) Exogenous stress proteins enhance the immunogenicity of apoptotic tumor cells and stimulate antitumor immunity. Blood 101:245–252
54. Fiebiger E, Meraner P, Weber E, Fang IF, Stingl G, Ploegh H, Maurer D (2001) Cytokines regulate proteolysis in major histocompatibility complex class II-dependent antigen presentation by dendritic cells. J Exp Med 193:881–892
55. Fonteneau JF, Larsson M, Bhardwaj N (2002) Interactions between dead cells and dendritic cells in the induction of antiviral CTL responses. Curr Opin Immunol 14:471–477
56. Fonteneau JF, Gilliet M, Larsson M, Dasilva I, Munz C, Liu YJ, Bhardwaj N (2003) Activation of influenza virus-specific CD4+ and CD8+ T cells: a new role for plasmacytoid dendritic cells in adaptive immunity. Blood 101:3520–3526
57. Fonteneau JF, Kavanagh DG, Lirvall M, Sanders C, Cover TL, Bhardwaj N, Larsson M (2003) Characterization of the MHC class I cross-presentation pathway for cell-associated antigens by human dendritic cells. Blood 102:4448–4455
58. Freigang S, Egger D, Bienz K, Hengartner H, Zinkernagel RM (2003) Endogenous neosynthesis vs. cross-presentation of viral antigens for cytotoxic T cell priming. Proc Natl Acad Sci USA 100:13477–13482
59. Gabrilovich DI, Patterson S, Timofeev AV, Harvey JJ, Knight SC (1996) Mechanism for dendritic cell dysfunction in retroviral infection of mice. Clin Immunol Immunopathol 80:139–146

60. Gallucci S, Lolkema M, Matzinger P (1999) Natural adjuvants: endogenous activators of dendritic cells. Nat Med 5:1249–1255
61. Garrett WS, Chen LM, Kroschewski R, Ebersold M, Turley S, Trombetta S, Galan JE, Mellman I (2000) Developmental control of endocytosis in dendritic cells by Cdc42. Cell 102:325–334
62. Geissmann F, Revy P, Regnault A, Lepelletier Y, Dy M, Brousse N, Amigorena S, Hermine O, Durandy A (1999) TGF-beta 1 prevents the noncognate maturation of human dendritic Langerhans cells. J Immunol 162:4567–4575
63. Gorak PM, Engwerda CR, Kaye PM (1998) Dendritic cells, but not macrophages, produce IL-12 immediately following *Leishmania donovani* infection. Eur J Immunol 28:687–695
64. Gregory CD (2000) CD14-dependent clearance of apoptotic cells: relevance to the immune system. Curr Opin Immunol 12:27–34
65. Grouard G, Rissoan MC, Filgueira L, Durand I, Banchereau J, Liu YJ (1997) The enigmatic plasmacytoid T cells develop into dendritic cells with interleukin (IL)-3 and CD40-ligand. J Exp Med 185:1101–1111
66. Guermonprez P, Saveanu L, Kleijmeer M, Davoust J, Van Endert P, Amigorena S (2003) ER-phagosome fusion defines an MHC class I cross-presentation compartment in dendritic cells. Nature 425:397–402
67. Hanayama R, Tanaka M, Miwa K, Shinohara A, Iwamatsu A, Nagata S (2002) Identification of a factor that links apoptotic cells to phagocytes. Nature 417:182–187
68. Harding CV, Song R (1994) Phagocytic processing of exogenous particulate antigens by macrophages for presentation by class I MHC molecules. J Immunol 153:4925–4933
69. Harshyne L, Watkins S, Gambotto A, Barratt-Boyes S (2001) Dendritic cells acquire antigens from live cells for cross-presentation to CTL. J Immunol 166:3717–3723
70. Harshyne LA, Zimmer MI, Watkins SC, Barratt-Boyes SM (2003) A role for class a scavenger receptor in dendritic cell nibbling from live cells. J Immunol 170:2302–2309
71. Hawiger D, Inaba K, Dorsett Y, Guo K, Mahnke K, Rivera M, Ravetch JV, Steinman RM, Nussenzweig MC (2001) Dendritic cells induce peripheral T cell unresponsiveness under steady state conditions in vivo. J Exp Med 194:769–780
72. Heath WR, Belz GT, Behrens GMN, Smith CM, Forehan SP, Parish IA, Davey GM, Wilson NS, Carbone FR, Villadangos JA (2004) Cross-presentation, dendritic cell subsets, and the generation of immunity to cellular antigens. Immunol Rev 199:9–26
73. Heil F, Hemmi H, Hochrein H, Ampenberger F, Kirschning C, Akira S, Lipford G, Wagner H, Bauer S (2004) Species-specific recognition of single-stranded RNA via Toll-like receptor 7 and 8. Science 303:1526–1529
74. Heit A, Maurer T, Hochrein H, Bauer S, Huster KM, Busch DH, Wagner H (2003) Cutting Edge: Toll-like receptor 9 expression is not required for CpG DNA-aided cross-presentation of DNA-conjugated antigens but essential for cross-priming of CD8 T cells. J Immunol 170:2802–2805
75. Henson PM, Bratton DL, Fadok VA (2001) The phosphatidylserine receptor: a crucial molecular switch? Nat Rev Mol Cell Biol 2:627–633

76. Herr W, Ranieri E, Olson W, Zarour H, Gesualdo L, Storkus WJ (2000) Mature dendritic cells pulsed with freeze-thaw cell lysates define an effective in vitro vaccine designed to elicit EBV-specific $CD4^+$ and $CD8^+$ T lymphocyte responses. Blood 96:1857–1864
77. Hochrein H, O'Keeffe M, Wagner H (2002) Human and mouse plasmacytoid dendritic cells. Hum Immunol 63:1103–1110
78. Hochweller K, Anderton SM (2004) Systemic administration of antigen-loaded CD40-deficient dendritic cells mimics soluble antigen administration. Eur J Immunol 34:990–998
79. Hoffmann PR, deCathelineau AM, Ogden CA, Leverrier Y, Bratton DL, Daleke DL, Ridley AJ, Fadok VA, Henson PM (2001) Phosphatidylserine (PS) induces PS receptor-mediated macropinocytosis and promotes clearance of apoptotic cells. J Cell Biol 155:649–659
80. Houde M, Bertholet S, Gagnon E, Brunet S, Goyette G, Laplante A, Princiotta MF, Thibault P, Sacks D, Desjardins M (2003) Phagosomes are competent organelles for antigen cross-presentation. Nature 425:402–406
81. Howard LM, Miga AJ, Vanderlugt CL, Dal Canto MC, Laman JD, Noelle RJ, Miller SD (1999) Mechanisms of immunotherapeutic intervention by anti-CD40L (CD154) antibody in an animal model of multiple sclerosis. J Clin Invest 103:281–290
82. Huang FP, Platt N, Wykes M, Major JR, Powell TJ, Jenkins CD, MacPherson GG (2000) A discrete subpopulation of dendritic cells transports apoptotic intestinal epithelial cells to T cell areas of mesenteric lymph nodes. J Exp Med 191:435–444
83. Hughes J, Liu Y, Van Damme J, Savill J (1997) Human glomerular mesangial cell phagocytosis of apoptotic neutrophils: mediation by a novel CD36-independent vitronectin receptor/thrombospondin recognition mechanism that is uncoupled from chemokine secretion. J Immunol 158:4389–4397
84. Hugues S, Mougneau E, Ferlin W, Jeske D, Hofman P, Homann D, Beaudoin L, Schrike C, Von Herrath M, Lehuen A, Glaichenhaus N (2002) Tolerance to islet antigens and prevention from diabetes induced by limited apoptosis of pancreatic β cells. Immunity 16:169–181
85. Huynh ML, Fadok VA, Henson PM (2002) Phosphatidylserine-dependent ingestion of apoptotic cells promotes TGF-beta1 secretion and the resolution of inflammation. J Clin Invest 109:41–50
86. Ichikawa HT, Williams LP, Segal BM (2002) Activation of APCs through CD40 or Toll-like receptor 9 overcomes tolerance and precipitates autoimmune disease. J Immunol 169:2781–2787
87. Ignatius R, Marovich M, Mehlhop E, Villamide L, Mahnke K, Cox WI, Isdell F, Frankel SS, Mascola JR, Steinman RM, Pope M (2000) Canarypox virus-induced maturation of dendritic cells is mediated by apoptotic cell death and tumor necrosis factor alpha secretion. J Virol 74:11329–11338
88. Inaba K, Turley S, Yamaide F, Iyoda T, Mahnke K, Inaba M, Pack M, Subklewe M, Sauter B, Sheff D, Albert M, Bhardwaj N, Mellman I, Steinman RM (1998) Efficient presentation of phagocytosed cellular fragments on the major histocompatibility complex class II products of dendritic cells. J Exp Med 188:2163–2173

89. Ishimoto Y, Ohashi K, Mizuno K, Nakano T (2000) Promotion of the uptake of PS liposomes and apoptotic cells by a product of growth arrest-specific gene, gas6. J Biochem (Tokyo) 127:411–417
90. Iyoda T, Shimoyama S, Liu K, Omatsu Y, Akiyama Y, Maeda Y, Takahara K, Steinman RM, Inaba K (2002) The $CD8^+$ dendritic cell subset selectively endocytoses dying cells in culture and in vivo. J Exp Med 195:1289–1302
91. Janda J, Schoneberger P, Skoberne M, Messerle M, Russmann H, Geginat G (2004) Cross-presentation of listeria-derived CD8 T cell epitopes requires unstable bacterial translation products. Immunol 173:5644–5651
92. Jarvelainen HA, Galmiche A, Zychlinsky A (2003) Caspase-1 activation by *Salmonella*. Trends Cell Biol 13:204–209
93. Jeannin P, Renno T, Goetsch L, Miconnet I, Aubry JP, Delneste Y, Herbault N, Baussant T, Magistrelli G, Soulas C, Romero P, Cerottini JC, Bonnefoy JY (2000) OmpA targets dendritic cells, induces their maturation and delivers antigen into the MHC class I presentation pathway. Nat Immunol 1:502–509
94. Ji JD, Tassiulas I, Park-Min KH, Aydin A, Mecklenbrauker I, Tarakhovsky A, Pricop L, Salmon JE, Ivashkiv LB (2003) Inhibition of interleukin 10 signaling after Fc receptor ligation and during rheumatoid arthritis. J Exp Med 197:1573–1583
95. Jung S, Unutmaz D, Wong P, Sano G-I, De los Santos K, Sparwasser T, Wu S, Vuthoori S, Ko K, Zavala F, Pamer EG, Littman DR, Lang RA (2002) In vivo depletion of $CD11c^+$ dendritic cells abrogation priming of $CD8^+$ T cells by exogenous cell-associated antigens. Immunity 17:211–220
96. Kabarowski JH, Zhu K, Le LQ, Witte ON, Xu Y (2001) Lysophosphatidylcholine as a ligand for the immunoregulatory receptor G2A. Science 293:702–705
97. Kaisho T, Akira S (2003) Regulation of dendritic cell function through Toll-like receptors. Curr Mol Med 3:373–385
98. Kariko K, Ni H, Capodici J, Lamphier M, Weissman D (2004) mRNA is an endogenous ligand for toll-like receptor 3. J Biol Chem 279:12542–12550
99. Kim SJ, Gershov D, Ma X, Brot N, Elkon KB (2002) I-PLA(2) Activation during apoptosis promotes the exposure of membrane lysophosphatidylcholine leading to binding by natural immunoglobulin M antibodies and complement activation. J Exp Med 196:655–665
100. Kobie JJ, Wu RS, Kurt RA, Lou S, Adelman MK, Whitesell LJ, Ramanathapuram LV, Arteaga CL, Akporiaye ET (2003) Transforming growth factor beta inhibits the antigen-presenting functions and antitumor activity of dendritic cell vaccines. Cancer Res 63:1860–1864
101. Korb LC, Ahearn JM (1997) C1q binds directly and specifically to surface blebs of apoptotic human keratinocytes: complement deficiency and systemic lupus erythematosus revisited. J Immunol 158:4525–4528
102. Kovacsovics-Bankowski M, Clark K, Benacerraf B, Rock KL (1993) Efficient major histocompatibility complex class I presentation of exogenous antigen upon phagocytosis by macrophages. Proc Natl Acad Sci USA 90:4942–4946
103. Kovacsovics-Bankowski M, Rock KL (1995) A phagosome-to-cytosol pathway for exogenous antigens presented on MHC class I molecules. Science 267:243–246

104. Krug A, Veeraswamy R, Pekosz A, Kanagawa O, Unanue ER, Colonna M, Cella M (2003) Interferon-producing cells fail to induce proliferation of naive T cells but can promote expansion and T helper 1 differentiation of antigen-experienced unpolarized T cells. J Exp Med 197:899–906
105. Larsson M, Messmer D, Somersan S, Fonteneau JF, Donahoe SM, Lee M, Dunbar PR, Cerundolo V, Julkunen I, Nixon DF, Bhardwaj N (2000) Requirement of mature dendritic cells for efficient activation of influenza A-specific memory CD8+ T cells. J Immunol 165:1182–1190
106. Larsson M, Fonteneau JF, Bhardwaj N (2001) Dendritic cells resurrect antigens from dead cells. Trends Immunol 22:141–148
107. Larsson M, Fonteneau JF, Somersan S, Sanders C, Bickham K, Thomas EK, Mahnke K, Bhardwaj N (2001) Efficiency of cross presentation of vaccinia virus-derived antigens by human dendritic cells. Eur J Immunol 31:3432–3442
108. Larsson M, Fonteneau JF, Lirvall M, Haslett P, Lifson JD, Bhardwaj N (2002) Activation of HIV-1 specific CD4 and CD8 T cells by human dendritic cells: roles for cross-presentation and non-infectious HIV-1 virus. AIDS 16:1319–1329
109. Lateef Z, Fleming S, Halliday G, Faulkner L, Mercer A, Baird M (2003) Orf virus-encoded interleukin-10 inhibits maturation, antigen presentation and migration of murine dendritic cells. J Gen Virol 84:1101–1109
110. Lauber K, Bohn E, Krober SM, Xiao YJ, Blumenthal SG, Lindemann RK, Marini P, Wiedig C, Zobywalski A, Baksh S, Xu Y, Autenrieth IB, Schulze-Osthoff K, Belka C, Stuhler G, Wesselborg S (2003) Apoptotic cells induce migration of phagocytes via caspase-3-mediated release of a lipid attraction signal. Cell 113:717–730
111. Lauber K, Blumenthal SG, Waibel M, Wesselborg S (2004) Clearance of apoptotic cells: getting rid of the corpses. Mol Cell 14:277–287
112. Le Bon A, Etchart N, Rossmann C, Ashton M, Hou S, Gewert D, Borrow P, Tough DF (2003) Cross-priming of CD8+ T cells stimulated by virus-induced type I interferon. Nat Immunol 4:1009–1015
113. Le LQ, Kabarowski JH, Weng Z, Satterthwaite AB, Harvill ET, Jensen ER, Miller JF, Witte ON (2001) Mice lacking the orphan G protein-coupled receptor G2A develop a late-onset autoimmune syndrome. Immunity 14:561–571
114. Leadbetter EA, Rifkin IR, Hohlbaum AM, Beaudette BC, Shlomchik MJ, Marshak-Rothstein A (2002) Chromatin-IgG complexes activate B cells by dual engagement of IgM and Toll-like receptors. Nature 416:603–607
115. Lee AW, Truong T, Bickham K, Fonteneau JF, Larsson M, Da Silva I, Somersan S, Thomas EK, Bhardwaj N (2002) A clinical grade cocktail of cytokines and PGE_2 results in uniform maturation of human monocyte-derived dendritic cells: implications for immunotherapy. Vaccine 20:A8–A22
116. Li M, Davey GM, Sutherland RM, Kurts C, Lew AM, Hirst C, Carbone FR, Heath WR (2001) Cell-associated ovalbumin is cross-presented much more efficiently than soluble ovalbumin in vivo. J Immunol 166:6099–6103
117. Li MO, Sarkisian MR, Mehal WZ, Rakic P, Flavell RA (2003) Phosphatidylserine receptor is required for clearance of apoptotic cells. Science 302:1560–1563
118. Liu K, Iyoda T, Saternus M, Kimura K, Inaba K, Steinman RM (2002) Immune tolerance after delivery of dying cells to dendritic cells in situ. J Exp Med 196:1091–1097

119. Liu Z, Geboes K, Colpaert S, Overbergh L, Mathieu C, Heremans H, de Boer M, Boon L, D'Haens G, Rutgeerts P, Ceuppens JL (2000) Prevention of experimental colitis in SCID mice reconstituted with CD45RBhigh $CD4^+$ T cells by blocking the CD40-CD154 interactions. J Immunol 164:6005–6014
120. Lizee G, Basha G, Tiong J, Julien JP, Tian M, Biron KE, Jefferies WA (2003) Control of dendritic cell cross-presentation by the major histocompatibility complex class I cytoplasmic domain. Nat Immunol 4:1065–1073
121. Lucas M, Stuart LM, Savill J, Lacy-Hulbert A (2003) Apoptotic cells and innate immune stimuli combine to regulate macrophage cytokine secretion. J Immunol 171:2610–2615
122. Luder CG, Gross U, Lopes MF (2001) Intracellular protozoan parasites and apoptosis: diverse strategies to modulate parasite-host interactions. Trends Parasitol 17:480–486
123. Marguet D, Luciani MF, Moynault A, Williamson P, Chimini G (1999) Engulfment of apoptotic cells involves the redistribution of membrane phosphatidylserine on phagocyte and prey. Nat Cell Biol 1:454–456
124. Marovich MA, McDowell MA, Thomas EK, Nutman TB (2000) IL-12p70 production by *Leishmania major*-harboring human dendritic cells is a CD40/CD40 ligand-dependent process. J Immunol 164:5858–5865
125. Marth T, Kelsall BL (1997) Regulation of interleukin-12 by complement receptor 3 signaling. J Exp Med 185:1987–1995
126. Matsui H, Tsuji S, Nishimura H, Nagasawa S (1994) Activation of the alternative pathway of complement by apoptotic Jurkat cells. FEBS Lett 351:419–422
127. Matzinger P (2002) An innate sense of danger. Ann NY Acad Sci 961:341–342
128. McDowell MA, Sacks DL (1999) Inhibition of host cell signal transduction by *Leishmania*: observations relevant to the selective impairment of IL-12 responses. Curr Opin Microbiol 2:438–443
129. McGuirk P, Mills KH (2000) Direct anti-inflammatory effect of a bacterial virulence factor: IL-10-dependent suppression of IL-12 production by filamentous hemagglutinin from *Bordetella pertussis*. Eur J Immunol 30:415–422
130. McPherson SW, Roberts JP, Gregerson DS (1999) Systemic expression of rat soluble retinal antigen induces resistance to experimental autoimmune uveoretinitis. J Immunol 163:4269–4276
131. Mevorach D, Mascarenhas JO, Gershov D, Elkon KB (1998) Complement-dependent clearance of apoptotic cells by human macrophages. J Exp Med 188:2313–2320
132. Moffatt OD, Devitt A, Bell ED, Simmons DL, Gregory CD (1999) Macrophage recognition of ICAM-3 on apoptotic leukocytes. J Immunol 162:6800–6810
133. Mold C, Morris CA (2001) Complement activation by apoptotic endothelial cells following hypoxia/reoxygenation. Immunology 102:359–364
134. Morelli AE, Larregina AT, Shufesky WJ, Zahorchak AF, Logar AJ, Papworth GD, Wang Z, Watkins SC, Falo LD Jr, Thomson AW (2003) Internalization of circulating apoptotic cells by splenic marginal zone dendritic cells: dependence on complement receptors and effect on cytokine production. Blood 101:611–620
135. Morelli AE, Larregina AT, Shufesky WJ, Sullivan ML, Stolz DB, Zahorchak AF, Logar AJ, Papworth GD, Wang Z, Watkins SC, Falo LD Jr, Thomson AW (2004)

Endocytosis, intracellular sorting, and processing of exosomes by dendritic cells. Blood 104:3257–3266
136. Morelli AE, Thomson AW (2003) Dendritic cells: regulators of alloimmunity and opportunities for tolerance induction. Immunol Rev 196:125–146
137. Moser M (2003) Dendritic cells in immunity and tolerance—Do they display opposite functions? Immunity 19:5–8
138. Newbold CI (1999) Antigenic variation in *Plasmodium falciparum*: mechanisms and consequences. Curr Opin Microbiol 2:420–425
139. Norbury CC, Basta S, Donohue KB, Tscharke DC, Princlotta MF, Berglund P, Gibbs J, Bennink JR, Yewdell JW (2004) D8+ T cell cross-priming a transfer of proteasome abstrates. Science 304:1318–1321
140. Norbury CC, Hewlett LJ, Prescott AR, Shastri N, Watts C (1995) Class I MHC presentation of exogenous soluble antigen via macropinocytosis in bone marrow macrophages. Immunity 3:783–791
141. Norsworthy PJ, Fossati-Jimack L, Cortes-Hernandez J, Taylor PR, Bygrave AE, Thompson RD, Nourshargh S, Walport MJ, Botto M (2004) Murine CD93 (C1qRp) contributes to the removal of apoptotic cells in vivo but is not required for C1q-mediated enhancement of phagocytosis. J Immunol 172:3406–3414
142. Ocana-Morgner C, Mota MM, Rodriguez A (2003) Malaria blood stage suppression of liver stage immunity by dendritic cells. J Exp Med 197:143–151
143. Ogden CA, de Cathelineau A, Hoffmann PR, Bratton D, Ghebrehiwet B, Fadok VA, Henson PM (2001) C1q and mannose binding lectin engagement of cell surface calreticulin and CD91 initiates macropinocytosis and uptake of apoptotic cells. J Exp Med 194:781–795
144. Ohashi K, Burkart V, Flohe S, Kolb H (2000) Cutting edge: heat shock protein 60 is a putative endogenous ligand of the toll-like receptor-4 complex. J Immunol 164:558–561
145. Oka K, Sawamura T, Kikuta K, Itokawa S, Kume N, Kita T, Masaki T (1998) Lectin-like oxidized low-density lipoprotein receptor 1 mediates phagocytosis of aged/apoptotic cells in endothelial cells. Proc Natl Acad Sci USA 95:9535–9540
146. Okamura Y, Watari M, Jerud ES, Young DW, Ishizaka ST, Rose J, Chow JC, Strauss JF 3rd (2001) The extra domain A of fibronectin activates Toll-like receptor 4. J Biol Chem 276:10229–10233
147. Oldenhove G, de Heusch M, Urbain-Vansanten G, Urbain J, Maliszewski C, Leo O, Moser M (2003) CD4+ CD25+ regulatory T cells control T helper cell type 1 responses to foreign antigens induced by mature dendritic cells in vivo. J Exp Med 198:259–266
148. Ota S, Ono T, Morita A, Uenaka A, Harada M, Nakayama E (2002) Cellular processing of a multibranched lysine core with tumor antigen peptides and presentation of peptide epitopes recognized by cytotoxic T lymphocytes on antigen-presenting cells. Cancer Res 62:1471–1476
149. Pangburn MK (2000) Host recognition and target differentiation by factor H, a regulator of the alternative pathway of complement. Immunopharmacology 49:149–157
150. Park JS, Svetkauskaite D, He Q, Kim JY, Strassheim D, Ishizaka A, Abraham E (2004) Involvement of Toll-like receptors 2 and 4 in cellular activation by high mobility group Box 1 protein. J Biol Chem 279:7370–7377

151. Pasare C, Medzhitov R (2003) Toll pathway-dependent blockade of CD4+CD25+ T cell-mediated suppression by dendritic cells. Science 299:1033–1036
152. Pfeifer JD, Wick MJ, Roberts RL, Findlay K, Normark SJ, Harding CV (1993) Phagocytic processing of bacterial antigens for class I MHC presentation to T cells. Nature 361:359–362
153. Platt N, Suzuki H, Kurihara Y, Kodama T, Gordon S (1996) Role for the class A macrophage scavenger receptor in the phagocytosis of apoptotic thymocytes in vitro. Proc Natl Acad Sci USA 93:12456–12460
154. Pollara G, Speidel K, Samady L, Rajpopat M, McGrath Y, Ledermann J, Coffin RS, Katz DR, Chain B (2003) Herpes simplex virus infection of dendritic cells: balance among activation, inhibition, and immunity. J Infect Dis 187:165–178
155. Pulendran B (2004) Modulating vaccine responses with dendritic cells and Toll-like receptors. Immunol Rev 199:227–250
156. Quezada SA, Jarvinen LZ, Lind EF, Noelle RJ (2004) CD40/CD154 interactions at the interface of tolerance and immunity. Annu Rev Immunol 22:307–328
157. Radic M, Marion T, Monestier M (2004) Nucleosomes are exposed at the cell surface in apoptosis. J Immunol 172:6692–6700
158. Ramirez MC, Sigal LJ (2002) Macrophages and dendritic cells use the cytosolic pathway to rapidly cross-present antigen from live, vaccinia-infected cells. J Immunol 169:6733–6742
159. Reis e Sousa C, Germain RN (1995) Major histocompatibility complex class I presentation of peptides derived from soluble exogenous antigen by a subset of cells engaged in phagocytosis. J Exp Med 182:841–851
160. Rodriguez A, Regnault A, Kleijmeer M, Ricciardi-Castagnoli P, Amigorena S (1999) Selective transport of internalized antigens to the cytosol for MHC class I presentation in dendritic cells. Nat Cell Biol 1:362–368
161. Roncarolo MG, Levings MK, Traversari C (2001) Differentiation of T regulatory cells by immature dendritic cells. J Exp Med 193:F5–9
162. Rosen A, Casciola-Rosen L, Ahearn J (1995) Novel packages of viral and self-antigens are generated during apoptosis. J Exp Med 181:1557–1561
163. Roth E, Schwartzkopff J, Pircher H (2002) CD40 ligation in the presence of self-reactive CD8 T cells leads to severe immunopathology. J Immunol 168:5124–5129
164. Rovere P, Vallinoto C, Bondanza A, Crosti MC, Rescigno M, Ricciardi-Castagnoli P, Rugarli C, Manfredi AA (1998) Bystander apoptosis triggers dendritic cell maturation and antigen-presenting function. J Immunol 161:4467–4471
165. Rubartelli A, Poggi A, Zocchi MR (1997) The selective engulfment of apoptotic bodies by dendritic cells is mediated by the $alpha_v beta_3$ integrin and requires intracellular and extracellular calcium. Eur J Immunol 27:1893–1900
166. Rubartelli A, Poggi A, Zocchi MR (1997) The selective engulfment of apoptotic bodies by dendritic cells is mediated by the $\alpha_v\beta_3$ integrin and requires intracellular and extracellular calcium. Eur J Immunol 27:1893–1900
167. Salio M, Palmowski MJ, Atzberger A, Hermans IF, Cerundolo V (2004) CpG-matured murine plasmacytoid dendritic cells are capable of in vivo priming of functional CD8 T cell responses to endogenous but not exogenous antigens. J Exp Med 199:567–579

168. Sambrano GR, Steinberg D (1995) Recognition of oxidatively damaged and apoptotic cells by an oxidized low density lipoprotein receptor on mouse peritoneal macrophages: role of membrane phosphatidylserine. Proc Natl Acad Sci USA 92:1396–1400
169. Sansonetti PJ, Phalipon A, Arondel J, Thirumalai K, Banerjee S, Akira S, Takeda K, Zychlinsky A (2000) Caspase-1 activation of IL-1beta and IL-18 are essential for *Shigella flexneri*-induced inflammation. Immunity 12:581–590
170. Sauter B, Albert ML, Francisco L, Larsson M, Somersan S, Bhardwaj N (2000) Consequences of cell death: exposure to necrotic tumor cells, but not primary tissue cells or apoptotic cells, induces the maturation of immunostimulatory dendritic cells. J Exp Med 191:423–434
171. Savill J, Hogg N, Ren Y, Hasslet C (1992) Thrombospondin cooperates with CD36 and the vitronectin receptor in macrophage recognition of neutrophils undergoing apoptosis. J Clin Invest 90:1513–1522
172. Savill J, Dransfield I, Gregory C, Haslett C (2002) A blast from the past: clearance of apoptotic cells regulates immune responses. Nat Rev Immunol 2:965–275
173. Savill JS, Dransfield I, Hogg N, Haslett C (1990) Vitronectin receptor-mediated phagocytosis of cells undergoing apoptosis. Nature 343:170–173
174. Scaffidi P, Misteli T, Bianchi ME (2002) Release of chromatin protein HMGB1 by necrotic cells triggers inflammation. Nature 418:191–195
175. Schagat TL, Wofford JA, Wright JR (2001) Surfactant protein A enhances alveolar macrophage phagocytosis of apoptotic neutrophils. J Immunol 166:2727–2733
176. Schaible UE, Winau F, Sieling PA, Fischer K, Collins HL, Hagens K, Modlin RL, Brinkmann V, Kaufmann SH (2003) Apoptosis facilitates antigen presentation to T lymphocytes through MHC-I and CD1 in tuberculosis. Nat Med 9:1039–1046
177. Scheinecker C, McHugh R, Shevach EM, Germain RN (2002) Constitutive presentation of a natural tissue autoantigen exclusively by dendritic cells in the draining lymph node. J Exp Med 196:1079–1090
178. Schneider-Schaulies S, ter Meulen V (2002) Triggering of and interference with immune activation: interactions of measles virus with monocytes and dendritic cells. Viral Immunol 15:417–428
179. Schulz O, Pennington DJ, Hodivala-Dilke K, Febbraio M, Reis e Sousa C (2002) CD36 or alphavbeta3 and alphavbeta5 integrins are not essential for MHC class I cross-presentation of cell-associated antigen by CD8 alpha+ murine dendritic cells. J Immunol 168:6057–6065
180. Scott RS, McMahon EJ, Pop SM, Reap EA, Caricchio R, Cohen PL, Earp HS, Matsushima GK (2001) Phagocytosis and clearance of apoptotic cells is mediated by MER. Nature 411:207–211
181. Servet-Delprat C, Vidalain PO, Valentin H, Rabourdin-Combe C (2003) Measles virus and dendritic cell functions: how specific response cohabits with immunosuppression. Curr Top Microbiol Immunol 276:103–123
182. Shen L, Sigal LJ, Boes M, Rock KL (2004) Important role of cathepsin S in generating peptides for TAP-independent MHC class I crosspresentation in vivo. Immunity 21:155–165

183. Shen L, Rock KL (2004) Cellular protein is the source of cross-priming antigen in vivo. Proc Natl Acad Sci USA 101:3035–3040
184. Shi Y, Evans JE, Rock KL (2003) Molecular identification of a danger signal that alerts the immune system to dying cells. Nature 425:516–521
185. Sigal LJ, Crotty S, Andino R, Rock KL (1999) Cytotoxic T-cell immunity to virus-infected non-haematopoietic cells requires presentation of exogenous antigen. Nature 398:77–80
186. Sigal LJ, Rock KL (2000) Bone marrow-derived antigen-presenting cells are required for the generation of cytotoxic T lymphocyte responses to viruses and use transporter associated with antigen presentation (TAP)-dependent and -independent pathways of antigen presentation. J Exp Med 192:1143–1150
187. Škoberne M, Schenk S, Hof H, Geginat G (2002) Cross-presentation of *Listeria monocytogenes*-derived CD4 T cell epitopes. J Immunol 169:1410–1418
188. Škoberne M, Beignon AS, Bhardwaj N (2004) Danger signals: a time and space continuum. Trends Mol Med 10:251–257
189. Smiley ST, King JA, Hancock WW (2001) Fibrinogen stimulates macrophage chemokine secretion through toll-like receptor 4. J Immunol 167:2887–2894
190. Sohn JH, Bora PS, Suk HJ, Molina H, Kaplan HJ, Bora NS (2003) Tolerance is dependent on complement C3 fragment iC3b binding to antigen-presenting cells. Nat Med 9:206–212
191. Somersan S, Bhardwaj N (2001) Tethering and tickling: a new role for the phosphatidylserine receptor. J Cell Biol 155:501–504
192. Somersan S, Larsson M, Fonteneau JF, Basu S, Srivastava P, Bhardwaj N (2001) Primary tumor tissue lysates are enriched in heat shock proteins and induce the maturation of human dendritic cells. J Immunol 167:4844–4852
193. Sorg RV, Kogler G, Wernet P (1999) Identification of cord blood dendritic cells as an immature CD11c− population. Blood 93:2302–2307
194. Steinbrink K, Wolfl M, Jonuleit H, Knop J, Enk AH (1997) Induction of tolerance by IL-10-treated dendritic cells. J Immunol 159:4772–4780
195. Steinbrink K, Jonuleit H, Muller G, Schuler G, Knop J, Enk AH (1999) Interleukin-10-treated human dendritic cells induce a melanoma-antigen-specific anergy in $CD8^+$ T cells resulting in a failure to lyse tumor cells. Blood 93:1634–1642
196. Steinbrink K, Graulich E, Kubsch S, Knop J, Enk AH (2002) $CD4^+$ and $CD8^+$ anergic T cells induced by interleukin-10-treated human dendritic cells display antigen-specific suppressor activity. Blood 99:2468–2476
197. Steinman RM, Turley S, Mellman I, Inaba K (2000) The induction of tolerance by dendritic cells that have captured apoptotic cells. J Exp Med 191:411–416
198. Steinman RM, Nussenzweig MC (2002) Avoiding horror autotoxicus: the importance of dendritic cells in peripheral T cell tolerance. Proc Natl Acad Sci US 99:351–358
199. Stent G, Reece JC, Baylis DC, Ivinson K, Paukovics G, Thomson M, Cameron PU (2002) Heterogeneity of freshly isolated human tonsil dendritic cells demonstrated by intracellular markers, phagocytosis, and membrane dye transfer. Cytometry 48:167–176
200. Stuart LM, Lucas M, Simpson C, Lamb J, Savill J, Lacy-Hulbert A (2002) Inhibitory effects of apoptotic cell ingestion upon endotoxin-driven myeloid dendritic cell maturation. J Immunol 168:1627–1635

201. Subklewe M, Paludan C, Tsang ML, Mahnke K, Steinman RM, Munz C (2001) Dendritic cells cross-present latency gene products from Epstein-Barr virus-transformed B cells and expand tumor-reactive $CD8^+$ killer T cells. J Exp Med 193:405–411
202. Tabi Z, Moutaftsi M, Borysiewicz LK (2001) Human cytomegalovirus pp65- and immediate early 1 antigen-specific HLA class I-restricted cytotoxic T cell responses induced by cross-presentation of viral antigens. J Immunol 166:5695–5703
203. Taylor PR, Carugati A, Fadok VA, Cook HT, Andrews M, Carroll MC, Savill JS, Henson PM, Botto M, Walport MJ (2000) A hierarchical role for classical pathway complement proteins in the clearance of apoptotic cells in vivo. J Exp Med 192:359–366
204. Termeer C, Benedix F, Sleeman J, Fieber C, Voith U, Ahrens T, Miyake K, Freudenberg M, Galanos C, Simon JC (2002) Oligosaccharides of hyaluronan activate dendritic cells via toll-like receptor 4. J Exp Med 195:99–111
205. Terpstra V, Bird DA, Steinberg D (1998) Evidence that the lipid moiety of oxidized low density lipoprotein plays a role in its interaction with macrophage receptors. Proc Natl Acad Sci USA 95:1806–1811
206. Travaglione S, Falzano L, Fabbri A, Stringaro A, Fais S, Fiorentini C (2002) Epithelial cells and expression of the phagocytic marker CD68: scavenging of apoptotic bodies following Rho activation. Toxicol In Vitro 16:405–411
207. Tsuji S, Kaji K, Nagasawa S (1994) Activation of the alternative pathway of human complement by apoptotic human umbilical vein endothelial cells. J Biochem (Tokyo) 116:794–800
208. Tvinnereim AR, Harty JT (2000) $CD8^+$ T-cell priming against a nonsecreted *Listeria monocytogenes* antigen is independent of the antimicrobial activities of gamma interferon. Infect Immun 68:2196–2204
209. Urban BC, Ferguson DJ, Pain A, Willcox N, Plebanski M, Austyn JM, Roberts DJ (1999) *Plasmodium falciparum*-infected erythrocytes modulate the maturation of dendritic cells. Nature 400:73–77
210. Urban BC, Willcox N, Roberts DJ (2001) A role for CD36 in the regulation of dendritic cell function. Proc Natl Acad Sci USA 98:8750–8755
211. Urban BC, Roberts DJ (2002) Malaria, monocytes, macrophages and myeloid dendritic cells: sticking of infected erythrocytes switches off host cells. Curr Opin Immunol 14:458–465
212. Vabulas RM, Ahmad-Nejad P, da Costa C, Miethke T, Kirschning CJ, Hacker H, Wagner H (2001) Endocytosed HSP60s use toll-like receptor 2 (TLR2) and TLR4 to activate the toll/interleukin-1 receptor signaling pathway in innate immune cells. J Biol Chem 276:31332–31339
213. Vabulas RM, Braedel S, Hilf N, Singh-Jasuja H, Herter S, Ahmad-Nejad P, Kirschning CJ, Da Costa C, Rammensee HG, Wagner H, Schild H (2002) The endoplasmic reticulum-resident heat shock protein Gp96 activates dendritic cells via the Toll-like receptor 2/4 pathway. J Biol Chem 277:20847–20853
214. Vabulas RM, Wagner H, Schild H (2002) Heat shock proteins as ligands of toll-like receptors. Curr Top Microbiol Immunol 270:169–184
215. Vandivier RW, Ogden CA, Fadok VA, Hoffmann PR, Brown KK, Botto M, Walport MJ, Fisher JH, Henson PM, Greene KE (2002) Role of surfactant proteins

A, D, and C1q in the clearance of apoptotic cells in vivo and in vitro: calreticulin and CD91 as a common collectin receptor complex. J Immunol 169:3978–3986
216. Verbovetski I, Bychkov H, Trahtemberg U, Shapira I, Hareuveni M, Ben-Tal O, Kutikov I, Gil O, Mevorach D (2002) Opsonization of apoptotic cells by autologous iC3b facilitates clearance by immature dendritic cells, down-regulates DR and CD86, and up-regulates CC chemokine receptor 7. J Exp Med 196:1553–1561
217. Verhoven B, Schlegel RA, Williamson P (1995) Mechanisms of phosphatidylserine exposure, a phagocyte recognition signal on apoptotic cells. J Exp Med 182:1597–1601
218. Verthelyi D, Zeuner RA (2003) Differential signaling by CpG DNA in DCs and B cells: not just TLR9. Trends Immunol 24:519–522
219. Voll RE, Herrmann M, Roth EA, Stach C, Kalden JR (1997) Immunosuppressive effects of apoptotic cells. Nature 390:350–351
220. Weinrauch Y, Zychlinsky A (1999) The induction of apoptosis by bacterial pathogens. Annu Rev Microbiol 53:155–187
221. West MA, Wallin RPA, Matthews SP, Svensson HG, Zaru R, Ljunggren HG, Prescott AR, Watts C (2004) Enchanced dendritic cell antigen capture via toll-like receptor-induced actin remodeling. Science 305:1153–1157
222. White S, Rosen A (2003) Apoptosis in systemic lupus erythematosus. Curr Opin Rheumatol 15:557–562
223. Yang Y, Huang CT, Huang X, Pardoll DM (2004) Persistent Toll-like receptor signals are required for reversal of regulatory T cell-mediated CD8 tolerance. Nat Immunol 5:508–515
224. Yewdell JW, Norbury CC, Bennink JR (1999) Mechanisms of exogenous antigen presentation by MHC class I molecules in vitro and in vivo: implications for generating CD8+ T cell responses to infectious agents, tumors, transplants, and vaccines. Adv Immunol 73:1–77
225. Yewdell JW, Hill AB (2002) Viral interference with antigen presentation. Nat Immunol 3:1019–1025
226. Yoshida Y, Kang K, Berger M, Chen G, Gilliam AC, Moser A, Wu L, Hammerberg C, Cooper KD (1998) Monocyte induction of IL-10 and down-regulation of IL-12 by iC3b deposited in ultraviolet-exposed human skin. J Immunol 161:5873–5879
227. Yrlid U, Wick MJ (2000) *Salmonella*-induced apoptosis of infected macrophages results in presentation of a bacteria-encoded antigen after uptake by bystander dendritic cells. J Exp Med 191:613–624
228. Zimmermann VS, Bondanza A, Monno A, Rovere-Querini P, Corti A, Manfredi AA (2004) TNF-alpha coupled to membrane of apoptotic cells favors the cross-priming to melanoma antigens. J Immunol 172:2643–2650

Subject Index

Current Topics in Microbiology and Immunology

Volumes published since 1989 (and still available)

Vol. 245/I: **Justement, Louis B.; Siminovitch, Katherine A. (Eds.):** Signal Transduction and the Coordination of B Lymphocyte Development and Function I. 2000. 22 figs. XVI, 274 pp. ISBN 3-540-66002-X

Vol. 245/II: **Justement, Louis B.; Siminovitch, Katherine A. (Eds.):** Signal Transduction on the Coordination of B Lymphocyte Development and Function II. 2000. 13 figs. XV, 172 pp. ISBN 3-540-66003-8

Vol. 246: **Melchers, Fritz; Potter, Michael (Eds.):** Mechanisms of B Cell Neoplasia 1998. 1999. 111 figs. XXIX, 415 pp. ISBN 3-540-65759-2

Vol. 247: **Wagner, Hermann (Ed.):** Immunobiology of Bacterial CpG-DNA. 2000. 34 figs. IX, 246 pp. ISBN 3-540-66400-9

Vol. 248: **du Pasquier, Louis; Litman, Gary W. (Eds.):** Origin and Evolution of the Vertebrate Immune System. 2000. 81 figs. IX, 324 pp. ISBN 3-540-66414-9

Vol. 249: **Jones, Peter A.; Vogt, Peter K. (Eds.):** DNA Methylation and Cancer. 2000. 16 figs. IX, 169 pp. ISBN 3-540-66608-7

Vol. 250: **Aktories, Klaus; Wilkins, Tracy, D. (Eds.):** Clostridium difficile. 2000. 20 figs. IX, 143 pp. ISBN 3-540-67291-5

Vol. 251: **Melchers, Fritz (Ed.):** Lymphoid Organogenesis. 2000. 62 figs. XII, 215 pp. ISBN 3-540-67569-8

Vol. 252: **Potter, Michael; Melchers, Fritz (Eds.):** B1 Lymphocytes in B Cell Neoplasia. 2000. XIII, 326 pp. ISBN 3-540-67567-1

Vol. 253: **Gosztonyi, Georg (Ed.):** The Mechanisms of Neuronal Damage in Virus Infections of the Nervous System. 2001. approx. XVI, 270 pp. ISBN 3-540-67617-1

Vol. 254: **Privalsky, Martin L. (Ed.):** Transcriptional Corepressors. 2001. 25 figs. XIV, 190 pp. ISBN 3-540-67569-8

Vol. 255: **Hirai, Kanji (Ed.):** Marek's Disease. 2001. 22 figs. XII, 294 pp. ISBN 3-540-67798-4

Vol. 256: **Schmaljohn, Connie S.; Nichol, Stuart T. (Eds.):** Hantaviruses. 2001, 24 figs. XI, 196 pp. ISBN 3-540-41045-7

Vol. 257: **van der Goot, Gisou (Ed.):** Pore-Forming Toxins, 2001. 19 figs. IX, 166 pp. ISBN 3-540-41386-3

Vol. 258: **Takada, Kenzo (Ed.):** Epstein-Barr Virus and Human Cancer. 2001. 38 figs. IX, 233 pp. ISBN 3-540-41506-8

Vol. 259: **Hauber, Joachim, Vogt, Peter K. (Eds.):** Nuclear Export of Viral RNAs. 2001. 19 figs. IX, 131 pp. ISBN 3-540-41278-6

Vol. 260: **Burton, Didier R. (Ed.):** Antibodies in Viral Infection. 2001. 51 figs. IX, 309 pp. ISBN 3-540-41611-0

Vol. 261: **Trono, Didier (Ed.):** Lentiviral Vectors. 2002. 32 figs. X, 258 pp. ISBN 3-540-42190-4

Vol. 262: **Oldstone, Michael B.A. (Ed.):** Arenaviruses I. 2002, 30 figs. XVIII, 197 pp. ISBN 3-540-42244-7

Vol. 263: **Oldstone, Michael B. A. (Ed.):** Arenaviruses II. 2002, 49 figs. XVIII, 268 pp. ISBN 3-540-42705-8

Vol. 264/I: **Hacker, Jörg; Kaper, James B. (Eds.):** Pathogenicity Islands and the Evolution of Microbes. 2002. 34 figs. XVIII, 232 pp. ISBN 3-540-42681-7

Vol. 264/II: **Hacker, Jörg; Kaper, James B. (Eds.):** Pathogenicity Islands and the Evolution of Microbes. 2002. 24 figs. XVIII, 228 pp. ISBN 3-540-42682-5

Vol. 265: **Dietzschold, Bernhard; Richt, Jürgen A. (Eds.):** Protective and Pathological Immune Responses in the CNS. 2002. 21 figs. X, 278 pp. ISBN 3-540-42668-X

Vol. 266: **Cooper, Koproski (Eds.):** The Interface Between Innate and Acquired Immunity, 2002, 15 figs. XIV, 116 pp. ISBN 3-540-42894-1

Vol. 267: **Mackenzie, John S.; Barrett, Alan D. T.; Deubel, Vincent (Eds.):** Japanese Encephalitis and West Nile Viruses. 2002. 66 figs. X, 418 pp. ISBN 3-540-42783-X

Vol. 268: **Zwickl, Peter; Baumeister, Wolfgang (Eds.):** The Proteasome-Ubiquitin Protein Degradation Pathway. 2002, 17 figs. X, 213 pp. ISBN 3-540-43096-2

Vol. 269: **Koszinowski, Ulrich H.; Hengel, Hartmut (Eds.):** Viral Proteins Counteracting Host Defenses. 2002, 47 figs. XII, 325 pp. ISBN 3-540-43261-2

Vol. 270: **Beutler, Bruce; Wagner, Hermann (Eds.):** Toll-Like Receptor Family Members and Their Ligands. 2002, 31 figs. X, 192 pp. ISBN 3-540-43560-3

Vol. 271: **Koehler, Theresa M. (Ed.):** Anthrax. 2002, 14 figs. X, 169 pp. ISBN 3-540-43497-6

Vol. 272: **Doerfler, Walter; Böhm, Petra (Eds.):** Adenoviruses: Model and Vectors in Virus-Host Interactions. Virion and Structure, Viral Replication, Host Cell Interactions. 2003, 63 figs., approx. 280 pp. ISBN 3-540-00154-9

Vol. 273: **Doerfler, Walter; Böhm, Petra (Eds.):** Adenoviruses: Model and Vectors in Virus-Host Interactions. Immune System, Oncogenesis, Gene Therapy. 2004, 35 figs., approx. 280 pp. ISBN 3-540-06851-1

Vol. 274: **Workman, Jerry L. (Ed.):** Protein Complexes that Modify Chromatin. 2003, 38 figs., XII, 296 pp. ISBN 3-540-44208-1

Vol. 275: **Fan, Hung (Ed.):** Jaagsiekte Sheep Retrovirus and Lung Cancer. 2003, 63 figs., XII, 252 pp. ISBN 3-540-44096-3

Vol. 276: **Steinkasserer, Alexander (Ed.):** Dendritic Cells and Virus Infection. 2003, 24 figs., X, 296 pp. ISBN 3-540-44290-1

Vol. 277: **Rethwilm, Axel (Ed.):** Foamy Viruses. 2003, 40 figs., X, 214 pp. ISBN 3-540-44388-6

Vol. 278: **Salomon, Daniel R.; Wilson, Carolyn (Eds.):** Xenotransplantation. 2003, 22 figs., IX, 254 pp.ISBN 3-540-00210-3

Vol. 279: **Thomas, George; Sabatini, David; Hall, Michael N. (Eds.):** TOR. 2004, 49 figs., X, 364 pp.ISBN 3-540-00534-X

Vol. 280: **Heber-Katz, Ellen (Ed.):** Regeneration: Stem Cells and Beyond. 2004, 42 figs., XII, 194 pp.ISBN 3-540-02238-4

Vol. 281: **Young, John A. T. (Ed.):** Cellular Factors Involved in Early Steps of Retroviral Replication. 2003, 21 figs., IX, 240 pp. ISBN 3-540-00844-6

Vol. 282: **Stenmark, Harald (Ed.):** Phosphoinositides in Subcellular Targeting and Enzyme Activation. 2003, 20 figs., X, 210 pp. ISBN 3-540-00950-7

Vol. 283: **Kawaoka, Yoshihiro (Ed.):** Biology of Negative Strand RNA Viruses: The Power of Reverse Genetics. 2004, 24 figs., IX, 350 pp. ISBN 3-540-40661-1

Vol. 284: **Harris, David (Ed.):** Mad Cow Disease and Related Spongiform Encephalopathies. 2004, 34 figs., IX, 219 pp. ISBN 3-540-20107-6

Vol. 285: **Marsh, Mark (Ed.):** Membrane Trafficking in Viral Replication. 2004, 19 figs., IX, 259 pp. ISBN 3-540-21430-5

Vol. 286: **Madshus, Inger H. (Ed.):** Signalling from Internalized Growth Factor Receptors. 2004, 19 figs., IX, 187 pp. ISBN 3-540-21038-5

Vol. 287: **Enjuanes, Luis (Ed.):** Coronavirus Replication and Reverse Genetics. 2005, 49 figs., XI, 257 pp. ISBN 3-540-21494-1

Vol. 288: **Mahy, Brain W. J. (Ed.):** Foot-and-Mouth Disease Virus. 2005, 16 figs., IX, 278 pp. ISBN 3-540-22419-X

Printing: Mercedes-Druck, Berlin
Binding: Stein+Lehmann, Berlin